PAT Applied in Biopharmaceutical Process Development and Manufacturing

An Enabling Tool for Quality-by-Design

BIOTECHNOLOGY AND BIOPROCESSING SERIES

Series Editor
Anurag Rathore

PAT Applied in Biopharmaceutical Process Development and Manufacturing

An Enabling Tool for Quality-by-Design

Edited by

Cenk Undey
Duncan Low
Jose C. Menezes
Mel Koch

CRC Press
Taylor & Francis Group
Boca Raton London New York

CRC Press is an imprint of the
Taylor & Francis Group, an **informa** business

CRC Press
Taylor & Francis Group
6000 Broken Sound Parkway NW, Suite 300
Boca Raton, FL 33487-2742

© 2012 by Taylor & Francis Group, LLC
CRC Press is an imprint of Taylor & Francis Group, an Informa business

No claim to original U.S. Government works

Printed in the United States of America on acid-free paper
Version Date: 20110810

International Standard Book Number: 978-1-4398-2945-5 (Hardback)

Library of Congress Cataloging-in-Publication Data

PAT applied in biopharmaceutical process development and manufacturing : an enabling tool for quality-by-design / edited by Cenk Undey... [et al.].
    p. cm. -- (Biotechnology and bioprocessing)
  Includes bibliographical references and index.
  ISBN 978-1-4398-2945-5 (hardback)
    1. Pharmaceutical biotechnology. 2. Biopharmaceutics. 3. Drug development. I. Undey, Cenk. II. Title. III. Series.

RS380.P37 2011
615.1'9--dc23   2011031862

Visit the Taylor & Francis Web site at
http://www.taylorandfrancis.com

and the CRC Press Web site at
http://www.crcpress.com

To all PAT and QbD practitioners coming from different backgrounds, united by the quest to advance process knowledge and understanding to deliver better processes and better products through better science.

Cenk Undey, Duncan Low, Jose Menezes, Mel Koch

Contents

Foreword

The publication of the draft of the guidance on Process Analytical Technology (PAT) on September 3, 2003 was a milestone for a much needed change for the pharmaceuticals and biopharmaceuticals industry. Energy and resources had been dedicated and applied to the process of drafting the guidance and finalizing it a year later in September 2004. The guidance was the culmination of collaboration between the regulators and the regulated, and the fruit of knowledge and experience of many years.

For the FDA the guidance was a first: it was historic; it was not prescriptive; it challenged the industry to re-evaluate its operations and knowledge of the manufacturing processes; it challenged the regulators world-wide as well as the industry to cease operating in silos and to collaborate across functions. A desired goal of PAT is to enhance understanding and control the manufacturing process, which is consistent with our current drug quality system: quality cannot be tested into products; it should be built-in or should be by design.*

Nearly a decade later its application is not universal for several reasons such as:

- Benefits of real-time control are neither appreciated nor understood;
- The regulatory process has not been explored sufficiently to facilitate and allow change;
- There is still confusion as to what the PAT framework is and how to implement it.

Continued success of the pharmaceutical industry requires innovation and efficiency. PAT is one framework capable of facilitating innovation and efficiency, thus assuring success of pharmaceuticals and biopharmaceuticals industries. As tenets of quality by design are consistent with PAT procedures, implementing the PAT framework can enable quality by design, can reduce the risk to quality and regulatory concerns whilst improving efficiency.

This book is a first comprehensive text on the PAT for biologics. It covers all aspects of the framework from different angles and perspectives. It is a rich offering for those who have not yet tried PAT in changing the research, development, and manufacturing within the industry. And it is useful for practitioners as a reference book.

Ali Afnan, PhD

* http://www.fda.gov/downloads/Drugs/GuidanceComplianceRegulatoryInformation/Guidances/ucm070305.pdf.

Introduction to Process Analytical Technology in Biopharmaceuticals

Mel Koch and Ray Chrisman

1 INTRODUCTION

In the past few years, the field of bioprocessing has experienced a major increase in interest, as products resulting from this field are having an impact on their respective markets. As a result, significant growth in research and development and a dramatic expansion of production capacity have occurred. The significant increase in research and development is principally in the pharmaceutical and biotechnology industries, which have recognized bioprocessing as a potential source of a whole new range of bioactive materials. The capacity expansion has been led by the rapid growth in bio-ethanol and other fuels-related production. Adding to the general increase in interest in bioprocessing is the fact that other industries such as chemicals are exploring the potential of a biological approach for a more sustainable source of materials.

However, bioprocessing brings a whole new range of challenges to the process development team. These may include constantly varying quality of raw materials, optimization, and control of organism-mediated molecular production and postproduction concentration and purification steps. In addition, most bioprocesses require sterile operating conditions.

Biopharmaceutical production often has additional challenges in that the desired component may be trapped inside the cell, requiring a lysing step and filtering. Protein therapeutics impose additional conditions such as working with large and heat- and shear-sensitive molecules. Moreover, there are various, sometimes subtle, factors that can alter the post-translational modifications of the proteins, which can have undesirable effects on efficacy.

As with all of pharmaceutical production, the regulatory environment for the production of therapeutics has been changing. This change is a direct result of the implementation of the U.S. Food and Drug Administration (FDA)–initiated quality by design (QbD) guidelines and corresponding activity from other regulatory agencies and the International Committee on Harmonization (ICH) activities, notably Q8, Q9, and Q10. Given the rapid growth in the biopharmaceutical area and the complexity of the molecules, the optimum utilization of these concepts is still being developed, which requires the team to be very proactive in their efforts to satisfy regulatory requirements during process development. Fortunately, the development and utilization of process analytical technology (PAT) has also been growing rapidly in this field. This book will offer many examples of the ways in which design teams are using PAT to not only speed up process development for biopharmaceutical production but also how it is being used to ensure high-quality production with enhanced regulatory compliance.

2   OVERVIEW OF BOOK STRUCTURE

The goal of this book, *PAT Applied in Biopharmaceutical Process Development and Manufacturing: An Enabling Tool for Quality-by-Design*, is to provide the reader with an up-to-date overview of the rapidly growing field of process analysis in bioprocessing. As such, Chapter 1, "Scientific and Regulatory Overview of Process Analytical Technology in Bioprocesses," begins the book by defining the concept of PAT from a regulatory point of view. Validation implications are outlined for the future use of PAT tools in biomanufacturing of drug substance and final dosage form.

Chapter 2, "Strategic Vision for Integrated Process Analytical Technology and Advanced Control in Biologics Manufacturing," describes how integrating PAT into biologics manufacturing requires a strategic vision and plan for advanced control. Experience with small-molecule production is a valuable starting point here. Achieving advanced control usually requires the implementation of multivariate analytical techniques (more than one variable per measurement, e.g., multispectral techniques like infrared or chromatography or combinations of measurements) during the process development phase. The chapter finishes with an example where PAT data are used for the control of glycosylation patterns of proteins.

Statistical treatment of the data from the design of experiments is used to predict a design space for process operation and is described in Chapter 3, "Multivariate Techniques in Biopharmaceutical Process Development." A variety of techniques, including fault detection and identification, predictive monitoring, partial least squares, principal component analysis, and spectral approaches, are reviewed. The chapter finishes with an example of the impact of PAT tools on effective production operations and a look at the future of process characterization.

Another facet of overall plant operation for process control is described in Chapter 4, "Analysis, Monitoring, Control, and Optimization of Batch Processes: Multivariate Dynamic Data Modeling." Several historical examples of successful modeling based on handling process data are described.

The focus of the book then shifts in Chapter 5, "Multivariate Data Analysis in Biopharmaceuticals," to the use of multivariate data analysis (MVDA) for problem solving and model development at the pilot scale and demonstrates how this knowledge can flow into uses for production control. The chapter also describes the various potential problems in modeling and how the appropriate use of MVDA can help solve them.

Chapter 6, "Process Analytical Technology Advances and Applications in Recombinant Protein Cell Culture Processes," focuses on the challenges in applying PAT to recombinant protein production processes. This involves regulatory perspectives, tool selection, and knowledge management. The chapter finishes with applications to recombinant protein production.

Chapter 7, "Process Analytical Technology Applied to Raw Materials," begins with the characterization of the extreme complexity of raw material feeds and how chemometrics and MVDA can help ensure product quality given the variations in the feed. These points are enhanced by the use of several examples, including yeast extracts and fermentation, insulin micronization design development, and the media influences on glycosylation in mammalian cell cultivation.

Given the importance of cleaning to ensure product quality and safety and reduce production holdups, the use of PAT for this production phase is reviewed in Chapter 8, "Process Analytical Technology for Enhanced Verification of Bioprocess System Cleaning." Its utilization to enhance product release and ensure compliance with regulatory requirements is described. The efforts required to ensure cell culture quality are explored in Chapter 9, "Cell Culture Process Analytical Technology Multiplexing Near-Infrared," via techniques of multiplexing near-infrared technology.

One of the most significant processing areas from a time and capital investment perspective is the whole area of bioseparations. The treatment of the subject in the next chapter demonstrates the value of PAT utilization in the area to ensure optimum operation. Thus, Chapter 10, "Process Analytical Technology for Bioseparation Unit Operations," reviews PAT applications in major harvest and purification operations, such as centrifugation, filtration, homogenization, and chromatography.

To give a broader understanding of the applications of PAT to the unique challenges of using biomaterials for production, an overview of its use in biofuels is described in Chapter 11, "Process Analytical Technology Use in Biofuels Manufacturing," to point out the many new and expanding applications to production processes in the bioprocessing field.

In Chapter 12, "Application of Microreactors for Innovative Bioprocess Development and Manufacturing," a major effort is made to help the reader understand the utility of the growing use of microscale technology to explore the broad parameter space covered by modern-day bioprocessing. The description demonstrates not only the utility of PAT for characterization of organism growth, but also serves as a reminder of the number and complexity of interactions of the parameters that affect their growth.

A key unit operation that is somewhat unique to bioprocessing is lyophilization. PAT tools useful for process control in this section of the process are described in Chapter 13, "Real-Time Monitoring and Controlling of Lyophilization Process Parameters through Process Analytical Technology Tools." This chapter demonstrates how the use of PAT can enhance the performance of the sometimes overlooked intricacies of lyophilization by describing the process parameters and the PAT tools available to monitor them.

Chapter 14, "Process Analytical Technology's Role in Operational Excellence" takes a broader view and evaluates all data flowing from the bioprocess to facilitate manufacturing excellence and lean manufacturing concepts. As an example, the idea of PAT-based predictive maintenance is presented.

The book then closes in Chapter 15, "Conclusions, Current Status, and Future Vision of Process Analytical Technology in Biologics," with a brief look to the future of PAT in the biopharmaceutical production area. It builds its vision of the future based on current uses of PAT in bioprocessing and then describes expected technology developments and how they may impact process design, monitoring, and control.

3 BRIEF BACKGROUND OF PAT IN BIOPHARMACEUTICAL PRODUCTION

There is growing global interest in the concepts of product and process optimization and quality assurance based on PAT to ensure public safety and product efficacy. The efforts are rapidly gaining support from both governments and industry as the various advantages of effective PAT utilization are more fully understood. Current implementation projects have clearly demonstrated improvements in not only quality but also cost of manufacture, environmental impact, energy efficiency, and the safety and security of operations.

The positive impacts of technological advances in PAT that have been seen in small-molecule drug development and manufacturing are a significant part of the reason that the concepts are now being applied to biopharmaceutical manufacturing. Thus, even though biomolecule production is one of the oldest areas of effective production, as evidenced by the use of fermentation through the ages, there are still significant improvements being discovered. These improvements are a result of new technology, enabling a more fundamental understanding of how bioprocesses work and what influences their efficiency.

What is now known is that there are many complexities in bioprocessing relative to the organism used, the media, the makeup and quality of the nutrients, the processing conditions, and the harvesting of the product. With this level of complexity, a key part of the advancement in understanding is based on the application of statistical data analysis to relate the multivariate analytical measurements to the broad parameter space to facilitate a much better process understanding. This powerful information analysis approach leads to actionable process understanding to advance the goal of process improvements mentioned above. This book will focus more on the methods and value of relating data to functional parameters throughout the process than it will on the specifics of the measurement science techniques used. However, because the appropriate measurements for the problem are a key part of analytical success, sufficient time will be spent in technique description to ensure that the reader understands and appreciates the justification for the approach chosen.

In addition to the traditional uses of bioprocessing in foods and beverages, there is rapid growth in its emerging use for biopharmaceuticals, biofuels, and bio-based production of chemicals. Though PAT for bioprocessing overall is still in its infancy, the rapid growth in several of the related fields has led to several new implementations of the technology. Bioprocessing, as it is related to bioenergy, has received a strong emphasis recently due to rising energy costs, largely due to increasing fossil fuel prices.

The use of biomaterials as a source for hydrocarbons has presented many challenges to the process engineer. As an example, new approaches to process separations are needed for both the removal of the water from the product and also in separating the product from other components in the mixture. Bioethanol is a good example where the cost of removing the water from the product is the most expensive cost in the process other than buying the raw material, corn. These separation unit operations need to have analytical measurements for monitoring purposes to be controlled for optimum operation.

Because some of the key unit operations and their control are common across these related bioprocessing fields, Chapter 11 is included to describe the use of PAT in the biofuels area. It is expected that some of the learnings developed for the use of PAT in the production of biofuels will be useful to broaden the general understanding of successful applications of the technology.

One of the key differences in the application of PAT to biopharmaceutical process development from its historical bioprocessing uses has been the utilization of the technology to develop a more fundamental understanding of the underlying organism growth. As part of the effort to apply the concepts of QbD and fully characterize the operating range of a bioprocess, the utilization of PAT is proving to be indispensable for the development of the needed understanding. This understanding is then being utilized to optimize the process to improve productivity, reduce environmental impact, increase energy efficiency, and enhance process safety.

One approach that is being utilized to much more efficiently explore the wide range of variables that can impact organism growth is the use of microscale reactors. This technology can be highly automated and parallelized to run a significant number of experiments to grow organisms under an extremely wide range of conditions. The coupling of the approach with PAT generates very large data sets that profit from the use of the statistical analysis to understand the impact of changes in operating parameters. Chapter 12 is devoted to examining this powerful new approach that may be new to many who are working in the bioprocessing field.

As mentioned above, the ability to do more organism growth experiments aids the new push for process understanding that is part of the QbD concept embraced by the U.S. FDA pharmaceutical regulatory agency and advocated by regulatory agencies around the world. A key part of the QbD concept that will be described in more detail later in this introduction is the need to understand process variability as a function of process parameters.

However, this enhanced understanding is necessary for all aspects of the manufacturing process and is not limited to the organism growth sections. Thus, in addition to the bioreactor, the other unit operations, such as lyophilization, separation, and cleaning, are each described in much more detail to help the reader appreciate the PAT needed for characterization of the corresponding unit operations. The descriptions are taken through data analysis to help build awareness of the approaches that lead to the desired process knowledge for optimization of the steps.

Just as composition variability in a barrel of oil can dramatically change the efficiency of operation of a refinery, the variability of the raw material feed of a bioprocess can have a significant impact on process performance. Chapter 7 presents the state-of-the-art in measuring this variability and then describes how to use the information for optimized process operation.

As with any process design, care must be taken to avoid suboptimization of the process. A review of the contributions of PAT to the development of operational excellence in the analysis of the whole process is included in Chapter 14. Part of the uses of the approach is to ensure that the understandings that are developed in one unit operation are compatible with needs of other processing steps.

Another key theme that is present throughout the book is that as the manufacturing process is being designed, there is a need to ensure the scalability of information. PAT is proving to be very important to facilitate effective scale-up. In fact the inability to move from one scale to another has been a real problem for quality production as new small-molecule processes are implemented. To solve this problem in the small-molecule drug area, the microreactor world uses the concept of numbering up. The idea is that instead of building bigger reactors, scale-up is achieved by adding more microscale reactors so that process conditions such as heat and mass transfer do not change (Hessel et al. 2009).

This number up concept has proven to work and alleviate many of the problems with scaling to meet product launch or rapidly expanding product demand. However, it is now being demonstrated that the microscale equipment coupled with PAT can be used to characterize the underlying chemistry to the point that first principles types of process models can now be built (Koch et al. 2007). The ability to develop these models that are deconvoluted from the effects of the processing equipment enables the choice of processing equipment based on cost-effective performance to optimize the underlying chemistry as opposed to having to do extensive and expensive testing of the reaction at each new scale. The ability to more precisely understand the chemistry is a relatively new concept that is still evolving and is expected to become more important as companies work to achieve more cost-effective manufacturing. Although it is unclear if these first principles types of models can be developed for biopharmaceutical manufacturing, it is clear that PAT will have a significant impact on better model development for process scale-up.

Finally, the effective use of PAT also provides data that can greatly improve process understanding and facilitate the regulatory filing process. As this is a significant part of any new drug development process, we begin the book with a detailed look at how PAT can be effectively used to help biomanufacturing in this needed effort.

4 OVERVIEW OF PROCESS ANALYTICAL TECHNOLOGY

For more than 70 years, PAT (McMahon and Wright 1996; Gregory et al. 1946) has been practiced to monitor materials production. Initially, PAT was used to follow the progress of chemical reactions, but it also proved useful for problem-solving purposes if the process ever went into an upset condition. The first analytical tools that were taken to the process environment were the instruments that historically were used in the laboratory to characterize the product being produced.

The instruments were taken out of the laboratory and put on-line due to the realization that taking samples and transporting them to a central analytical laboratory was not only dangerous and costly, but it also resulted in inaccurate representation of the process—as the dynamics of the process were often missed because of the time required to make the measurement. It rapidly became clear that the same instrument that gave data on the parameters of the final product when used in the lab could, if successfully placed on-line, represent process intermediates, many of which could predict product quality.

As PAT has matured, sampling techniques have improved to enable many measurement tools to be used in an at-line or in-line mode, which has reduced the need for manual sampling. However, sampling still remains the largest source of problems with on-line installations. This sampling problem is magnified with the often multiphase character of bioprocesses.

As the need for PAT grows, the types of instrumentation being implemented on-line continue to expand the breadth of analytical tools being utilized. It spans an increasing range of optical spectroscopy, various new approaches in separation science, a growing list of tools such as mass spectrometry, chemical sensors, acoustics, chemical imaging, and light scattering (Workman et al. 2009). Moreover, developments in PAT are coming from a wide range of laboratories, including the traditional analytical chemistry labs as well as many significant advancements coming from engineering, biological sciences, and computer science.

A key problem with this wide-ranging PAT development is that it is increasingly difficult for any one technical organization to follow all of the innovations in PAT. The developments are occurring across the whole range of processing industries as well as in government, and academic laboratories. This has become an increasingly more important problem as the need for PAT in a broader range of unit operations has risen and the implementing organizations continue to have limited resources of funding, man power, and time.

As such, most organizations are forced to look outside their company for help in following and implementing developments in PAT and QbD. This is evident in the increasing participation in symposium-based forums such as the International Foundation for Process Analytical Chemistry (IFPAC) and in a growing number of workshops and technical presentations in the fields of PAT and QbD that are being offered on a global basis by professional societies (including ISPE, PDA, and AAPS), standard setting groups (such as ASTM and ASME), and other commercial organizations (such as IBC, CHI, and IQPC).

A powerful additional strategy to stay abreast of the broad-ranging PAT field is the concept of leveraging activity via an industrial consortium. In this approach, members of a consortium are able to follow developments in related industries and leverage their resources to fund research projects in selected areas for the development of new PAT tools.

The Center for Process Analysis and Control (CPAC) at the University of Washington in Seattle, Washington (www.cpac.washington.edu), is one of the earliest and long standing examples of industrial-academic consortia from which several concepts in PAT emerged and were further developed.

The reason an academic consortium of industrial organizations survives is the ability to leverage scarce resources. In the pharmaceutical area, it has proved to serve as a benchmarking tool to understand how QbD can be a cost-effective approach to accomplish process and product optimization. Leveraging also occurs in a consortium because industry is exposed to additional members of a broad academic research team which is needed to develop modern PAT.

5   PAT CHALLENGES THAT ARE UNIQUE TO BIOPHARMACEUTICAL PRODUCTION

In a general sense, there are a series of fairly straightforward, simple steps that are taken in the chemical and related product areas that have proven to be quite useful to gauge the feasibility of a PAT implementation on a selected unit operation. The first step in a successful PAT implementation is finding the appropriate measurement tools that can characterize the components in the selected unit operation. This requires matching the speed of analysis, the specificity, and the quantitative precision and accuracy of the chosen technique to the rate of change and the compositional complexity of the unit.

Presently in biopharmaceutical operations, many conventional sensors (pH, conductivity, UV, etc.) are used but they have limitations related to fouling and nonspecificity for the desired product. Understanding more complex signals from advanced measurement instruments (Raman, NIR, IR, etc.) is challenging, but their use shows promise (see Chapters 6, 7, and 9). Work continues to show underlying correlations to 'cause and effect' in an attempt to achieve process understanding. As well, there is a challenge to develop methods for rapid microbiological testing to determine viability of a batch at an early step of processing.

It is worth mentioning at this point that in addition to the traditional approach just mentioned, newer data analysis approaches enable correlating instrument response to a broader range of process parameters such as product quality, process performance, and even equipment performance. Thus, the information from an in-process measurement, when properly analyzed, can provide a wealth of information leading to higher levels of product quality, enhanced process control, and even information leading to advanced strategies for preventative maintenance.

Once the needs are understood, the next step is developing a sampling strategy. This in many respects is the hardest part of the preliminary evaluation and as mentioned earlier is the source of failure in most implementations. This can be even more of an issue in biopharmaceutical applications as the streams can be far more complex than in the more traditional PAT applications in the small-molecule area.

The seemly rather straightforward role of the sampling system is to take a representative sample from the process stream and prepare it or condition it for analysis by the analytical device. The sample conditioning may include an initial separation to remove liquids from a solid sample or solids from a liquid sample. It must then adjust the temperature, pressure, and concentration to be compatible with the requirements of the analysis device. It may also need to adjust pH, change solvents, add additional components such as standards or other reagents, and possibly do an initial molecular or salt separation to reduce interferences. These steps all need to be done reproducibly as well as fast enough for needed control requirements. They must also be automated with minimal and preferably no operator intervention. It must also do these steps day and night, summer, and winter for extended periods of time with no loss of precision of the measurement.

Of course there are other issues like materials of construction and external environment. The environmental conditions such as temperature, power quality, and other external feeds can have a major impact on long-term system drift. Other issues such as dust, humidity/rain, insect and animal infestations, and mechanical vibrations have also been known to cause problems with system performance. The point is that the list of potential problems is long and being constantly expanded. The key to handling the long list of potential problems is to develop a rugged reliable design based on the wealth of existing experience in the field.

Unfortunately, the design problem for biopharmaceutical production processes can be even more complicated by issues such as multiple phases in, for example, a bioreactor. Not only are there the organisms but there may be solids in the nutrient feed and potentially gas bubbles. Something as simple as media saturated with air in the bioreactor can cause flow inconsistencies in tubing and filters or bubble buildup on optical surfaces. In addition, the molecule of interest may still be inside the cells, which means that a cell lysis step must be added. All of these problems are solvable, but they do require a careful design to ensure that the on-line system produces the required data quality in a timely fashion. A system must be proven to be reliable to be useful and believed by the operating staff.

In order to improve sampling systems reliability, standardized flow path approaches were developed by a working group facilitated by CPAC that enabled the use of modularized and miniaturized sampling components in the petrochemical world (collective effort called New Sampling and Sensor Initiative, NeSSI) (CPAC 2011).

Although the NeSSI sampling system approach is viable for biopharmaceutical production, the implementation is currently not compatible with some important though unique concerns to pharmaceutical production processes. Even though solutions are known that would solve the problems, it has so far not been commercially reasonable to retool equipment designs to provide sampling system components that are qualified for use in pharmaceutical production. It is hoped that this problem will be resolved in a timely fashion to enable more rapid and reliable implementations of PAT in biopharmaceutical production processes.

Finally, to complete the design phase for a process analyzer, the system must be designed to ensure that process sterilization concerns are met both during operation and also during any system maintenance. If the sampling system design seems to be able to meet the various operational needs and the analysis system can perform the needed analysis, then it is reasonable to begin the process of system development. This description is meant to demonstrate that there is much more to a successful implementation than just running a test sample in the lab, or in other words, the name PAT is no misnomer.

It should be pointed out that each unit operation has its own challenges when it comes to on-line analyzers such as physical properties, analysis time, and sample complexity. It should also be pointed out that the operation of the particular step must also be understood to ensure that the data from the analyzer is of value for control of the unit.

Another aspect that is often mentioned in technical meetings by control engineers is that process analytical chemists will develop a system that provides information on component concentrations and assume that the problem is solved. However, the process often has no control point for directly changing product concentration and thus a process model needs to be developed to relate concentrations to controllable parameters like valve settings and motor speeds. The process model may suggest that reducing/raising the temperature may improve yield or that changing residence times or agitation rate or some feed flow rate or all of the above may change the concentration but a model must be built to understand what to change at what rate to alter the concentration. Fortunately, if the data are available, it becomes much more possible to do but it is not necessarily easy to do.

6   REGULATORY ASPECTS OF PAT RELATIVE TO QBD

With the FDA's recent emphasis on quality, cost, and productivity, the use of PAT has emerged as a key resource to accomplish QbD. The general description of PAT encompasses the system that surrounds the measurement tools. It involves the sampling of the process, the sample conditioning needed for the effective measurement, the measurement, the data pretreatment, the data handling, and connection to the process control system. This has been described in an FDA Guidance.

The FDA has also worked with the ICH to outline an approach for describing QbD for the small-molecule area. In fact, this outline is an excellent template for all industries to follow in their process development and optimization activities.

The sample QbD approach (as incorporated in the International Harmonization document Q8R) was presented by a FDA representative (Nasr 2008) in Siena, Italy, in 2008. It involved:

1. Targeting the product profile
2. Determining critical quality attributes (CQAs)
3. Linking raw material attributes and process parameters to CQAs and performing risk assessment
4. Developing a design space for operating the process
5. Designing and implementing a process control strategy
6. Managing product life cycle, including continual improvement

It was emphasized that the application of PAT tools is useful for optimizing these steps in operating an effective process.

Large-molecule (biological) production can learn from successes in small-molecule manufacturing where "design space" was calculated based on designed variable experimentation. This often involved varying raw material quality, stoichiometry, and process conditions. Moreover, the value of monitoring was seen in establishing process control systems from effective process models. Monitoring was also the key to continuous improvement activities. In addition, as mentioned earlier, there are challenges for manufacturing of large (biological) molecules that include: diverse reaction media, sterilization concerns, equipment cleaning, size of unit operations, and process development activities, all of which can be aided by appropriate PAT.

7   INDUSTRY PERSPECTIVE ON THE NEEDS AND FUTURE OF PAT

Much of what is being pursued presently in the biopharmaceutical industry regarding areas around PAT is summarized in Figure 1 (Thomas 2009). There is a need for improved fundamental scientific understanding to improve process productivity. Much more of this knowledge can be gained by designing studies that are more

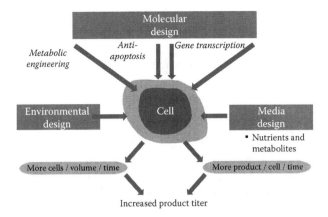

FIGURE 1 Greater productivity requires improved fundamental scientific understanding (Thomas 2009).

productive with the utilization of PAT in the process development phase (Thomas 2009).

Various processing steps are involved in the process development, and improvements can be made throughout the sequence of steps to bring a protein therapeutic into production. Although it is clear that these are very complex materials to manufacture, it is also clear that significant improvements have been made and are continuing to be made in their production (Thomas 2009).

Bioreactor operation is one of the most critical steps in producing proteins, and improvements are needed in molecular design, as well as in the characterization of the media and bioreactor conditions to improve productivity. Again, these can be facilitated by PAT (Thomas 2009).

However, there are significant barriers in measuring the products of bioprocessing as reflected in Thomas' presentation. Challenges and trends in measuring protein therapeutics include, as summarized by Thomas (2009),

1. Greater sensitivity and more specificity
   a. Peak profile (i.e., cation exchange chromatography (CEX)) does not identify specific chemical changes
   b. Move toward more informative assays
2. Elimination of subjectivity
   a. Visual inspection, presence of particles, color
   b. Move toward automated instrument-based inspection
3. Focus on CQAs
   a. Move toward specific chemical change and the site of modification
   b. Move toward understanding the biological importance of chemical changes

In conclusion, Thomas (2009) stated that technical opportunities for improving bioprocessing include the ability to more specifically measure the post-translational modifications of proteins and to more rapidly understand their significance. In addition, given the large number of parameters associated with cell growth, PAT methods need to be able to support high-throughput experimentation for exploring parameter space to understand the impact of parameter variation. There are significant opportunities for PAT to provide a much better understanding of the quality of raw materials and the impact of variations in them (Thomas 2009).

8 CONCLUSION

In response to the need for improved measurement approaches for bioprocessing, a significant amount of research is underway in this area, though it is spread over a broad range of organizations. For enhancing the ability to assess chemical modifications of proteins which can occur in a process and their biological relevancies, new measurement platforms are being considered. There is no doubt that, based on continued successful advances in miniaturization technology, most of the new measurement needs will be solved.

In addition, novel microfabrication approaches are now being demonstrated in the production of devices, which can reduce measurement time and cost. With these advancements now being incorporated into devices, manufacturing new high-throughput tools for process development can be considered. These developments will have broad-based effects on the scope of PAT, which includes sampling, sensing, control, and eventual data communication.

It is expected that those practicing in the bioprocessing field will find the contents of this book to be very timely and relevant to their efforts as it explores the wide range of applications of PAT to bioprocessing.

REFERENCES

CPAC. 2011. Center for Process Analysis and Control, University of Washington. Available on-line at http://www.cpac.washington.edu/NeSSI/. Accessed on March 13, 2011.

Gregory, C.H. (Team Leader), H.B. Appleton, A.P. Lowes, and F.C. Whalen. 1946. Instrumentation & Control in the German Chemical Industry, British Intelligence Operations Subcommittee Report #1007 (12 June 1946) (per discussion with Terry McMahon).

Hessel, V., A. Renken, J.C. Schouten, and J. Yoshida (Editors). 2009. *Micro Process Engineering: A Comprehensive Handbook*. Weinheim, Germany: Wiley-VCH.

Nasr, M. 2008. Siena Workshop on Product and Process Optimization, October 5–8, 2008. Available on-line at http://ifpac.com/siena/. Accessed on March 13, 2011.

Koch, M.V., K.M. VandenBussche, and R.W. Chrisman (Editors). 2007. *Micro Instrumentation for High Throughput Experimentation and Process Intensification—A Tool for PAT*. Weinheim, Germany: Wiley-VCH.

McMahon, T. and E.L. Wright. 1996. *Analytical Instrumentation: A Practical Guide for Measurement and Control*, edited by R.E. Sherman and L.J. Rhodes. Research Triangle Park, NC: Instrument Society of America.

Thomas, J. 2009. *Needs in Bio-Processing*, CPAC Summer Institute, Seattle, WA, July 2009.

Workman, Jr., J., M.V. Koch, B. Lavine, and R. Chrisman. 2009. Process analytical chemistry. *Anal. Chem.* 81: 4623–4643. Available on-line at www.infoscience.com. Accessed on March 13, 2011.

Contributors

Ali Afnan
Step Change Pharma, Inc.
Olney, Maryland

Keith Bader
Hyde Engineering + Consulting, Inc.
Boulder, Colorado

Rahul Bhambure
Department of Chemical Engineering
Indian Institute of Technology
New Delhi, India

Kurt A. Brorson
Office of Pharmaceutical Sciences
Center for Drug Evaluation and
    Research
U.S. Food and Drug Administration
Silver Spring, Maryland

Ana Veronica Carvalhal
Genentech, Inc.
South San Francisco, California

Ray Chrisman
University of Washington
Seattle, Washington

Jeremy S. Conner
Amgen, Inc.
Thousand Oaks, California

Fetanet Ceylan Erzen
Amgen, Inc.
West Greenwich, Rhode Island

Mariana L. Fazenda
University of Strathclyde
Glasgow, Scotland, UK

Pedro Felizardo
Instituto Superior Técnico
Technical University of Lisbon
Lisbon, Portugal

Joydeep Ganguly
Biogen-Idec Corp.
Research Triangle Park, North Carolina

Vishal Ghare
Department of Chemical Engineering
Indian Institute of Technology
New Delhi, India

Linda M. Harvey
University of Strathclyde
Glasgow, Scotland, UK

John P. Higgins
Merck Sharp and Dohme Corp.
West Point, Pennsylvania

John M. Hyde
Hyde Engineering + Consulting, Inc.
Boulder, Colorado

Feroz Jameel
Amgen, Inc.
Thousand Oaks, California

William J. Kessler
Physical Sciences Inc.
Andover, Massachusetts

Mansoor A. Khan
Office of Pharmaceutical Sciences
Center for Drug Evaluation and
    Research
U.S. Food and Drug Administration
Silver Spring, Maryland

Mel Koch
University of Washington
Seattle, Washington

Theodora Kourti
GlaxoSmithKline
Global Manufacturing and Supply
Global Functions

Duncan Low
Amgen, Inc.
Thousand Oaks, California

Meghan M. McCabe
Department of Chemical Engineering
University of Delaware
Newark, Delaware

Brian McNeil
University of Strathclyde
Glasgow, Scotland, UK

José C. Menezes
Instituto Superior Técnico
Technical University of Lisbon
Lisbon, Portugal

Peter G. Millili
Department of Chemical Engineering
University of Delaware
Newark, Delaware

M. Joana Neiva-Correia
Instituto Superior Técnico
Technical University of Lisbon
Lisbon, Portugal

Louis Obando
Merck Sharp and Dohme Corp.
West Point, Pennsylvania

Babatunde A. Ogunnaike
Department of Chemical Engineering
University of Delaware
Newark, Delaware

Jun T. Park
Office of Pharmaceutical Sciences
Center for Drug Evaluation and
    Research
U.S. Food and Drug Administration
Silver Spring, Maryland

Anurag S. Rathore
Department of Chemical Engineering
Indian Institute of Technology
New Delhi, India

Erik K. Read
Office of Pharmaceutical Sciences
Center for Drug Evaluation and Research
U.S. Food and Drug Administration
Silver Spring, Maryland

Seth T. Rodgers
Seahorse Bioscience
San Francisco, California

A. Peter Russo
Seahorse Bioscience
San Francisco, California

Victor M. Saucedo
Genentech, Inc.
South San Francisco, California

Rakhi B. Shah
Office of Pharmaceutical Sciences
Center for Drug Evaluation and
    Research
U.S. Food and Drug Administration
Silver Spring, Maryland

Erik Skibsted
Novo Nordisk
Bagsværd, Denmark

Manbir Sodhi
Department of Mechanical, Systems
    and Industrial Engineering
University of Rhode Island
Kingston, Rhode Island

Melissa M. St. Amand
Department of Chemical Engineering
University of Delaware
Newark, Delaware

Peter K. Watler
Hyde Engineering + Consulting, Inc.
Boulder, Colorado

Christopher Watts
Office of Pharmaceutical Sciences
Center for Drug Evaluation and Research
U.S. Food and Drug Administration
Silver Spring, Maryland

Shuichi Yamamoto
Yamaguchi University
Ube, Japan

Seongkyu Yoon
Department of Chemical Engineering
University of Massachusetts
Lowell, Massachusetts

1 Scientific and Regulatory Overview of Process Analytical Technology in Bioprocesses

Rakhi B. Shah, Kurt A. Brorson, Erik K. Read,
Jun T. Park, Christopher Watts, and
Mansoor A. Khan

CONTENTS

1.1 INTRODUCTION

Process analytical technology (PAT) is a system for designing, analyzing, and controlling manufacturing processes based on (1) an understanding of the scientific and engineering principles involved and (2) identification of variables that affect product quality. In the pharmaceutical world, PAT concepts were adopted early on by the small molecule industry; the progress in biotech community may be more incremental and gradual but optimistic. The PAT approach should give the biotech industry an incentive to explore various preexisting or novel analytical tools for measurements during, rather than at the end of, a process to get more information about the process and control it in real time. With biopharmaceuticals, testing is more complex; one cannot test for everything. However, we believe that the targeted research and development (R&D) can make PAT approaches evolve even for complex protein properties such as secondary structure and glycosylation patterns. We envision that PAT evolves from process control based on real-time measurement of (1) parameters that confirm that a unit operation/piece of equipment continues to be fit for purpose, moving on to (2) those parameters that directly correlate with critical quality

1

attributes (CQAs), and then finally to (3) actual product (or raw material) CQAs (i.e., the traditional conception of PAT). As stated above, achievement of (3) requires surmounting significant technology barriers by intense and purposeful R&D. The present chapter gives an overview of application of PAT in various bioprocesses and briefly provides the current regulatory experience of PAT by CDER's (Center for Drug Evaluation and Research) Office of Biotechnology Products.

The PAT initiative is with the belief that quality cannot be tested into products, but should be built-in or by design. According to the Food and Drug Administration (FDA) guidance, the desired state of pharmaceutical manufacturing is that

1. Product quality and performance are ensured through the design of effective and efficient manufacturing processes.
2. Product and process specifications are based on a mechanistic understanding of how formulation and process factors affect product performance.
3. Quality assurance is continuous and in real time.
4. Relevant regulatory policies and procedures are tailored to accommodate the most current level of scientific knowledge.
5. Risk-based regulatory approaches recognize both the level of scientific understanding and the capability of process control related to product quality and performance.

The primary goal of PAT is to provide processes that consistently generate products of predetermined quality. Improved quality and efficiency are expected from (1) reduction of cycle times using on-, in-, or at-line measurements and controls, (2) prevention of waste of raw materials (e.g., chemicals and high-quality water), (3) real-time product release, and (4) increased use of automation.

Biopharmaceuticals are inherently complex proteins that can display some isoform heterogeneity. Starting materials (i.e., animal-derived raw materials and cell culture broths) are complex mixtures of the protein drug and extraneous biologically derived materials and, thus, pose unique challenges for real-time measurement of material attributes of the actual product quality (and consequently *in vivo* performance). Current analytical techniques are limited in their capabilities by (1) the complexity of the technology required to provide real-time characterization of product quality (e.g., conformational changes and tertiary structure) and (2) the need to filter out signals from the extraneous material described above. These limitations do not promote the direct measurement and control of such attributes. Because of these limitations, current control strategies are based on inferential techniques that correlate with product quality (e.g., pH, dissolved oxygen in cell culture). As measurement systems provide greater understanding of product and process quality, strategies for process control will advance, providing timely means of characterizing and assuring product quality.

As has been previously described by the FDA, the expectation is that product quality assurance will gradually evolve from a strategy founded on fixed control (i.e., of process parameters) to one of adaptive control (process parameters are adjusted based on timely measurements of material attributes). The goal of such a strategy would be the more efficient and reliable attainment of desired material attributes of the end product (drug substance and/or drug product).

1.2 SCIENTIFIC OVERVIEW OF PAT IN BIOPROCESSES

The benefit of PAT is to help introduce well-designed and well-understood processes into manufacturing, which allows product quality to move toward the desired state. Although adoption of PAT in regulatory submissions has been relatively slow, there are numerous scientific reports detailing its development for bioprocesses (summarized in Read et al. 2010a,b). The current challenge is to develop and implement on-line and in-line measurement sensors that can measure actual product attributes during the process. In this section, we provide a few examples of how the use of PAT has enabled better process control in drug substance and drug product manufacturing for biotech products. Readers are encouraged to survey the scientific literature and other chapters of this book to gain in-depth knowledge on this topic.

1.2.1 APPLICATION OF PAT IN BIOTECH DRUG SUBSTANCE MANUFACTURING

The manufacture of biotechnology products is a complex process that evolves throughout the development of each product. Heterogeneity in a biotechnology product arises from inherent variation within living cells (i.e., the production substrate) and within subsequent purification. The entire manufacturing process encompasses upstream cell culture and downstream purification steps through the final fill-finished drug product process. This complex scheme generates various post-translational modifications that occur during cell culture as well as process-related (i.e., host-cell proteins, host-cell DNA, or leachables) and product-related impurities (i.e., aggregates or product fragments) that should largely be removed during purification. We have previously reviewed various strategies currently under development to implement PAT in bioprocessing (Read et al. 2010a,b).

Production culture represents unique challenges for the application of PAT because cell culture processes are inherently complex, and direct measurement and control of critical quality attributes are challenging and often not applied. Instead, the instruments traditionally used in cell culture measure and control important operation parameters such as temperature, agitation rate, and pH, or culture parameters such as dissolved oxygen, cell density, and glucose. These have enabled process control but do not fit the classical definition of PAT because they do not directly measure a product attribute. It is believed that application of PAT to this complex process could yield better process controls and improvements in biotech active pharmaceutical ingredient (API) quality and purity. As an example, the presence of essential media components during culture is critical for cell culture process success. Generally, cell culture media are composed of dozens of ingredients and although some are obviously important (e.g., glucose and glutamine), the relative importance of each on culture growth and production is not fully understood. Moreover, the interaction of these components is not known in detail (i.e., Is there synergy? Which are limiting?). Monitoring systems have been designed to track multiple components one at a time

(Speciner et al. 2007); these could allow process control, for example, by implementing specialized feeds should one fall below a certain level. This would not only help in understanding the importance of such components but also would minimize waste if a particular component is deemed to be 'unnecessary' for growth, and hence a feed could be avoided or at least minimized. Measurement could be accomplished by sampling followed by analysis (at-line) or by sensors (on-line). Near-infrared spectroscopy is one of the PAT tools widely applied during small-molecule production. Hypothetically, it has a potential to be used as a PAT tool for on-line cell culture processes also. As an alternative, a cell culture state-based monitor/control strategy has been proposed for nutrient supplementation (Sitton and Srienc 2008). The proposed system involves at-line, automated flow cytometry-based monitoring for an increase in nonviable cells or entry into stationary phase. This presumably heralds the depletion of media nutrients, which are then generically supplemented to sustain drug substance production. In summary, there are various tools that in theory could be used on-line (Raman spectroscopy and UV spectroscopy) or at-line (capillary electrophoresis, high-performance liquid chromatography, surface plasmon resonance, and flow cytometry). Many of these have been used to monitor cell culture, at least in R&D studies, with intent to develop PAT applications (Read et al. 2010a).

Control of downstream purification processes usually starts with control of the input, that is, ensuring consistent upstream cell culture. This is because the harvest material condition impacts the performance of at least upstream purification. The output of each step impacts the subsequent step, all the way to the final drug product fill-finished processes. The downstream purification process aims to purify a biotechnology product from a complex feedstock to the point where desired product quality and formulation buffer conditions are achieved. Typical downstream purification steps used for biotechnology product production include diafiltration/ultrafiltration, virus filtration and/or inactivation, column chromatography, mixing, inclusion body solubilization/refolding step, and conjugation reactions (e.g., PEGylation or antibody–drug conjugation). These downstream purification steps are monitored and controlled using various control strategies, which may include real-time methods. However, process understanding is often derived from indirect measures because there are fundamental scientific barriers to the direct/in-line measurement of critical quality attributes of the product and key process parameters (Lee et al. 2008).

It is believed that the implementation of PAT will result in more consistent product quality, reduce time to product/process development and time-to-market, and improve patient safety (FDA 2004). However, unit operations used for production of biotechnology products are complex and present significant challenges to employment of PAT tools that monitor CQAs directly and control the processes. Control of the production process for biotechnology products is further complicated by variability, complexity, and relatively limited understanding of the raw and feed materials used (Read et al. 2010a). To overcome these issues, the 2008 International Forum and Exhibition of Process Analytical Technology (Process Analysis & Control, IFPAC) identified specific targets for targeted R&D: (1) rapid glycosylation measurements during cell culture processes, (2) on-line monitoring of cell metabolism, (3) on-line protein aggregation detection, and (4) developing active-control feedback systems for monitoring and controlling upstream fermentation processes (Lee et al. 2008).

1.2.2 Application of PAT in Biotech Drug Product Manufacturing

Drug product manufacturing for biotech involves various operations where process knowledge and process control are important such as filtration, sterile filling, freeze-drying, and finally inspection of vials for defects or particulate matter.

In filtration, for instance, PAT can play a role in gaining process knowledge or as an assessment tool, with respect to protein adsorption, protein degradation, and filtration yield. For process control, PAT can be used to monitor filtration pressure and flow rate. In sterile filtration, microbiological burden before and after filling can be verified by use of various PAT tools (Read et al. 2010b).

In the filling operation, the target is to optimize fill accuracy. The process knowledge needed to achieve this is filling speed, filling yield, filling duration, and the product temperature. Wireless temperature measurement is a PAT tool. PAT in filling requires a fill-weight check, which is done by balance and nuclear magnetic resonance.

In the freeze-drying unit operation, also known as lyophilization unit operation, the optimum process relies on the knowledge of critical properties of the formulation and process parameters, and then applies this information to process design. Freeze-drying is a complex, time-consuming, and hence expensive multiple-step process during which the starting material (solution) undergoes several transformations leading to the end product (dry cake). To improve process efficiency and to guarantee the final product quality, methods based on in-line and real-time measurements using suitable process analyzers must be developed for continuous control of all critical process aspects. During lyophilization, it is crucial to ensure that the endpoint of all intermediate process steps is reached before the next process step is initiated and to monitor the solid state of the freeze-dried product. Process analyzers allowing the continuous monitoring of all chemical and physical phenomena occurring during freeze-drying can help to increase process understanding and process knowledge. Attempts to use PAT tools in the freeze-drying process with information on sensors and their application at various stages of freeze-drying were captured in a recent literature review (Read et al. 2010b). The list includes temperature sensors, pressure rise analysis, manometric temperature measurements, calorimetry, microscopy, and various spectroscopic techniques such as near-infrared, Raman, infrared, laser Doppler shift, and frequency modulation.

Manufacture of sterile biopharmaceuticals requires visual inspection of the final drug product filled in sealed containers to ensure that there is no contamination from foreign particulates. Although done manually in the past, more automated systems are being utilized in biotech industries. Automated visual inspections are used as PAT tools to achieve this goal. Machine parameters, product properties, and fill configuration are all important factors that determine the performance of such systems (Knapp and Abramson 1990; Knapp 2007).

In summary, to fulfill the PAT objectives in a bioprocess, it is necessary to use and implement a combination of PAT tools (chemometric tools, process analyzers, endpoint monitoring tools, and knowledge management tools). Process and product measurement, understanding, and control give rise to a better management of knowledge tools and can revolutionize how we achieve consistent quality biotech products.

1.3   REGULATORY OVERVIEW OF PAT IN BIOPROCESSES

Discussion of the PAT concept has been gaining momentum in the biotech community due to the potential for improved operational control and compliance. So far, tangible results from this buzz have been confined to R&D and early process development applications in the areas of upstream and downstream processing. Rather than a step change revolution, PAT is likely to be adopted in an evolutionary, piecemeal basis, with simpler and more indirect PAT applications first.

Over time, numerous PAT approaches have been envisioned by the wider biotech community; they coalesce in three broad categories (Read et al. 2010a): (1) Process control based on real-time, direct measurement of product (or raw material) CQAs; (2) process control based on real-time measurement of parameters that directly correlate with a CQA; or (3) process control based on real-time measurement of parameters that confirm that a unit operation/piece of equipment continues to be fit for purpose (e.g., filter back pressure, column integrity, and feedback from a vibrational monitor on a pump).

Implementation of PAT in bioprocessing is likely to follow the traditional incremental pattern of technology development in the bioprocess world (i.e., not a step transition such as those propounded by technology gurus in other industries). Type 3 approaches described above are standard, built-in feedbacks for many pieces of equipment and were common long before the introduction of the PAT guidance. Type 1 is "canonical PAT," which is the ultimate goal for process control. Based on practicalities and the traditional pattern of technology evolution (i.e., R&D-based, incremental, deliberate, and gradual), we believe that the PAT in biotechnology is likely to be gradual and follow a path like this: Type 3 (always there) → type 2 (being implemented now) → type 1 (now in R&D stage, being evaluated for implementation on a case-specific basis).

To illustrate this point, the authors searched through the process development section of a random sampling of electronic antibody biological license applications and found very few examples of type 1 PAT, that is, process control based on real-time, direct measurement of product (or raw material) CQAs. No measurement and control strategies conforming to the definition of PAT were found to be currently implemented. One developing license application did include a controlled bioreactor nutrient feeding bolus triggered when the culture reached sufficient viable cell density. However, further process development demonstrated insignificant benefit from the control strategy, and instead a new feeding strategy based only on time since inoculation was adopted. In contrast, there were several examples of type 2 PAT process control. For example, some firms monitor and control virus filter flux decay because this directly correlates with virus breakthrough. Thus, although there are examples of type 1 bioprocessing PAT in the scientific literature and case studies in process development labs in industry, we have not yet reached the stage where they are transitioning into routine manufacturing.

The goal of PAT is to "enhance understanding and control the manufacturing process, which is consistent with our current drug quality system: quality cannot be tested into products; it should be built-in or should be by design" (FDA 2004). We are currently at a stage where the accumulation of this process understanding is accelerating. Part of this acceleration results from knowledge-based capitalization by platform approaches, where process and product understanding gained on Product 1 is applicable and can be utilized for development and commercialization of Product 2 and other platform products in the pipeline.

Adoption of PAT control strategies will require a changed approach toward process and analytical development, manufacturing, quality assurance, and regulatory filings. Challenges for implementing PAT in biopharmaceutical processes sometimes limit its use within an organization. Putting new concepts into manufacturing can involve culture change and requires management support for a multidisciplinary approach to development and training.

The potential benefits are great, for example, real-time release testing leading to a reduction in end-product release testing. Real-time release testing refers to the ability to evaluate and ensure the acceptable quality of in-process and/or final product based on process data. PAT could also result in elimination of superfluous monitoring and control at the manufacturing scale (Dean and Bruttin 2000).

1.4 CONCLUSIONS

The complicated nature of biotech products makes understanding and controlling their production process not just necessary but often times difficult. Recent emphasis within the FDA on manufacturing sciences and PAT encourages the biotechnological industry to further optimize and improve current bioprocesses and to design new cycles that are robust and economical from the very beginning. One way to achieve process efficiency and to ensure the final product quality is the use of methods based on in-line, on-line, at-line, or real-time measurements using suitable process analyzers for continuous control of critical and other process parameters. For example, process analyzers can allow continuous monitoring of important chemical and physical phenomena occurring in bioprocesses that can help to increase process understanding and process knowledge. Combination of PAT tools such as chemometric tools, process analyzers, endpoint monitoring tools, and knowledge management tools will enable active and adaptive control over the production process.

REFERENCES

Dean, D. and F. Bruttin. 2000. The risks and economics of regulatory compliance. *PDA J. Pharm. Sci. Technol.* 54 (3): 253–263.

FDA. 2004. *PAT—A Framework for Innovative Pharmaceutical Development, Manufacturing, and Quality Assurance.* Rockville, MD: U.S. Department of Health and Human Services Food and Drug Administration Center for Biologics Evaluation and Research.

Knapp, J.Z. and L.R. Abramson. 1990. Automated particulate inspection systems: Strategies and implications. *J. Parenter. Sci. Technol.* 44 (2): 74–107.

Knapp, J.Z. 2007. Overview of the forthcoming PDA task force report on the inspection for visible particles in parenteral products: Practical answers to present problems. *PDA J. Pharm. Sci. Technol.* 75 (2): 131–147.

Lee, A., K.O. Webber, D. Ripple, M.J. Tarlov, and E.J. Bruce. 2008. Future technology needs for biomanufacturing: Special session at IFPAC 2008 identifies goals and gaps. *Biopharm. Int.* 21 (3). http://biopharminternational.findpharma.com/biopharm/News/

Future-Technology-Needs-for-Biomanufacturing-Speci/ArticleStandard/detail/502426 March 12.

Read, E.K., J.T. Park, R.B. Shah et al. 2010a. Process analytical technology (PAT) for biopharmaceutical products: Concepts and applications—Part I. *Biotechnol. Bioeng.* 105: 276–284.

Read, E.K., J.T. Park, R.B. Shah et al. 2010b. Process analytical technology (PAT) for biopharmaceutical products: Concepts and applications—Part II. *Biotechnol. Bioeng.* 105: 285–295.

Sitton, G. and F. Srienc. 2008. Mammalian cell culture scale-up and fed-batch control using automated flow cytometry. *J. Biotechnol.* 135: 174.

Speciner, L., G. Barringer, J. Perez, C. Grimaldi, R. Reineke, and A. Arroyo. 2007. Implementing an automated reactor sampling system for monitoring cell culture bioreactors. 234th ACS National Meeting, Boston.

2 Strategic Vision for Integrated Process Analytical Technology and Advanced Control in Biologics Manufacturing

Melissa M. St. Amand, Peter G. Millili, Meghan M. McCabe, and Babatunde A. Ogunnaike

CONTENTS

2.1 INTRODUCTION

With a U.S. market of approximately $50 billion in 2005 and projected sales near $100 billion by the end of 2010, biopharmaceuticals represent the largest growing class of therapeutics (Biostrategy 2005; Pavlou and Reichert 2004; Walsh 2006). Over one third of drugs currently in clinical trials are recombinant protein therapeutics, and the success of antibody therapies alone will increase that percentage. Due to the therapeutic nature of these products, global regulatory agencies aim to protect consumers by requiring manufacturers to follow stringent quality guidelines that are meant to ensure the safety and efficacy of biopharmaceutical products. Unfortunately, the quality control practices used to meet these quality guidelines are largely inefficient and often ineffective. Until recently, biopharmaceutical manufacturers could remain profitable in spite of significant inefficiencies in their manufacturing processes. However, with rapidly expanding product portfolios, the emergence of generics, and tighter regulatory requirements, to remain competitive, biopharmaceutical manufacturers will need to develop and implement novel methods for assuring product quality consistently and efficiently.

Process analytical technology (PAT), a general methodology predicated on using a comprehensive understanding of the drug product and its manufacturing process for improving product quality, safety, and production efficiency, will prove beneficial for biopharmaceutical manufacturing if implemented properly. However, the true spirit behind PAT and how it is to be implemented properly in practice remain unclear to many in the biopharmaceutical industry. This chapter presents a strategic vision of how PAT should be implemented for biopharmaceutical manufacturing processes, especially emphasizing the need to integrate process and product quality control as an integral component. The principles and concepts are explained and then illustrated with an actual experimental case study.

The rest of the chapter is organized as follows: in Section 2.2, we provide some background on quality and quality control in the biopharmaceutical industry, along with a brief introduction to PAT as conceived by the Food and Drug Administration. Section 2.3 is devoted to an overview of the tasks involved in achieving the objectives of PAT in practice before presenting, in Section 2.4, our strategic vision of how these tasks are to be accomplished in practice for manufacturing biologics. In Section 2.5, we illustrate the concepts presented in Sections 2.3 and 2.4 with an actual implementation of PAT on a laboratory-scale process developed at the University of Delaware, where the ultimate goal is on-line, real-time control of glycosylation during monoclonal antibody (MAb) production. The discussion includes results obtained to date along with an overview of future work before closing with some concluding remarks in Section 2.6.

2.2 BACKGROUND

2.2.1 QUALITY IN THE BIOPHARMACEUTICAL INDUSTRY

As defined by the "International Conference on Harmonization," quality in pharmaceutical products is the ". . . suitability of a drug substance or drug product for its intended use" (ICH 2000). In practical terms, such "suitability" is determined objectively on the basis of such attributes as product identity, bioactivity, purity, toxicity, and potency. In an effort to ensure that consumers receive high-quality biopharmaceuticals, regulatory agencies such as FDA and European Medicines Agency require therapeutics manufacturers to determine the critical quality attributes (CQAs) of their products and establish a set of acceptance criteria, or specifications, to which the drug substance or drug product must conform to be considered acceptable for its intended use.

Formally, a CQA is defined as any ". . . physical, chemical, biological, or microbiological property or characteristic that should be maintained within an appropriate limit, range, or distribution to ensure the desired product quality" (ICH 2009). These CQAs and their appropriate specifications are typically determined during early stage development after a series of characterization studies of the physicochemical properties, biological activity, immunochemical properties, and impurities of the product. Once established and accepted, the CQA specifications become the critical quality standards endorsed by regulatory authorities as conditions of approval for the product (FDA 1999).

2.2.2 LIMITATIONS OF CURRENT QUALITY CONTROL PRACTICES IN THE BIOPHARMACEUTICAL INDUSTRY AND PAT

Unfortunately, most CQAs can only be characterized with off-line, post-production assays that require extensive sample preparation. As a result, the quality control practices currently used by the biopharmaceutical industry, to ensure that a product batch has adhered to the established specifications, typically involve performing a series of off-line product quality control assays such as those listed in Table 2.1.

As sales steadily increase and product portfolios expand, the longstanding biopharmaceutical industry standard of using off-line, post-production assays as the primary method for product quality control faces severe limitations. First, some established specifications may not truly result in a quality product especially when these specifications are established through univariate experimentation, which neglects effects of interactions that may exist between process variables and/or CQAs. For example, consider a product whose quality attributes are influenced by two process variables y_1 and y_2. When considered independently, acceptable product quality is achieved when each process variable is maintained within its individual, established acceptable range. Consider now that the interaction between the two variables has an unknown effect on a CQA. With univariate specifications, it is possible for the process variables y_1 and y_2 to fall within their respective acceptable ranges, but the product still fails to meet the associated CQA specification. This is because, as illustrated in Figure 2.1, an important consequence of the interactions between y_1 and y_2 is that the actual acceptable range for the product's CQA is in fact the much tighter ellipse, not the rectangle.

Defining the acceptable ranges for variables of interest when multivariate effects are taken into consideration is known as "establishing the design space," and the FDA has recently issued guidelines for establishing an effective design space (FDA 2009). It is expected that operating within the design space will result in the product meeting the defined CQA specifications. However, the use of off-line, post-production assays to ensure that a product meets the established specifications of the design

TABLE 2.1

Examples of Biopharmaceutical Quality Control Assays

Assay	Description of Purpose
SDS-PAGE (reduced and unreduced, Coomassie blue, and silver stain)	Identify protein identity and occurrence of proteolysis, dissociation, cross-linking Determine presence of impurities such as residual amounts of extraneous animal proteins
Isoelectric focusing	Identify heterogeneity in the protein structure such as charge isoforms, deamidation, deglycosylation, and oxidation
Size exclusion HPLC	Determine presence of aggregation, dissociation, proteolysis, and protein concentration
Absorbance (280 nm)	Verify protein concentration
Amino acid analysis	Determine protein's amino acid coefficient, extinction coefficient
N-terminal sequencing	Confirm protein identity
Peptide mapping	Determine protein identity, primary structure, and detect protein impurities such as deamidation, substitution, oxidation, proteolysis, cross-linking
Mass spectroscopy	Determine protein molecular mass, primary structure, and glycosylation
Biophysical assays using NMR, CD, intrinsic fluorescence, etc.	Determine secondary and tertiary structure, degree of unfolding and denaturation
Immunoassays	Determine antigen-binding activity, identify specific containments, isotyping
Binding assay	Determine protein potency, antigen binding activity, neutralization activity
Western blotting/antisera probes	Identify presence of specific protein contaminants

Source: Brass J.M., K. Krummen, and C. Moll-Kaufmann, *Pharm. Acta Helv.*, 71, 395–403, 1996.
Note: SDS-PAGE, sodium dodecyl sulfate polyacrylamide gel electrophoresis; HPLC, high-performance liquid chromatography; NMR, nuclear magnetic resonance; CD, circular dichroism.

space severely limits the effectiveness of current quality control practices in the biopharmaceutical industry. A post-production, off-line quality control strategy makes biopharmaceutical manufacturing processes extremely inefficient because the only reasonable course of action when a CQA is found to be out of range is to discard the entire product batch (due to regulatory concerns). Thus, even when the design space has been properly established, meeting CQA specifications consistently (and efficiently) with current industrial quality control procedures remains a major challenge.

On-line product quality control, which has been a standard in the petroleum and chemical process industries for decades (Schuyten 1967; Ogunnaike 1994), significantly improves process efficiency because it makes "in-production" corrections possible, thereby preventing out-of-specification conditions from persisting during production.

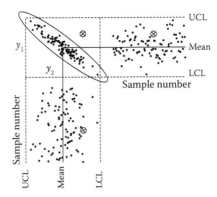

FIGURE 2.1 Quality control of two variables indicates misleading nature of univariate charts. (From MacGregor, J.F. and T. Kourti., *Control Eng. Pract.*, 3, 403–414, 1995. With permission.)

Such on-line product quality control strategies not only improve product yields and reduce production costs but also result in higher quality products that meet customer specifications (and government regulations) more easily and more consistently.

In this regard, the biopharmaceutical industry, with its off-line and post-production quality assurance methods, clearly lags behind other industries in the development and implementation of innovative strategies to ensure product quality. This state of affairs in biopharmaceutical manufacturing is usually attributed to the fact that bioprocessing, in general, involves highly complex processes with many production and purification steps, in addition to long holding times of product intermediates between subsequent process steps (hold times that can range from 1 to 9 months). As a result, manufacturers often do not fully understand the effects of product components and process characteristics on final product quality. Without this kind of comprehensive understanding, the potential for seemingly minor process alterations to compromise product quality and safety originally prompted regulatory agencies to enforce strict regulations on biopharmaceutical manufacturing to prevent process alterations from causing fatal consequences. However, these strict regulatory guidelines have also dissuaded most biopharmaceutical manufacturers from implementing innovative on-line control strategies that would otherwise actually improve product quality.

In 2004, with the goal of supporting and promoting innovation and efficiency in pharmaceutical development, manufacturing, and quality assurance, the FDA introduced a set of PAT guidelines (FDA 2004). According to this FDA guidance for industry, PAT is a ". . . system for designing, analyzing, and controlling manufacturing through timely measurements of critical quality and performance attributes of raw and in-process materials and processes, with the goal of ensuring final product quality." The intention of PAT is to promote a comprehensive understanding of the drug product and its manufacturing process, which will help manufacturers better understand how their products are degraded, released, and absorbed; determine what major factors affect product quality; and identify the major sources of variability. The FDA guidance suggests that with this kind of comprehensive understanding, biologics manufacturers will be able to design and develop processes that consistently meet predefined end product quality specifications, thereby significantly improving product quality and safety, as well as production efficiency. The FDA guidance states clearly that PAT will benefit both the biopharmaceutical manufacturer and consumer through higher quality products; however, precisely how biopharmaceutical manufacturers are to achieve the objectives of PAT in practice remains unclear.

In the next section, we present an overview of the tasks required in achieving the objectives of PAT, justifying each one before presenting in Section 2.4 a recommendation of the component systems required for accomplishing these tasks in practice.

2.3   ACHIEVING PAT OBJECTIVES IN PRACTICE

Achieving the objectives of PAT in practice fundamentally requires carrying out the following four tasks.

1. *Measure*. Measure key process variables and CQAs to acquire timely data about the process and the product; communicate the data between process equipment and the data acquisition and control system.
2. *Analyze*. Analyze the data obtained from measurements to extract useful information about the process conditions and product attributes; convert such information to process understanding and knowledge.
3. *Control*. On the basis of acquired information and process knowledge (and using an appropriate control strategy), manipulate appropriate variables to keep the process under control and maintain the product quality attributes within specification limits.
4. *Optimize*. Continuously improve process understanding by monitoring process performance over time and comparing it with design objectives; utilize improved understanding to optimize process operation to ensure consistent attainment of approval specifications for the CQAs.

Before discussing how these tasks should be accomplished in practice, it is important to place all four of them within the context of the spirit of and organizing principles behind PAT.

For the products of any manufacturing process to meet predefined quality specifications consistently, there must be a comprehensive measurement system in place for acquiring appropriate measurements of key process variables and of raw materials; there must also be a reliable system determining critical quality and performance attributes of end products sufficiently frequently. Biopharmaceutical manufacturing should be no exception. In all cases, as a necessary precursor to the establishment of the measurement system, the key process variables and CQAs of interest must have already been identified—an exercise that, in the biopharmaceutical industry, requires detailed characterization studies of the physicochemical properties, biological activity, immunochemical properties, purity, and impurities of the product. Appropriate sensors and analyzers must then be acquired for the identified key process variables, with preference for those sensors capable of providing real-time measurements on-line where possible. It should be clear, therefore, why the first step in the implementation of PAT must be the establishment of an effective measurement system: without appropriate and adequate process and product data, the manufacturing process simply cannot be operated efficiently, and the product quality cannot be assured.

Finally, because the mere acquisition of data from a manufacturing process is just a means to an end and should not be considered as an end in itself, there must be, as an integral part of an effective measurement system, an infrastructure for transmitting data back and forth between process equipment and the data acquisition and control system (as discussed shortly).

Next, we note that the true goals of PAT extend well beyond the mere installation of (on-line) measurement devices and the gathering of raw data. To be useful, raw process data sets must be analyzed appropriately to extract the process information they contain. Following data acquisition, therefore, the next important task is that of data analysis for information extraction. As will be discussed shortly, such information processing can be carried out in a variety of ways, ranging from basic data analysis (e.g., the computation of sample statistics) and more advanced multivariable data analysis [e.g., principal component analysis (PCA)], to capturing the information in the form of explicit mathematical models of the process characteristics.

However, even such process information is not meant to be acquired as an end in itself; process information must be used actively to adapt process operation to compensate for, as necessary, the effects of unavoidable variability in raw materials, ambient conditions, etc., to which all manufacturing processes are subject. Without such judicious process adaptation, the product attribute targets cannot be met consistently in the face of such persistent and unavoidable "disturbances." Accomplishing this task requires the design and implementation of an effective control system whose purpose is to adjust strategic manipulated variables (MVs) judiciously, based on an appropriate control strategy that uses basic process understanding and recently acquired information about process conditions. The primary objective of such a control system is to ensure that the critical product quality attributes are maintained within the required specification limits consistently. This is one of the least understood—and hence least practiced—aspects of PAT implementation.

Finally, the process must be monitored continuously over extended periods of operation, and periodically, the archived historical process and product data analyzed to update and improve process knowledge/understanding. Such updated process knowledge should then be used to determine operating conditions that will lead to optimal process performance.

With this perspective of the tasks that must be accomplished in meeting the objectives of PAT, we now turn to the central question of this chapter: How can these tasks be accomplished specifically for a biopharmaceutical manufacturing process?

2.4 STRATEGIC VISION FOR BIOPHARMACEUTICAL MANUFACTURING

A specific configuration of the component systems required for successfully carrying out the tasks discussed in the previous section for biopharmaceutical manufacturing is shown in Figure 2.2.

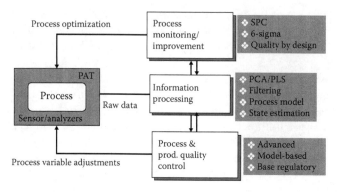

FIGURE 2.2 Strategic vision of configuration of biopharmaceutical processing incorporating PAT.

2.4.1 PROCESS AND DATA ACQUISITION

The foundational and central component in this configuration is the biopharmaceutical process that produces and purifies the therapeutic product of interest. In our strategic vision for the biopharmaceutical industry, we include, as an integral part of this manufacturing process, all other ancillary equipment that might be used during manufacturing, particularly the sensors and analyzers that are used to acquire measurements.

In establishing such a process in practice, it is essential to determine the key process variables that are central to the process operation, in addition to the critical quality and performance attributes for the raw materials, in-process variables, and especially for the end products. For example, the key process variables for most bioreactor processes used to produce therapeutic proteins are pH, dissolved oxygen (DO), biomass concentration, as well as nutrient and metabolite concentrations in the media (Hopkins et al. 2010). Typical product end-use performance attributes include protein activity and purity that are determined by such product quality characteristics as concentration, glycosylation pattern, degree of proteolysis, and protein aggregation.

These variables, attributes, and characteristics are typically identified during comprehensive preliminary studies designed specifically for these purposes, following which appropriate hardware/software (i.e., sensors, electrodes, and analyzers) for acquiring the necessary raw data must then be installed.

"On-line" sensors are exposed directly to the process fluid and provide measurements in real time; "at-line" measurements are obtained locally from samples that have been aseptically withdrawn from the bioreactor. "Off-line" measurements are usually performed on aseptically stored supernatant within hours or days of sampling. Ideally, data should be collected with on-line or at-line measurement devices, but some measurements can only be determined by off-line measurements. With current bioreactor automation technology, on-line measurements of such process variables as pH, temperature, and DO have been made possible with the advent of steam sterilizable probes for these variables (Johnson et al. 1964). At-line measurements are available for biomass, nutrient, and metabolite concentrations through automated sampling systems coupled to such bioanalyzers as the BioFlex (Nova Biomedical, Waltham, MA). The majority of quality attributes, if they can be measured at all, must be measured by off-line assays.

In terms of the infrastructure for transmitting data back and forth between process equipment and the data acquisition and control system, OPC has become the industrial communication standard for pharmaceutical manufacturing. This protocol enables real-time transmission of data between computer systems among different instruments. The interoperability of the software between instruments of different manufactures is assured through the creation and maintenance of open standards specifications that are regulated by the OPC Foundation (http://www.opcfoundation .org). Originally introduced in 1996 as an acronym for "object linking and embedding (OLE) for process control," OPC is no longer considered an acronym due to the wide applicability of the technology outside the area of process control.

OPC is configured with a client/server architecture: instruments and process systems that generate data require an OPC server, whose sole function is to distribute stored process data to the OPC client software. Process systems and industrial software applications that utilize the process data to carry out various functions use OPC client software to call OPC servers and acquire the desired process data. The communication between the OPC clients and servers is typically achieved via proper configuration of Microsoft Window's COM or DCOM settings within a network, or by establishing an OPC tunnel between the client and server over the Internet.

2.4.2 INFORMATION PROCESSING

As noted in Section 2.3, following the implementation of an appropriate measurement system (consisting of appropriate sensors, electrodes, and analyzers), it is not enough merely to collect and archive the raw operating process and product data. The acquired data sets must be converted to useful form, processed, and analyzed to extract useful information about the process.

The necessity for data processing arises from a variety of reasons. For example, on one end of the spectrum of reasons is the simple fact that each measurement, $y_i(t_k)$, obtained for variable i at time instant t_k consists of the true value, $\eta_i(t_k)$, and an associated measurement noise component, $\varepsilon_i(t_k)$, such that

$$y_i(t_k) = \eta_i(t_k) + \varepsilon_i(t_k).$$

Because the true state of the process variable is thus always and unavoidably corrupted by the measurement noise component, consequently, depending on how large or small $\varepsilon_i(t_k)$ is in relation to the measurement, it is often necessary for the acquired data to be "filtered" first to obtain a value $\tilde{y}_i(t_k)$ that is a "better" (in the sense of being less variable) estimate of $\eta_i(t_k)$ than the raw measurement. On the other end of the spectrum is a more complicated reason: the data set, consisting of measurements for $i = 1, 2, \ldots n_y$ variables, obtained at times $t_k, k = 1, 2, \ldots N$, encodes information about the process over the indicated time horizon, but such information is neither obvious nor explicit merely from the acquired raw measurement values. Such data will be multivariate (for $n_y > 1$) and is, therefore, likely to be correlated across the variables, as well as "dynamic," indicating joint, as well as individual, correlation in time. Extracting the encoded process information requires the use of techniques calibrated for such multivariate and dynamic data.

Finally, product quality measurements will only be available infrequently through off-line assays, and in some cases will not be available at all. Moreover, by definition, the product's final end-use performance attribute cannot be determined until the product has been tested in end use. However, these quality characteristics and end-use attributes are being incorporated into the product as it is being manufactured; as such, information about these product characteristics and end-use attributes are available in the process data if only we knew how to access them and how to use such information appropriately to devise operating procedures for the process.

As a result, the process information contained in the acquired data set must be extracted via a variety of techniques depending on the nature of the data set and the required information. These analysis techniques include, but are not limited to, data filtering, simple statistical summaries (means, standard deviations, correlation coefficients, and histograms), and such multivariate analysis as PCA and projection to latent structures also known as partial least squares (PLS). More sophisticated applications require the development, from the acquired data (and/or augmented by fundamental first principles), of mathematical models that represent the process behavior. Such models could then be used for "state estimation," the inference of unmeasured states such as product characteristics and end-use attributes, which are not available directly by measurement frequently enough.

2.4.3 PROCESS/PRODUCT QUALITY CONTROL

Although it is important to analyze the raw data to extract useful process information, the resulting process information is only beneficial to the extent that it is used actively to adapt process operation appropriately to meet the product quality specifications consistently, in spite of unavoidable and persistent variability to which all manufacturing processes are subject. The process control system is charged with the responsibility of using process measurements (acquired directly or inferred) to determine how best to adjust process conditions to maintain process and product quality variables within desired specification limits.

Most modern manufacturing processes are equipped with process control systems that are used to implement process control strategies of varying degrees of sophistication. At the most basic level is base regulatory control, where a single process variable known as a controlled variable (CV) (e.g., reactor temperature) is to be controlled to a desired set point by judiciously adjusting an MV (e.g., cooling water flow rate) that is capable of changing the value of the CV in question. The most common implementation strategy is feedback control using standard proportional–integral–derivative (PID) controllers and measurements of the CV to determine values of the MV required to maintain the CV at its desired set point (Ogunnaike and Ray 1994). Such a strategy can be used successfully within bioreactor automation systems to maintain pH, DO, temperature, and agitation rate at their respective desired values.

Advanced control strategies are required when the manipulation of a single MV results in changes to multiple CVs simultaneously and, for other cases, where the dynamic behavior of the process in question is complex. Such problems arise, for example, in controlling nutrient composition where changes to a single nutrient feed rate change the composition of the entire bioreactor content, not just the composition of a single nutrient, and the observed response arises as a result of complex biochemical and biophysical transformations. Under these circumstances, it is customary to use multivariable control strategies or other model-based strategies whereby the control action decision is made with the aid of a process model (see, for example, Ogunnaike and Ray 1994).

Effective process control strategies facilitate the maintenance of process variables close to desired set-point values, but this does not guarantee that product quality objectives will be met. An effective product quality control system ensures consistent attainment of desired end-use attributes in the manufactured product, although actual product quality and end-use attribute measurements are neither available immediately nor frequently enough for direct use in a classic feedback control configuration. Because unavailable or infrequently available product quality measurements must be inferred from available process measurements and augmented with appropriate process models, such control strategies will succeed to the extent that the "inference scheme" is effective in encapsulating the relationship between measurable process variables and product characteristics in a manner that can be exploited (1) directly for providing reliable estimates of the unavailable process and product information and (2) as a basis for determining appropriate control action required to ensure consistent attainment of desired end-use product characteristics (see Ogunnaike 1994; Ogunnaike et al. 2010).

2.4.4 PROCESS MONITORING AND IMPROVEMENTS

A supervisory system is required for monitoring and assessing the performance of the entire manufacturing process and its ancillary systems to ensure robust performance, improve process understanding, and optimize process performance. Such a system involves process and sensor fault detection and identification, as well as the detection of long-term trends in the trajectories of MVs in maintaining the process CVs and the product quality attributes at desired values.

Multivariate statistical process control (SPC) techniques are typically used for monitoring process performance and for identifying long-term deviations from normal operation characteristics (MacGregor and Kourti 1995); these techniques are also sometimes used to determine what is required to restore the process to optimal operation if necessary. The "Improve" portion of the Six Sigma methodology may also be co-opted for systematically improving process performance on the basis of long-term information acquired by the monitoring system.

The implementation of such a monitoring and optimization system should be driven by the principles of quality by design by which processes are operated around such operating conditions as are most conducive to the consistent attainment of high-quality product consistently with the minimum amount of variability.

The implementation of this ideal configuration for biopharmaceutical manufacturing will result in a robust process whose characteristics are well understood, whose product consistently meets the quality objectives, and whose performance is continuously being improved. Preliminary results of an actual experimental demonstration of these concepts are presented next.

2.5   EXAMPLE OF PAT IMPLEMENTATION: ON-LINE CONTROL OF GLYCOSYLATION DURING MONOCLONAL ANTIBODY PRODUCTION

To illustrate how the strategic vision discussed in the previous section may be implemented in practice, we now present some preliminary results from an experimental case study devoted to on-line, real-time control of glycosylation patterns on MAbs produced with Chinese hamster ovary (CHO) cells. Specifically, this presentation focuses on demonstrating the design and implementation of nutrient control and proposes a multiscale model-based control strategy for explicit control of glycan distribution.

MAb production has become the fastest growing sector of the biopharmaceutical industry with 23 MAbs currently on the market and more than 200 in development pipelines. As with other biopharmaceuticals, the therapeutic potential of MAbs requires manufacturers to establish quality criteria ranges for product attributes such as bioactivity, potency, and purity to assure consumers of the product's efficacy. Many factors affect the quality and bioactivity of MAbs, but arguably, the most important is glycosylation, a post-translational modification in which a carbohydrate chain termed a glycan is added to a protein (Beck et al. 2008; Butler 2006; Geyer and Geyer 2006; Goochee et al. 1991). The glycosylation patterns become especially critical when the MAb is intended to affect antibody-dependent cellular cytotoxicity and/or complement-dependent cytotoxicity function *in vivo*. Under such circumstances, the glycosylation pattern is therefore a CQA (Berridge et al. 2009), with the requirement that the glycosylation pattern formed on the manufactured MAb match the desired pattern as closely as possible to ensure that the MAb will function as intended *in vivo*. However, unlike other cellular processes, glycosylation has no master template so that glycan formation and attachment are subject to variability and are often nonuniform (Butler 2006). The challenge to meeting today's demands on quality assurance and effectiveness is making MAb products with consistent glycosylation patterns. Unfortunately, with current technology, glycosylation patterns can only be determined post-production so that if a batch fails to meet the quality benchmarks for glycosylation patterns, the entire lot must be discarded due to regulatory concerns. Although on-line control of glycosylation during production is to be preferred, strategies for ensuring consistent product quality during production are yet to be implemented broadly in the pharmaceutical industry in general. Especially for bioprocesses, the reasons are mostly attributable to the complexity of these processes, strict government regulations on making changes to validated processes, the nonavailability of on-line measurements, and the lack of comprehensive and effective control strategies.

This case study is concerned with the development of a comprehensive strategy for effective on-line control of protein glycosylation during the manufacture of MAbs, using a combination of multiscale modeling, hierarchical control, and state estimation. The experimental system is used to illustrate how to implement the strategic vision for biopharmaceutical manufacturing presented in Section 2.4, presenting results to date on the development—and experimental validation—of portions of the proposed strategy.

2.5.1 PROCESS AND DATA ACQUISITION

When establishing a MAb production process that will consistently produce protein with a desired glycosylation pattern, it is important to identify first the key process variables affecting this CQA. Unfortunately, much of a cell's internal regulatory mechanisms for glycosylation remain unknown. However, the final glycan structure is known to depend on glycosylation reaction site accessibility, glycosylation enzymes, intracellular sugar nucleotide donor availability, and the residence time in the Golgi apparatus (Butler 2006). Although these internal cellular conditions (which exert the most significant effect on glycosylation) are impossible to control directly on-line with the currently available technology, many studies have shown that glycosylation can be influenced indirectly by such external culture conditions as pH, temperature, DO, agitation rate, and nutrient and metabolite media concentrations (see Table 2.2).

Consequently, probes and analyzers for measuring these key process variables (ideally in real time) are essential to the successful implementation of PAT. Fortunately, steam sterilizable probes are currently available for on-line measurements of DO, temperature, and pH; nutrient and metabolite concentrations are available at-line through automated sampling systems connected to bioanalyzers.

In addition, a method must be developed for characterizing the glycosylation profiles of the produced MAbs—the primary product quality attribute of interest. Unfortunately, because current glycosylation characterization methods usually involve either high-performance liquid chromatography or liquid chromatography coupled with mass spectroscopy—methods that require extensive sample preparation—glycosylation characterization measurements can only be available at best off-line and at best once a day.

The equipment used for this case study involves a parallel bioreactor system (DASGIP, Jülich, Germany) with capability for temperature, pH, DO, feed rate, and

---

TABLE 2.2
Effects of Culture Conditions on Glycosylation

Process Variable	Effect on Glycosylation
Low glucose concentration	Reduced glycan site occupancy (Hayter et al. 1992)
Low glutamine concentration	Decreased sialylation; increased hybrid and high mannose glycans (Wong et al. 2005)
Ammonia accumulation	Reduced glycan site occupancy (Borys et al. 1994); decreased terminal sialylation (Yang and Butler 2002)
pH	Variations in degree of galactosylation (Muthing et al. 2003)
Low temperature	Increased glycan site occupancy (Gawlitzek et al. 2009)
Low dissolved oxygen	Reduced galactosylation levels (Kunkel et al. 2000)
High agitation rate	Reduced glycan site occupancy (Senger and Karim 2003)

agitation control. The bioreactor is integrated through an OPC interface with a bio-analyzer (Nova BioProfile 100+) that provides measurements of the metabolites. (The overall system, discussed in more detail in Section 2.5.3, is shown in Figure 2.5.)

Once the appropriate probes, analyzers, and assays are in place for obtaining measurements of DO, pH, temperature, agitation, glucose, and glutamine media concentrations, and for glycosylation characterization, appropriate bioreactor operating conditions must subsequently be determined. To do this systematically requires the performance of carefully designed experiments explicitly for quantifying the relationship between the key process variables and glycosylation patterns, from which one can then deduce what process conditions will lead to the production of a MAb with the desired glycosylation pattern. The recommended approach is as follows: first, the most critical factors affecting glycosylation should be identified with a screening design such as the Plackett–Burman or a fractional factorial design (Ogunnaike 2009, Chapter 19). Once the most critical factors have been identified, a response surface design, such as the Box–Behnken design, should be used to quantify the relationship between the critical factors and glycosylation. The resulting quantification can then be used to determine the optimal combination of variables required to achieve the target glycosylation pattern.

2.5.2   Information Processing

Other than the need to filter raw measurements to reduce the effects of noise, the defining issue with this case study is how to deal with the infrequent availability of glycosylation measurements. With the ultimate goal of ensuring the consistent production of MAbs with a desired glycosylation pattern, it is essential to obtain information about glycosylation sufficiently frequently for feedback control to be effective. Unfortunately, glycosylation measurements that are only available once a day will not be sufficient for traditional feedback control to be effective. The alternative approach we advocate is based on the following premise. Because culture conditions are known to affect glycosylation significantly, important information about the glycosylation pattern can be obtained by analyzing the available process measurements properly. A mathematical model representative of how these bioreactor conditions affect glycosylation pattern may therefore be combined with available process data to provide robust predictions of the glycosylation pattern in real time.

To achieve its objective with adequate fidelity, the mathematical model must capture the multiscale nature of glycosylation: the macroscale, where cells are grown in the bioreactor; the mesoscale, where the cells utilize the nutrients in the feed to carry out metabolic reactions and manufacture the protein; and finally, the microscale, where the glycosylation reactions occur in the ER and Golgi apparatus (see Figure 2.3).

Because no model is perfect, the predictions of this high-fidelity model will have to be augmented (and updated) judiciously with the delayed and infrequent measurements of glycosylation whenever they become available. This is accomplished by the design and implementation of a state estimator (Ogunnaike and Ray 1994) that will provide robust estimates of glycan distribution, using the process model, and all available process measurements, including the infrequent and delayed glycosylation measurements themselves. The estimated glycosylation characteristics can then be used in a model-based control scheme discussed in more detail in Section 2.5.3.

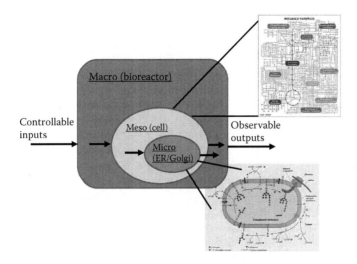

FIGURE 2.3 Overview of multiscale model for the prediction and ultimately control of glycosylation patterns on monoclonal antibodies with external culture conditions. The model includes macro (bioreactor), meso (cell), and micro (ER/Golgi apparatus) levels.

Work to date has produced a version of the macroscale and mesoscale aspects of the model, where data on culture conditions are used to predict the metabolic behavior of CHO cells during the growth phase. The model, based on one proposed by Provost and Bastin (2004), involves two initial substrates (glucose and glutamine), four terminal products (lactate, NH_4^+, CO_2, and alanine), twelve internal metabolites, and two terminal intracellular metabolites. The model has been validated against experimental data obtained from our DASGIP/Nova bioreactor system with CHO K1 cells grown in batch mode with serum-free media at 30% DO and pH 7.3. Figure 2.4 shows that the model prediction of growth, transition, and death phase consumption of glucose and glutamine, in addition to the production of lactate, NH_4^+, and biomass, is in good agreement with experimental data. Future work on the model will involve incorporating the micro (ER/Golgi) model.

2.5.3 PROCESS/PRODUCT QUALITY CONTROL

A major rationale for acquiring process information is to use such information for active control. How to implement effective control is illustrated in this section for the bioreactor system in Figure 2.5.

The DASGIP parallel bioreactor system (DASGIP) is equipped with an incubator for temperature control, along with capability for parallel monitoring of two pH and two DO sensors with temperature compensation; a gassing system for N_2, O_2, and CO_2 gassing of two vessels in parallel; a feeding system that allows up to four feeds per reactor vessel; and magnetic stir plates for agitation control. This is integrated with a Nova BioProfile 100+ with autosampler (Nova Biomedical), which is used to measure glucose (gluc), glutamine (gln), glutamate (glu), lactate (lac), ammonia (NH_4^+), sodium (Na^+), and potassium (K^+) concentrations in the media automatically at specified time

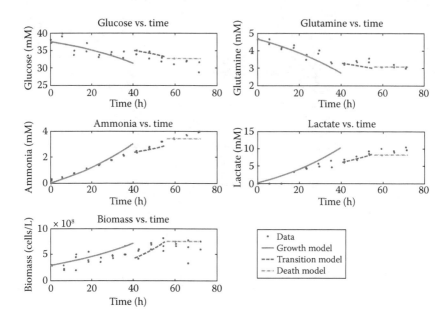

FIGURE 2.4 Macro and meso (bioreactor and cell) models validated with reactor data attained from the DASGIP/Nova bioreactor system growing CHO K1 cells under batch conditions in serum-free media at 30% DO and pH 7.3. Model fits data well for the growth phase, which ends after ~40 h runtime.

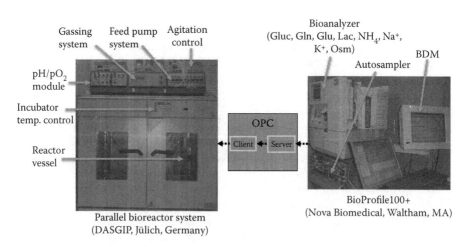

FIGURE 2.5 Overview of the bioreactor system with the ability to control nutrient levels and regularly monitor key cellular metabolites. The system consists of a DASGIP parallel bioreactor system linked to a Nova BioProfile 100+ with autosampler. BDM, bioprofile data management system.

intervals. The two systems are interfaced via OPC so that metabolite data from the Nova BioProfile 100+ are transferred to the DASGIP parallel bioreactor system, facilitating implementation of feedback control of the nutrients and metabolites.

At a minimum, base regulatory control must be established for the key process variables, in this case, DO, pH, temperature, and agitation rate, as well as nutrient and metabolite media concentrations. The current state of the art in industrial bioreactor control is limited to pH, DO, temperature, and agitation rate control (Hopkins et al. 2010). Precise control of glucose and glutamine concentrations in the media is not commonplace in either industry or academia, and metabolite measurements such as lactate and ammonia are usually performed off-line via kit assays or blood/gas analyzers and are, therefore, not available for automatic feedback control.

To demonstrate base regulatory control for the comprehensive bioreactor system (the bioreactor itself, the measurement devices, and the OPC linkage between them), CHO K1 cells were cultured in serum-free suspension culture with 30% DO and pH 7.3 and glucose and glutamine media concentrations measured at 3-h intervals over the course of a single 80-h experiment. The typical decrease in glucose and glutamine media concentration with cells grown under batch conditions is shown in Figures 2.6a and 2.6b, respectively (gray circles).

Simple single-input–single-output integral controllers were used to maintain media nutrient concentrations at desired set-points one at a time. Based on the Nova BioProfile readings of media concentration, the controller adjusted either the feed rate of a 50-g/L stock solution of glucose or a 29.3-g/L solution of glutamine. Figure 2.6 shows that even with such a rudimentary control algorithm, the system maintained glucose within 4% of the set-point and glutamine within 25% of the set-point. (The control action, feedstock pump output, is not shown.)

Although base regulatory control allows for precise control of the key process variables, this is not sufficient for effective control of the glycosylation pattern. Figure 2.7 shows a block diagram for the complete multiloop control scheme proposed for effective on-line control of glycosylation.

The two inner feedback loops of the glycosylation control scheme represent base regulatory control of the identified key process variables, which have been demonstrated on the DASGIP/Nova bioreactor system as discussed previously. The indicated outer loop is required for achieving explicit control of the glycosylation pattern using the multivariable controller C_2 designed to manipulate the set-points to the inner loop on the basis of continuous estimates of glycosylation determined from (1) the fundamental multiscale model of glycosylation discussed earlier, and process and bioreactor contents measurements, combined with (2) actual periodic measurements of glycosylation whenever available.

It is important to emphasize that glycosylation by nature is a heterogeneous process. As such, if left uncontrolled, the result will be a batch of manufactured MAb with an unacceptably broad range of glycans that may fail to meet regulatory requirements. The goal of the proposed control strategy is to narrow the glycan distribution, thereby increasing the likelihood that the desired glycosylation pattern is produced. Future work will involve completing the implementation of this comprehensive control scheme and subsequently developing and demonstrating the final component in Figure 2.2 for carrying out process monitoring and improvement.

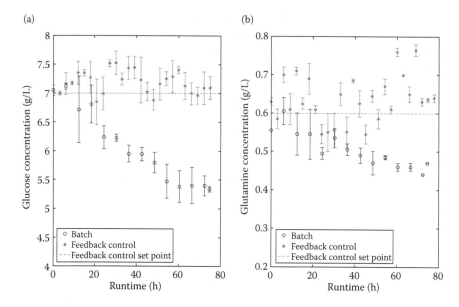

FIGURE 2.6   (a) Glucose and (b) glutamine media concentrations throughout a single 80-h passage with and without feedback control. Glucose ($n = 3$) and glutamine ($n = 2$) are from reactor runs with CHO K1 cells in serum-free suspension culture with 30% DO and pH 7.3. Without control, glucose and glutamine concentrations decrease throughout the 80-h runtime. With control, glucose media concentration is maintained within 4% of the set-point value, on average, whereas glutamine is maintained on average within 10% of the set-point value on average. Data shown are the averages of data obtained from two bioreactor runs and the error bars show the range of data obtained.

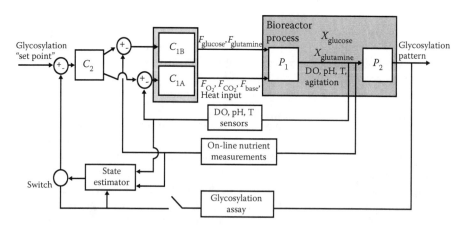

FIGURE 2.7   Proposed strategy for on-line glycosylation control.

2.6 CONCLUSION

In this chapter, we have presented a strategic vision for PAT implementation in biopharmaceutical manufacturing, and illustrated the application of these concepts with a case study involving the manufacture of MAbs in a laboratory-scale bioreactor. We have presented preliminary results demonstrating the establishment of a process integrated with a measurement system, on which base regulatory control has been successfully implemented as part of an overall plan for achieving on-line real-time control of glycosylation in MAbs produced in CHO cells.

It is important to note that on-line glycosylation control is merely one example of how the strategy presented in this chapter can be applied in the biopharmaceutical industry. The same principles apply to any other process and its quality attributes: implementing PAT effectively requires judicious data acquisition, appropriate information processing, effective control, and continuous process improvement.

REFERENCES

Beck, A., E. Wagner-Rousset, M.C. Bussat, M. Lokteff, C. Klinguer-Hamour, J.F. Haeuw, L. Goetsch, T. Wurch, A. Van Dorsselaer, and N. Corvaia. 2008. Trends in glycosylation, glycoanalysis and glycoengineering of therapeutic antibodies and Fc-fusion proteins. *Curr. Pharm. Biotechnol.* 9: 482–501.

Berridge, J., K. Seamon, and S. Venugopal. 2009. A-MAb: A case study in bioprocess development, pp. 1–278. CASSS and ISPE, CMC Biotech Working Group.

Biostrategy. 2005. Biopharmaceuticals—Current market dynamics and future outlook, pp. 1–66.

Borys, M.C., D.I.H. Linzer, and E.T. Papoutsakis. 1994. Ammonia affects the glycosylation patterns of recombinant mouse placental lactogen-1 by Chinese-hamster ovary cells in a pH-dependent manner. *Biotechnol. Bioeng.* 43: 505–514.

Brass, J.M., K. Krummen, and C. Moll-Kaufmann. 1996. Quality assurance after process changes of the production of a therapeutic antibody. *Pharm. Acta Helv.* 71: 395–403.

Butler, M. 2006. Optimisation of the cellular metabolism of glycosylation for recombinant proteins produced by mammalian cell systems. *Cytotechnology* 50: 57–76.

ICH. 2000. International Conference on Harmonisation, Guidance on Q6A specifications: *Test Procedures and Acceptance Criteria for New Drug Substances and New Drug Products: Chemical Substances*, edited by D.O.H.A.H. SERVICES, pp. 83041–83063.

FDA, U.S. Department of Health and Human Services. 2004. Guidance for Industry: PAT—A Framework for Innovative Pharmaceutical Development, Manufacturing, and Quality Assurance. Available online at http://www.fda.gov/downloads/Drugs/GuidanceComplianceRegulatoryInformation/Guidances/ucm070305.pdf.

ICH. 2009. Q8(R2): Pharmaceutical Development International Conference on Harmonization of Technical Requirements for the Registration of Pharmaceuticals for Human Use. Geneva, Switzerland.

FDA. 2009. *Q8(R1) Pharmaceutical Development*, revision 1, edited by C. CDER. Rockville, MD. http://www.fda.gov/RegulatoryInformation/Guidances/ucm128003.htm.

FDA, U.F.a.D.A. 1999. *Q6B Specifications: Test Procedures and Acceptance Criteria for Biotechnological/Biological Products*, edited by C. CDER. Rockville, MD. http://www.fda.gov/downloads/Drugs/guidanceComplianceRegulatoryInformation/Guidances/UCM073488.pdf.

Gawlitzek, M., M. Estacio, T. Furch, and R. Kiss. 2009. Identification of cell culture conditions to control N-glycosylation site-occupancy of recombinant glycoproteins expressed in CHO cells. *Biotechnol. Bioeng.* 103: 1164–1175.

Geyer, H. and R. Geyer. 2006. Strategies for analysis of glycoprotein glycosylation. *Biochim. Biophys. Acta, Proteins Proteomics* 1764: 1853–1869.

Goochee, C.F., M.J. Gramer, D.C. Andersen, J.B. Bahr, and J.R. Rasmussen. 1991. The oligosaccharides of glycoproteins—Bioprocess factors affecting oligosaccharide structure and their effect on glycoprotein properties. *Bio-Technology* 9: 1347–1355.

Hayter, P.M., E.M.A. Curling, A.J. Baines, N. Jenkins, I. Salmon, P.G. Strange, J.M. Tong, and A.T. Bull. 1992. Glucose-limited chemostat culture of Chinese-hamster ovary cells producing recombinant human interferon-gamma. *Biotechnol. Bioeng.* 39: 327–335.

Hopkins, D., M. St. Amand, and J. Prior. 2010. Bioreactor automation. In *Manual of Industrial Microbiology and Biotechnology*, edited by A.L.D.R. Baltz and J.E. Davies, pp. 719–730. Washington, DC: ASM Press.

ICH. 2009. Q8(R2): Pharmaceutical Development. International Conference on Harmonization of Technical Requirement for the Registration of Pharmaceuticals for Human Use. Geneva, Switzerland.

Johnson, M.J., C. Engblom, and J. Borkowsk. 1964. Steam sterilizable probes for dissolved oxygen measurement. *Biotechnol. Bioeng.* 6: 457.

Kunkel, J.P., D.C.H. Jan, M. Butler, and J.C. Jamieson. 2000. Comparisons of the glycosylation of a monoclonal antibody produced under nominally identical cell culture conditions in two different bioreactors. *Biotechnol. Prog.* 16: 462–470.

MacGregor, J.F. and T. Kourti. 1995. Statistical process-control of multivariate processes. *Control Eng. Pract.* 3: 403–414.

Muthing, J., S.E. Kemminer, H.S. Conradt, D. Sagi, M. Nimtz, U. Karst, and J. Peter-Katalinic. 2003. Effects of buffering conditions and culture pH on production rates and glycosylation of clinical phase I anti-melanoma mouse IgG3 monoclonal antibody R24. *Biotechnol. Bioeng.* 83: 321–334.

Ogunnaike, B.A. 1994. Online modeling and predictive control of an industrial terpolymerization reactor. *Int. J. Control.* 59: 711–729.

Ogunnaike, B.A. 2009. *Random Phenomena: Fundamentals of Probability & Statistics for Engineers.* Boca Raton, FL: CRC Press.

Ogunnaike, B.A. and W.H. Ray. 1994. *Process Dynamics, Modeling, and Control.* New York: Oxford University Press.

Ogunnaike, B.A., G. Francois, M. Soroush, and D. Bonvin. 2010. Control of Polymerization Processes, Chapter 12 in *The Control Handbook*, edited by W. Levine, 2nd edition. Boca Raton, FL: CRC Press.

Pavlou, A.K. and J.M. Reichert. 2004. Recombinant protein therapeutics—Success rates, market trends and values to 2010. *Nat. Biotechnol.* 22: 1513–1519.

Provost, A. and G. Bastin. 2004. Dynamic metabolic modelling under the balanced growth condition. *J. Process Control* 14: 717–728.

Schuyten, H. 1967. On-line digital computer control of refinery processes. In 7th World Petroleum Congress, Mexico City, Mexico.

Senger, R.S. and M.N. Karim. 2003. Neural-network-based identification of tissue-type plasminogen activator protein production and glycosylation in CHO cell culture under shear environment. *Biotechnol. Prog.* 19: 1828–1836.

Walsh, G. 2006. Biopharmaceuticals: Approval trends in 2005. *Biopharm. Int.* 19: 58.

Wong, D.C.F., K.T.K. Wong, L.T. Goh, C.K. Heng, and M.G.S. Yap. 2005. Impact of dynamic online fed-batch strategies on metabolism, productivity and *N*-glycosylation quality in CHO cell cultures. *Biotechnol. Bioeng.* 89: 164–177.

Yang, M. and M. Butler. 2002. Effects of ammonia and glucosamine on the heterogeneity of erythropoietin glycoforms. *Biotechnol. Prog.* 18: 129–138.

3 Multivariate Techniques in Biopharmaceutical Process Development

Jeremy S. Conner

CONTENTS

3.1 INTRODUCTION

Multivariate data analysis (MVDA) and process analytical technology (PAT) have been introduced into the process development organizations of the biopharmaceutical industry. There are several purposes for the introduction of these methods into the research and development (R&D) environment. First, the U.S. Food and Drug Administration's PAT initiative (U.S. Department of Health and Human Services et al. 2004) has provided an incentive to include advanced analytical techniques in a manufacturing environment. Management would like data-driven results before entertaining any changes to existing processes, and solid proof of performance must exist before such advanced techniques will be allowed into new processes. The process development organization is an obvious test bed for such techniques, as the implementation details can evolve along with the process while it is still in development. The development of appropriate control strategies for PAT continues to be an activity that is driven in the process development laboratory. Recent industry reviews (Gnoth et al. 2008; Read et al. 2010a) have indicated that the biotechnology industry is making advances in real-time measurement and control of critical quality attributes (CQAs).

Second, it is important to realize that the most important products to be delivered by the process development organization are the data produced from scale-down experiments. A wealth of data is generated from experimental designs, characterizing process performance over a range of conditions. The multivariate analysis techniques that are integral to PAT can be effectively applied to condense a large data package to manageable pieces, allowing visualization by scientists, engineers, and management.

Third, instrumentation that may be valuable in PAT applications can provide deeper insight into the process than has been possible using traditional process development and characterization protocols. Spectral analysis allows scientists to quantify species *in situ*, which were not possible to measure just a few years ago. The improved knowledge leads to better process modeling and process design. System architectures that can provide access to the information available from such instrumentation can be prototyped in the development laboratory (Cimander et al. 2003).

Additionally, multivariate analysis techniques allow batch-to-batch comparisons in a process development environment (Gunther et al. 2008). Although batch-to-batch comparison is a key aspect of manufacturing, it has a different purpose and meaning in an R&D environment, where key process parameters may be changed intentionally. Many traditional batch comparison techniques would immediately recognize these set-point changes as differences and identify the altered parameter as the assignable cause. Process development requires model-based techniques, where batch-to-batch comparisons can be made on the identified input–output relationships defined by the process model. Such comparisons provide for a level of fault detection and diagnostic capability to process development laboratories, leading to improved consistency. It is important to understand the typical development of a scale-down model and the use of this model for process development and characterization. A well-understood process model is key to any PAT implementation. The goal is to obtain a suitable small-scale system that accurately predicts the key process phenomena observed in large-scale systems. Typically, scale-down is performed by changing scale-dependent parameters by a volume ratio and leaving scale-independent parameters unchanged. Oftentimes, the scale-down reactor, which may be as small as 1 L, has similar geometry to the production-scale reactor, but this is not always the case. In cases where the geometries differ, parameters may be scaled based on the understanding of the differences. For example, it may be important to scale gas flow rates based on mass transfer coefficients rather than simply the bioreactor volume. Tescione et al. (2009) presented several methods for scale-down model development and evaluated them using an MVDA approach.

3.2   STATISTICAL TREATMENT OF DESIGN OF EXPERIMENT RESULTS

The goal of a properly designed experimental campaign is to maximize the generated information content. Experiments can be designed to maximize the power of statistical analysis to answer a specific question or to find an optimal operating condition in a minimal number of experiments. Mandenius and Brundin (2008) offer an excellent review of design of experiment (DOE) methodologies that are useful to the biotech industry. A good scale-down model should include sensitivity and uncertainty analysis. The inclusion of sensitivity analysis and the impact of uncertainty are considered to be good modeling practice and provide a means to evaluate the model's limits and usefulness (Sin et al. 2009).

In process characterization activities, data are usually generated at laboratory scale, and the results from these DOE campaigns are verified at pilot scale. A typical cell culture process characterization campaign would involve DOE at 2-L scale, followed by pilot runs at 2000-L scale. Kirdar et al. (2007) showed that MVDA can leverage small-scale mammalian cell culture DOE data to guide process characterization. Importantly, the analysis was shown to be useful in predicting pilot-scale results using laboratory-scale experiments.

During characterization campaigns, abnormal operation must be identified even though only a small number of calibration runs at the center point operating condition may be available (Gunther et al. 2007). Small-scale process development laboratories create a unique problem in fault detection and diagnosis. It is important that any pattern matching technique or attempt at fault detection does not pick up false positives from DOE data that were intentionally designed to operate at different set points. Gunther et al. showed that DOE experiments can be rapidly screened using partial least squares (PLS) and principal component analysis (PCA) modeling techniques to determine outliers (Gunther et al. 2008, 2009a). After generating models of input–output relationships for the process, these models can be compared to determine which experiments are exhibiting markedly different behavior.

3.2.1 Multivariate Design Space

The multivariate design space should be chosen such that the end effect on product CQAs can be determined. Rathore and Winkle (2009) provide a summary of how the design space can be determined. Working backward from the target product profile, the allowable variation (or permissible ranges) of CQAs can be determined. Once the CQAs are identified, and relevant specifications have been established, the product design space can be defined. Finally, key process parameter operating ranges can be defined using DOE methods in a process development laboratory. Figure 3.1 shows a two-dimensional representation of design space considerations. The normal operating range is defined by only a small variation in parameters. However, approval to stray from this normal range may be acceptable with the support of process development data over the characterized range. Such a claim can only be made with a full understanding of the process over the characterized range. There may be equipment constraints limiting some regions of the characterized range, as represented in Figure 3.1.

Although the process design space may allow some variation of parameters affecting CQAs, the optimum operating condition should still be investigated using factorial design techniques. Such techniques can be used to investigate interdependencies of controlled variables and CQAs (Mandenius and Brundin 2008). Harms et al. (2008) describe the use of failure modes and effect analysis to determine key

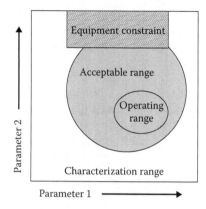

FIGURE 3.1   Multivariate design space showing constraints.

parameters and variables to be tested in scale-down models. Full factorial or fractional factorial DOEs are used to test the effect of indicated variables on CQAs.

Quality by design (QbD) implementation based on the concept of a design space provides opportunity for process improvements within the scope of the original product design (Rathore and Winkle 2009). According to the ICH Q8R2 guidance (U.S. Department of Health and Human Services et al. 2009), process alterations within the design space are not considered a change. However, regulatory post-approval change procedures could be triggered by alterations that deviate outside the design space. In the context of QbD, it is up to the applicant to propose the design space, which is then assessed and potentially approved by the appropriate regulatory body.

3.3   FAULT DETECTION AND IDENTIFICATION

Once a suitable process model has been identified, an obvious application is process fault detection and identification. Process fault detection could be considered a basic form of PAT with manual intervention at the process control point. The intent is to use plant data in combination with some model of expected process behavior to provide plant operators with information concerning the suitability of the current process to meet expectations. Major biopharmaceutical manufacturers including Genentech, Amgen, Wyeth, and MedImmune have reported using multivariate statistical techniques for in-process fault detection in manufacturing processes (Rathore et al. 2007). The focus in the next section will be on the PCA approach, which may be used in process development to identify important correlations among process data.

3.3.1   PCA MODELING APPROACHES

PCA is a valuable tool for analysis of multivariate data. It is well suited to large data sets such as those found in bioprocess industries. PCA, as formulated by Pearson (1901), seeks to find "lines and planes of closest fit to systems of points in space." In effect, PCA is performing a dimensionality reduction by transforming the data set into a model of smaller dimensionality by defining the correlation structure. PCA techniques have been successful in many scientific fields of study. A common use of PCA as a tool is to separate systematic structure in a large data set from the inherent noise, or as a data reduction technique (Wold et al. 1987).

Application of PCA to bioprocess data has been reported in many sources (Chiang et al. 2006). A basic modification is necessary to account for multiple batches of data, and the technique is commonly called multiway PCA (MPCA). The specific applications in process development often revolve around fault detection and diagnosis. Gunther et al. (2006) developed PCA models and identified faults using Hotelling's T^2 statistic for fed-batch fermentation (Gunther et al. 2006) and mammalian cell culture (Gunther et al. 2007) processes. Several process faults were detected and identified using the technique. A commercially available software package, SIMCA-P+ (Umetrics 2005), has proven useful for MVDA of process development data (Kirdar et al. 2007). The SIMCA-P+ approach to MVDA has proven useful for root cause analysis in a process development environment (Kirdar et al. 2008).

3.3.2 BATCH-TO-BATCH COMPARISONS

It is of particular interest to identify batches that appear to be performing abnormally in the laboratory. Early detection of these batches can either lead to corrective action or provide data about the source of the abnormality. Thus, batch-to-batch comparisons are as important in R&D as they are for obvious reasons in a manufacturing setting. Identification and control of batch-to-batch variability can contribute greatly to process optimization. Combining specialized instrumentation such as on-line autosamplers or spectral analyzers, which are described in a later section, with the powerful technique of PCA allows researchers to identify batch-to-batch variability and predict the productivity of the culture (Cimander et al. 2002).

Particular care must be taken when comparing batches across scales. Geometry differences, heat and mass transfer limitations, and operational differences can all indicate major variability in batch data across scales. When scaling up a process, it is vitally important to scale on the correct variables such that a suitable, representative multivariate model is valid across scales (Tescione et al. 2009). Developments of multivariate models that are suitable across scales are invaluable for troubleshooting and enabling R&D scientists to assist in manufacturing investigations (Kirdar et al. 2008).

A novel method for detecting differences among bioprocess batches using PCA similarity factors was presented by Gunther et al. (2009a). The PCA similarity factor compares the angles between loadings of two different PCA models (Krzanowski 1979). Graphically, if we have two PCA models, they are similar if the directions of the principal components are nearly the same, and they are different if the model's principal components are markedly different, as illustrated in Figure 3.2. It is important to note the advantages offered by individual batch comparisons, as large reference data sets are not necessary.

The PCA similarity factor is conveniently bounded between 0 and 1. Allowing $\theta_{a_1 a_2}$ to represent the angle between loading a_1 of model 1 and loading a_2 of model 2,

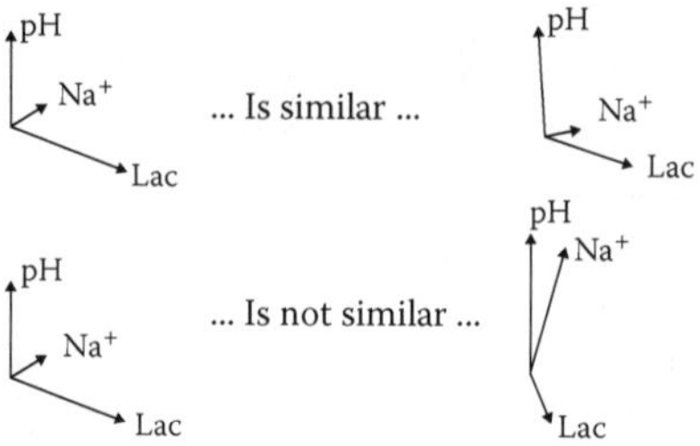

FIGURE 3.2 PCA model similarity comparisons.

the PCA similarity factor is defined by Equation 3.1. Each model has A principal components.

$$S_{\text{PCA1,2}} = \frac{1}{A} \sum_{a_1=1}^{A} \sum_{a_2=1}^{A} \cos^2 \theta_{a_1 a_2} \tag{3.1}$$

A matrix formulation of the PCA similarity factor is presented in Equation 3.2. P_1 and P_2 are the loading matrices of models 1 and 2, respectively.

$$S_{\text{PCA1,2}} = \text{trace} \frac{P_1' P_2 P_2' P_1}{A} \tag{3.2}$$

One problem with the standard similarity factor is that all principal components are weighted equally, even though the last few components may not account for much of the variability in the process data. A modification of the standard similarity factor was proposed by Johannesmeyer (1999) and applied by Singhal and Seborg (2002), where each loading is weighted by its respective eigenvalue, thus leveling the variances in the comparison. The modified PCA similarity factor is shown in Equation 3.3 and again in matrix form in Equation 3.4. Here, λ_{1,a_1} represents the eigenvalue for the a_1th principal component of model 1

$$S_{\text{PCA1,2}}^{\lambda} = \frac{\displaystyle\sum_{a_1=1}^{A} \sum_{a_2=1}^{A} \lambda_{1,a_1} \lambda_{2,a_2} \cos^2 \theta_{a_1 a_2}}{\displaystyle\sum_{a=1}^{A} \lambda_{1,a} \lambda_{2,a}} \tag{3.3}$$

$$S_{\text{PCA1,2}}^{\lambda} = \text{trace} \frac{\left(P_1^W\right)' \left(P_2^W\right)\left(P_2^W\right)' \left(P_1^W\right)}{\displaystyle\sum_{a=1}^{A} \lambda_{1,a} \lambda_{2,a}}, \tag{3.4}$$

where P_i^W are the weighted loadings for model i, defined by Equations 3.5 and 3.6.

$$\Lambda_i = \begin{bmatrix} \sqrt{\lambda_{i,1}} & 0 & 0 & 0 \\ 0 & \sqrt{\lambda_{i,2}} & 0 & 0 \\ 0 & 0 & \ddots & 0 \\ 0 & 0 & 0 & \sqrt{\lambda_{i,A}} \end{bmatrix} \tag{3.5}$$

$$P_i^W = P_i \Lambda_i . \tag{3.6}$$

Batch-to-batch comparisons using S_{PCA}^{λ} have proven quite useful in the process development environment for detecting process differences (Gunther et al. 2009a). Through the use of different portions of the entire data set, the technique can identify certain process behaviors by clustering, which can lead to a starting point for process troubleshooting during investigations (Conner et al. 2008).

3.4 PREDICTIVE MONITORING IN PROCESS DEVELOPMENT ENVIRONMENT

Another key research topic in process development organizations is the development of predictive monitoring techniques. Estimation algorithms for analytes and properties that cannot be readily measured are quite valuable in assessing the state of the process. The term "soft sensor" has been used to describe such methods, where software analyses of physical measurements are used to infer the value of another measurement. One of the most common methods in use for this application is the PLS algorithm. PLS methods and applications will be described in the next section.

3.4.1 PLS TECHNIQUES

PLS models have a definite input–output structure, in contrast to PCA models as described above. PLS models can be quite useful when estimating a particular measurement from a large data set. The PLS model defines correlations in the model structure, and with this information, a model for the unmeasured component may be developed. PLS application in the bioprocess industry is likely more widespread than PCA application.

As was considered in Section 3.3.1, data for multiple batches are often combined to form a reference model. The multiway PLS approach was developed to deal with batch processes (Nomikos and MacGregor 1995). Ündey et al. (2003) consider several different methods for structuring the batch data and develop a framework for batch performance monitoring using PLS. In this framework, on-line quality prediction and identification of contributing factors can be achieved (Ündey et al. 2003). Quantitative values for fermentation CQAs can be estimated using PLS. Successful prediction of end-of-batch product concentration using on-line process measurements and PLS modeling has been reported (Lopes and Menezes 2003; Gunther et al. 2009b). Cimander et al. (2002) used process data from an electronic nose device to estimate quantitative predictions of *Escherichia coli* culture quality using PLS modeling. Process variables such as time of phosphate limitation and tryptophan yield coefficient, which would be unmeasurable by traditional on-line measurement techniques, could be estimated in this manner and have proven to have an impact on the culture quality (Cimander et al. 2002).

From a PAT perspective, the PLS technique can be quite useful. PLS modeling of specific batch components could allow a suitable control strategy to be designed. Multivariate control strategies designed around a PLS model of the process have demonstrated success at pilot scale (Cimander and Mandenius 2004).

In a similar manner, as was presented in Section 3.3.2, similarity factors can be defined for PLS models. These similarity factors can be used to detect and diagnose abnormal batches, even in a process development environment (Gunther et al. 2008). Rapid model-based identification of similarities and dissimilarities among large experimental campaigns is helpful in determining the validity of individual batches, in the diagnosis of potential sources of variability, and in process troubleshooting.

3.4.2 Spectral Techniques

Recent development of at-line sampling and spectral analysis instrumentation has led to an entirely different range of multivariate analysis techniques than those described above. Here, we find use for some of the more traditional chemometrics techniques that until now have proven useful only in analytical laboratory settings. It is possible to monitor individual components within a complex solution using these techniques. Near-infrared spectroscopy (NIR), high-performance liquid chromatography (HPLC), and traditional single-analyte measurements have the capability to provide a wealth of information concerning culture metabolites or product composition and quality. Placed in a rapid at-line or on-line analyzer, these tools are powerful in determining the state of the process at any given time. In a process development environment, such knowledge is particularly valuable when optimizing a process or developing techniques to control the process state or trajectory.

Given a model to estimate process CQAs from spectral data, these attributes can be monitored over time. Development and testing of a model and estimation technique for CQAs could require significant process development resources. Process spectral signatures are expected to change over time, as the underlying process progresses toward completion, as shown in Figure 3.3. If a robust model has been developed, and the spectral data are suitably rich, then the progression of CQAs can be tracked and potentially tied into a feedback control scheme.

Although these techniques are ideally suited to measure components in complex solutions, a common problem still tends to be the sampling and measurement environment. Changing culture conditions such as temperature, pH, and agitation can lead to unexpected quantitative changes in the measurements (Skibsted et al. 2001). Additional complexity in the model or using separate models for different phases of the process may help to alleviate some of these issues (Ödman et al. 2009). As a practical matter, the on-line sampling mechanism (i.e., probe) can present problems with some fermentation cultures. On-line spectral measurements show much promise for

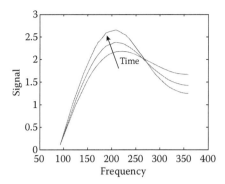

FIGURE 3.3   Spectrum progression in time.

the future of R&D for understanding culture conditions and development of novel techniques for controlling the culture state.

Marose et al. (1998) showed the utility of two-dimensional fluorescence spectroscopy in a wide range of cultures. Skibsted et al. (2001) showed that it is possible to measure a variety of culture properties in a *Pseudomonas fluorescens* culture using multiwavelength fluorescence spectroscopy and PLS models calibrated to the properties of interest. They were able to estimate CO_2 and O_2 composition in the exhaust gas, succinate, protein, optical density, and nitrate in the culture medium, even though these analytes themselves may not possess a fluorescence characteristic directly.

NIR is one of the most widely studied techniques for on-line analysis of the fermentation process. Recent reviews by Scarff et al. (2006) and Cervera et al. (2009) survey the potential applications for this promising technology. NIR sampling suffers from many of the same problems mentioned previously, including the impact of poor signal quality in some spectral regions (Arnold et al. 2002; Ge et al. 1994; Tamburini et al. 2003). NIR spectra are especially sensitive to temperature variation, and care must be taken to avoid misinterpretation of the signal due to temperature fluctuations (Cozzolino et al. 2007).

Development of on-line spectral-based monitoring and control techniques continues to be important for the industry. NIR-based feedback control has been demonstrated (Cimander and Mandenius 2004) and shows much promise, but work remains to be done in process development laboratories to overcome the difficulties around signal-to-noise, sampling, model development, and transfer. Potential future applications exist in all segments of the biopharmaceutical production process. In addition to the primarily upstream cell culture examples indicated above, advanced multivariate techniques could greatly benefit the downstream purification process, the formulation process, as well as the fill/finish process.

3.5   PAT APPLICATIONS

Application of PAT-enabling technology is quite widespread throughout the industry. PAT-enabling technologies include many of the multivariate analysis techniques described above, as well as specialized instrumentation designed to measure a specific

CQA. However, closed-loop control utilizing advanced measurement or analysis of the controlled variable is still a relatively new concept. For almost 25 years, various researchers in biotechnology process development laboratories have investigated the utility of PAT-like feedback control. Read et al. (2010a) specify three classes of PAT applications that have been generally acceptable in the biotech industry

1. Feedback control of a CQA
2. Feedback control of parameters directly correlating to a CQA
3. Feedback control of parameters that indicate a particular unit operation is fit for its intended purpose

Early implementations of rapid chromatographic methods to make decisions concerning cell culture harvest and fraction analysis in purification were reported by Low (1986) and Cooley and Stevenson (1992). Figure 3.4 shows the process and information flow for a PAT application in a typical chromatographic separation process. It is important to collect the portion of the eluent peak that is determined to meet a certain quality specification. Using a rapid on-line analyzer, the distributed control system (DCS) can be triggered to divert the process outlet flow as appropriate, ensuring collection of eluent that meets the quality specification. The particular technology selected for the analytical instrument is dependent on the needs of the process. Oftentimes, process development laboratories will be able to test such devices long before they are accepted in a manufacturing environment. The extensive testing in process development laboratories allows scientists and engineers to develop complex models, explore the operating range, and develop confidence in the measurement and analysis.

PAT applications have been proposed for a variety of downstream unit operations in a biotech R&D laboratory (Rathore et al. 2006). In a recent series of studies, several different devices were investigated for the application shown in Figure 3.4. The traditional techniques of capturing the eluent by absorbance at 280 nm and off-line post-analysis of fractionated samples were compared with on-line HPLC (Rathore et al. 2008a). It was found that the PAT implementation was likely to result in more consistent quality than the prior art. However, operational complexity on the manufacturing plant floor would increase, as may the variability in step recovery. In a follow-up study, the authors investigated an alternate technology, ultra-performance

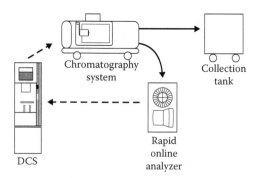

FIGURE 3.4   Chromatography PAT application equipment setup.

liquid chromatography (UPLC) (Rathore et al. 2008b). It was found that the UPLC accuracy compared favorably to the prior HPLC implementation, and the sample analysis time was faster. In applications such as Figure 3.4, sample time is of the essence because the product eluent must be slowed to the point that the diversion valve is reached with a delay equal to the sampling and analysis time.

In the upstream area, PAT applications usually focus around controlling the state of the cell culture or fermentation process, or determining the optimal timing for transfers between units. Controlling the state of the process does not always require specialized instrumentation. An application using generic model control to control a culture of *E. coli* along its optimal specific growth rate profile (Jenzsch et al. 2006) could be considered a PAT application. Here, the controlled variable is the specific growth rate of the process (μ), which is estimated using an extended Kalman filter. The $\mu(t)$ profile is certainly a CQA for the fermentation and, thus, the application meets the criteria for a closed-loop PAT example. Similarly, an application that controls fermentation total carbon dioxide production along a desired profile has been reported (Jenzsch et al. 2007).

Gross environmental conditions in the bioreactor have typically been the controlled variables in upstream applications, but it is becoming possible to measure, predict, and control specific metabolites. Technologies such as on-line HPLC and MPCA have been used in R&D laboratories to investigate amino acid metabolism (Larson et al. 2002). Even with sample intervals significantly longer than what is possible today, an early application by Lenas et al. (1997) used a fuzzy logic controller and on-line HPLC to control glucose and glutamine concentrations in mammalian cell culture. More recently, automated flow cytometry has been used to determine the state of the cell culture to optimize the scale-up process (Sitton and Srienc 2008). NIR and spectroscopic methods as mentioned in Section 3.4.2 have proven useful as a monitoring tool, but model transfer and scale-up issues have rendered elusive the full PAT application using these techniques. Modern chemometric techniques such as MPCA and PLS modeling have brought these applications closer to reality. A closed-loop application where NIR data were correlated to biomass has been reported (Cimander and Mandenius 2004). Ultimately, these techniques may be capable of robustly resolving measurements of individual metabolites, thus enabling multivariate PAT control of specific metabolic pathways.

3.6 FUTURE OF PROCESS CHARACTERIZATION AND TECHNOLOGY TRANSFER

Process characterization will be based on the concepts of QbD and further definition of the process design space. A thoroughly explored process design space, along with the supporting data showing an impact on product design space and CQAs, will be the goal of process characterization. Post-approval regulatory filings will need to be supported by process characterization work showing the effect of the proposed change on CQAs. One potential outcome of the QbD program is that process changes that do not affect CQAs would require less regulatory scrutiny than has been commonplace in the past (Rathore and Winkle 2009).

The future of biopharmaceutical manufacturing will be defined by much tighter cost containment measures than have been seen in the past. In such an environment, manufacturers cannot afford to have cell culture lots go to waste due to process trajectory excursions that cannot be explained. Investment in process monitoring and control technologies will become increasingly important to avoid potential losses (Gnoth et al. 2008). Further research into development of suitable NIR modeling strategies will help the technology to gain acceptance (Scarff et al. 2006).

PAT application is a component of QbD, whereby a means is provided for feedback control of the process based on measurements of CQAs. Process transfer between sites may be facilitated by PAT implementation, as the resulting CQAs can be monitored and controlled (Read et al. 2010b). Small-scale development of process models and standardized implementation of PAT control technology at manufacturing sites will likely reduce the costs of process transfer. Low-cost, flexible manufacturing technologies of the future will require detailed process knowledge derived from models developed in R&D laboratories.

REFERENCES

Arnold, S.A., R. Gaensakoo, L.M. Harvey, and B. McNeil. 2002. Use of at-line and in-situ near-infrared spectroscopy to monitor biomass in an industrial fed-batch *Escherichia coli* process. *Biotechnol. Bioeng.* 80 (4): 405–413.

Cervera, A.E., N. Petersen, A.E. Lantz, A. Larsen, and K.V. Gernaey. 2009. Application of near-infrared spectroscopy for monitoring and control of cell culture and fermentation. *Biotechnol. Prog.* 25 (6): 1561–1581.

Chiang, L.H., R. Leardi, R.J. Pell, and M.B. Seasholtz. 2006. Industrial experiences with multivariate statistical analysis of batch process data. *Chemom. Intell. Lab. Syst.* 81 (2): 109–119.

Cimander, C., T. Bachinger, and C.F. Mandenius. 2002. Assessment of the performance of a fed-batch cultivation from the preculture quality using an electronic nose. *Biotechnol. Prog.* 18: 380–386.

Cimander, C., T. Bachinger, and C.F. Mandenius. 2003. Integration of distributed multianalyzer monitoring and control in bioprocessing based on a real-time expert system. *J. Biotechnol.* 103: 237–248.

Cimander, C. and C.F. Mandenius. 2004. Bioprocess control from a multivariate process trajectory. *Bioprocess Biosyst. Eng.* 26: 401–411.

Conner, J., J. Gunther, and D.E. Seborg. 2008. *PCA/PLS Similarity Factors for Batch-to-Batch Comparisons*. Anaheim, CA: BioProcess International.

Cooley, R.E. and C.E. Stevenson. 1992. On-line HPLC as a process monitor in biotechnology. *Process Control Qual.* 2: 43–53.

Cozzolino, D., L. Liu, W.U. Cynkar, R.G. Dambergs, L. Janik, C.B. Colby, and M. Gishen. 2007. Effect of temperature variation on the visible and near infrared spectra of wine and the consequences on the partial least square calibrations developed to measure chemical composition. *Anal. Chim. Acta* 588: 224–230.

Ge, Z., A.G. Cavinato, and J.B. Callis. 1994. Noninvasive spectroscopy for monitoring cell density in a fermentation process. *Anal. Chem.* 60: 1354–1362.

Gnoth, S., M. Jenzsch, R. Simutis, and A. Lübbert. 2008. Control of cultivation processes for recombinant protein production: A review. *Bioprocess Biosyst. Eng.* 31: 21–29.

Gunther, J.C., J.B. Baclaski, D.E. Seborg, and J.S. Conner. 2009a. Pattern matching in batch bioprocesses—Comparisons across multiple products and operating conditions. *Comput. Chem. Eng.* 33: 88–96.

Gunther, J.C., J.S. Conner, and D.E. Seborg. 2007. Fault detection and diagnosis in an industrial fed-batch cell culture process. *Biotechnol. Prog.* 23: 851–857.

Gunther, J.C., J.S. Conner, and D.E. Seborg. 2008. PLS pattern matching in design of experiment, batch process data. *Chemom. Intell. Lab. Syst.* 94 (1): 43–50.

Gunther, J.C., J.S. Conner, and D.E. Seborg. 2009b. Process monitoring and quality variable prediction utilizing PLS in industrial fed-batch cell culture. *J. Proc. Control* 19 (5): 914–921.

Gunther, J.C., D.E. Seborg, and J.B. Baclaski. 2006. Fault detection and diagnosis in industrial fed-batch fermentation. Proceedings of American Control Conference, Minneapolis, MN, pp. 5511–5516.

Harms, J., X. Wang, T. Kim, X. Yang, and A.S. Rathore. 2008. Defining process design space for biotech products: Case study of *Pichia pastoris* fermentation. *Biotechnol. Prog.* 24: 655–662.

Jenzsch, M., S. Gnoth, M. Kleinschmidt, R. Simutis, and A. Lübbert. 2007. Improving the batch-to-batch reproducibility of microbial cultures during recombinant protein production by regulation of the total carbon dioxide production. *J. Biotechnol.* 128: 858–867.

Jenzsch, M., R. Simutis, and A. Lübbert. 2006. Generic model control of the specific growth rate in recombinant *Escherichia coli* cultivations. *J. Biotechnol.* 122: 483–493.

Johannesmeyer, M.C. 1999. Abnormal situation analysis using pattern recognition techniques and historical data. Master's thesis, University of California, Santa Barbara.

Kirdar, A.O., J.S. Conner, J. Baclaski, and A.S. Rathore. 2007. Application of multivariate analysis toward biotech processes: Case study of a cell-culture unit operation. *Biotechnol. Prog.* 23: 61–67.

Kirdar, A.O., K.D. Green, and A.S. Rathore. 2008. Application of multivariate data analysis for identification and successful resolution of a root cause for a bioprocessing application. *Biotechnol. Prog.* 24: 720–726.

Krzanowski, W.J. 1979. Between-groups comparison of principal components. *J. Am. Stat. Assoc.* 74: 703–707.

Larson, T.M., M. Gawlitzek, H. Evans, U. Albers, and J. Cacia. 2002. Chemometric evaluation of on-line high-pressure liquid chromatography in mammalian cell cultures: Analysis of amino acids and glucose. *Biotechnol. Bioeng.* 77 (5): 553–563.

Lenas, P., T. Kitade, H. Watanabe, H. Honda, and T. Kobayashi. 1997. Adaptive fuzzy control of nutrients concentration in fed-batch culture of mammalian cells. *Cytotechnology* 25: 9–15.

Lopes, J.A. and J.C. Menezes. 2003. Industrial fermentation end-product modelling with multilinear PLS. *Chemom. Intell. Lab. Syst.* 68 (1–2): 75–81.

Low, D.K.R. 1986. The use of the FPLC system in method development and process monitoring for industrial protein chromatography. *J. Chem. Technol. Biotechnol.* 36: 345–350.

Mandenius, C.F. and A. Brundin. 2008. Bioprocess optimization using design-of-experiments methodology. *Biotechnol. Prog.* 24: 1191–1203.

Marose, S., C. Lindemann, and T. Scheper. 1998. Two-dimensional fluorescence spectroscopy: A new tool for on-line bioprocess monitoring. *Biotechnol. Prog.* 14: 63–74.

Nomikos, P. and J.F. MacGregor. 1995. Multi-way partial least squares in monitoring batch processes. *Chemom. Intell. Lab. Syst.* 30 (1): 97–108.

Ödman, P., C.L. Johansen, L. Olsson, K.V. Gernaey, and A.E. Lantz. 2009. On-line estimation of biomass, glucose and ethanol in *Saccharomyces cerevisiae* cultivations using in situ multi–wavelength fluorescence and software sensors. *J. Biotechnol.* 144: 102–112.

Pearson, K. 1901. On lines and planes of closest fit to systems of points in space. *Philos. Mag.* 6 (2): 559–572.

Rathore, A.S., R. Johnson, O. Yu, A.O. Kirdar, A. Annamalai, S. Ahuja, and K. Ram. 2007. Applications of multivariate data analysis in biotech processing. *Biopharm. Int.* 20 (10): 130–144.

Rathore, A.S., A. Sharma, and D. Chilin. 2006. Applying process analytical technology to biotech unit operations. *Biopharm. Int.* 19 (8): 48–57.

Rathore, A.S. and H. Winkle. 2009. Quality by design for biopharmaceuticals. *Nat. Biotechnol.* 27 (1): 26–34.

Rathore, A.S., R. Wood, A. Sharma, and S. Dermawan. 2008b. Case study and application of process analytical technology (PAT) towards bioprocessing: II. Use of ultra-performance liquid chromatography (UPLC) for making real-time pooling decisions for process chromatography. *Biotechnol. Bioeng.* 101 (6): 1366–1374.

Rathore, A.S., M. Yu, S. Yeboah, and A. Sharma. 2008a. Case study and application of process analytical technology (PAT) towards bioprocessing: Use of on-line high-performance liquid chromatography (HPLC) for making real-time pooling decisions for process chromatography. *Biotechnol. Bioeng.* 100 (2): 306–316.

Read, E.K., J.T. Park, R.B. Shah, B.S. Riley, K.A. Brorson, and A.S. Rathore. 2010a. Process analytical technology (PAT) for biopharmaceutical products: Part I. Concepts and applications. *Biotechnol. Bioeng.* 105 (2): 276–284.

Read, E.K., R.B. Shah, B.S. Riley, J.T. Park, K.A. Brorson, and A.S. Rathore. 2010b. Process analytical technology (PAT) for biopharmaceutical products: Part II. Concepts and applications. *Biotechnol. Bioeng.* 105 (2): 285–295.

Scarff, M., S.A. Arnold, L.M. Harvey, and B. McNiel. 2006. Near infrared spectroscopy for bioprocess monitoring and control: Current status and future trends. *Crit. Rev. Biotechnol.* 26: 17–39.

Sin, G., K.V. Gernaey, and A.E. Lantz. 2009. Good modeling practice for PAT applications: Propagation of input uncertainty and sensitivity analysis. *Biotechnol. Prog.* 25 (4): 1043–1053.

Singhal, A. and D.E. Seborg. 2002. Pattern matching in historical batch data using PCA. *IEEE Control Syst. Mag.* 22 (5): 53–63.

Sitton, G. and F. Srienc. 2008. Mammalian cell culture scale-up and fed-batch control using automated flow cytometry. *J. Biotechnol.* 135: 174–180.

Skibsted, E., C. Lindemann, C. Roca, and L. Olsson. 2001. On-line bioprocess monitoring with a multi-wavelength fluorescence sensor using multivariate calibration. *J. Biotechnol.* 88: 47–57.

Tamburini, E., G. Vaccari, S. Tosi, and A. Trilli. 2003. Near-infrared spectroscopy: A tool for monitoring submerged fermentation processes using an immersion optical-fiber probe. *Appl. Spectrosc.* 57: 132–138.

Tescione, L., J. Lambropoulos, K. McElearney, R. Paranadi, R. Kshirsagar, H. Yusuf-Makagiansar, and T. Ryll. 2009. Qualification of a scale down cell culture model for process characterization. AIChE National Meeting, Nashville, TN.

Umetrics. 2005. *SIMCA-P and SIMCA-P+ 11 User Guide and Tutorial, Version 11.0.* Umea, Sweden: Umetrics.

Ündey, C., S. Ertunç, and A. Çinar. 2003. Online batch/fed-batch process performance monitoring, quality prediction, and variable-contribution analysis for diagnosis. *Ind. Eng. Chem. Res.* 42: 4645–4658.

U.S. Department of Health and Human Services, Food and Drug Administration, Center for Drug Evaluation and Research (CDER), Center for Veterinary Medicine (CVM) and Office of Regulatory Affairs (ORA). 2004. *PAT—A Framework for Innovative Pharmaceutical Development, Manufacturing, and Quality Assurance. Guidance for Industry.*

U.S. Department of Health and Human Services, Food and Drug Administration, Center for Drug Evaluation and Research (CDER) and Center for Biologics Evaluation and Research (CBER). 2009. *Q8(R2) Pharmaceutical Development. Guidance for Industry.*

Wold, S., K. Esbensen, and P. Geladi. 1987. Principal component analysis. *Chemom. Intell. Lab. Syst.* 2: 37–52.

4 Analysis, Monitoring, Control, and Optimization of Batch Processes: Multivariate Dynamic Data Modeling

Theodora Kourti

CONTENTS

4.1 INTRODUCTION

The pharmaceutical and biopharmaceutical industry is changing. Terms that reflect the current state are quality by design (QbD), design space, control strategy, process analytical technology (PAT), and process signature. Process modeling is an integral part of the QbD framework. Models for process understanding, statistical process monitoring, and process control are required during the life cycle of the product. Batch or semi-batch processes are dynamic, nonlinear, and of finite duration. Batch process variables are both autocorrelated and cross correlated. These characteristics (due to the nature of the batch process operation itself) should be considered when modeling batch operations for process understanding, statistical process control (SPC), real-time process control, and optimization. Such models may be mechanistic (based on first principles), or empirical (based on appropriate data), or hybrid. Batch unit operations may be described with first principles models when the chemical, biochemical, and physical processes that are taking place in the batch vessel are well understood (e.g., Cinar et al. 2003). However, when this is not the case or it is time-consuming and appropriate data are available, empirical models can be developed.

Models may be developed to address a large number of objectives: process understanding, SPC, real-time control actions, and process optimization. Empirical models can be used to analyze available historical data from past batch runs for process understanding and troubleshooting. Other models may be used for monitoring to establish that the process is in a state of SPC. In this case, a batch is checked against expected behavior. This may be possible in real time as the batch evolves or in "post-run analysis" immediately after the batch has been completed and before being released to market or mixed with another batch. Different types of models are required for each objective, and in the case of data-based models, different types of data are used to derive these models. Therefore, under the very general expression of "multivariate statistical analysis for batch processes," there are vast numbers of problems that can be addressed with a corresponding large number of methods available to address each problem. Table 4.1 gives a list of some of the problems that can be addressed. This chapter covers the foundations for empirical models based on latent variable methods for process understanding, SPC, fault detection and diagnosis, process control, optimization, and product transfer.

TABLE 4.1

Multivariate Statistical Analysis for Batch Processes

Historical data analysis
- Process understanding
- Troubleshooting

Inferential modeling
- Infer final product quality from process conditions during production at the end of the run (no lab testing).
- Infer final quality from process conditions during the run at given time (to calculate control action).
- Infer quality from process measurements in real time as process evolves (soft sensors).

Monitoring: check current batch against expected behavior
– In real time
- Establish an overall "process signature" and monitor it to determine that the batch progresses in a similar fashion with previous batches.
- Check that the process is in a state of SPC.
– Once a batch has been completed
- Use process data and other data to establish that the completed batch is similar to previous ones.

Fault detection
- Detect unusual/abnormal behavior in process, equipment, or product (either in real-time or post-run analysis). Consider appropriate modeling and control limits to detect faults (variable weights, choice of components, type of model).

Fault diagnosis
- Use appropriate methodology (and process knowledge) to diagnose the source of the fault.

Product transfer
- Establish operational knowledge and build appropriate models that can be used for product transfer and scale-up.

Process control
- Decide on mid-course correction of variable trajectories to control final quality.
- Explore feed-forward control possibilities.

Optimization
- Establish best operating conditions to satisfy certain criteria (quality, cost, safety, environmental requirements, etc.).

Multivariate statistical analysis has played an integral part in several industries, enabling process understanding, process monitoring, utilization of real-time analyzers, and real-time product release (Kourti 2005a,b). It is, therefore, appropriate to see it as an integral part in the efforts to address issues such as design space, control strategy, real-time process signature monitoring, process understanding, and correct technology transfer. Multivariate, data-based statistical methods play a critical role in providing solutions to these issues. Several applications in the biopharmaceutical industry have already been reported (Lennox et al. 2001; Ündey et al. 2003, 2004; Larson et al. 2003; Lopes et al. 2004) with recent ones in the PAT and QbD framework by Menezes et al. (2009), Ündey et al. (2010), Ganguly et al. (2011), and Kundu et al. (2011). From determining the acceptability of raw material entering the plant to ensuring quality of the product that leaves the plant, the multivariate analysis philosophy should govern all the operations that take that raw material and convert it to a final product in a cost-efficient way, while meeting safety and environmental constraints, from development to manufacturing to site transfer.

4.2 NATURE OF DATA FOR BATCH AND SEMI-BATCH PROCESSES

The final product in a manufacturing process is the result of several unit operations. For the purpose of this chapter, the expression "batch process" refers to a batch or semi-batch unit operation. Data collected from batch processes come from a variety of sources and cover a range of different formats, such as, for example, properties of raw materials entering the batch process, process variable trajectories, analyzer spectral data collected for the duration of the batch run, and batch product quality data; furthermore, other data such as preprocessing information as well as data indicating the performance of the batch output to subsequent unit operations and to final quality may be available. As a result, very large data sets are accumulated for each unit operation. Furthermore, these data sets, or subsets of them, may be used in different ways to analyze process and product performance behavior. Different types of data (and different types of models) are required to address the problems described in Table 4.1. The types of data typically available are summarized below, and their structure is shown in Table 4.2.

TABLE 4.2
Some Process Data Formats[a]

Matrix Symbol	Dimensions	Explanation
$\mathbf{X}$	$(n \times k)$; two-way matrix	Data from a continuous process, at given instant in time; or summary data from a batch (max T, min T, length of batch run, etc.)
	n observations in time or n batches; k process variable measurements	
$\mathbf{Y}$	$(n \times m)$; two-way matrix	Quality data from a continuous process corresponding to the process measurements in $\mathbf{X}$, properly lagged; or quality data at the end of a batch
	n observations in time or n batches; m product quality values	
$\underline{\mathbf{X}}$	$(I \times J \times K)$; three-way matrix	Data collected from batch process at several time intervals during production
	I batches; J process variables, measured at K time intervals for each batch	
$\mathbf{Z}$	$(n \times r)$; two-way matrix	Raw material; total cycle times; length between processes; preprocessing information
	n observations in time or n batches; r other variable measurements	

[a] Spectral data (NIR, etc.) are not included in this table.

4.2.1 Trajectories

Batch processes have finite duration. Some process variables are being recorded for the duration of the batch at different time intervals, such as, for example, agitation rate (RPM), pH, cooling agent flow (F), and temperatures in different locations in the reactor (T_1, T_2, and T_3). Historical data collected from a batch process for I batches of same duration (same length of time or same number of aligned observations), where all J process variables are measured at K time intervals, or K aligned observation numbers (AONs), can be represented by a three-dimensional data matrix $\mathbf{X}$ ($I \times J \times K$). Including trajectory information, introduces a complex data structure that brings very rich information about the autocorrelation and cross correlation of variables for the duration of the batch, and leads to detailed understanding of the process.

4.2.2 Summary Data

Sometimes, despite the fact that full trajectories are recorded, only summary data such as the minimum, maximum, or the average of the trajectory are reported for the full run or for different phases of the run. Examples of summary data are maximum temperature, average agitation rate, time duration of batch, and slopes of some variable trajectories. With summary data, the detailed information on the trajectories is lost; however, by capturing key characteristics, one might be able to use this information for certain types of simpler models to address certain problems with less modeling effort.

4.2.3 Other Data Related to the Batch Process

Most of the time, the product quality properties ($\mathbf{y}_{it}$) are measured at the end of the batch process at time t. However, these properties are a function of the process conditions at time t, but also a function of the process conditions prevailing several lags before, and in most cases, a function of the conditions prevailing during the entire batch. This is a point that one should consider in the modeling phase. Other data related to a batch process may be preprocessing information, as well as information on raw material properties and impurities. Information may also be available on recipe, other preprocessing data, or even the vessels involved as well as the operators involved. Data may also be collected in the form of spectra from real-time analyzers such as near infrared (NIR).

4.3 INTRODUCTION TO LATENT VARIABLE MODELING (TWO-WAY MATRICES)

Latent variables are most suitable for modeling data retrieved from process databases that consist of measurements on large numbers of variables that are highly correlated. Latent variables exploit the main characteristic of these correlated variables, that is, the effective dimension of the space in which they move is very small. Typically, only a few process disturbances or independent process changes routinely occur, and the dozens of measurements on the process variables are only different reflections of these few underlying events. For a historical process data set consisting of an $(n \times k)$ matrix of process variable measurements $\mathbf{X}$ and a corresponding $(n \times m)$ matrix of product quality data $\mathbf{Y}$, for linear spaces, latent variable models have the following common framework (Burnham et al. 1996):

$$\mathbf{X} = \mathbf{TP}^T + \mathbf{E} \tag{4.1}$$

$$\mathbf{Y} = \mathbf{TQ}^T + \mathbf{F} \tag{4.2}$$

where $\mathbf{E}$ and $\mathbf{F}$ are error terms, $\mathbf{T}$ is an $(n \times A)$ matrix of latent variable scores, and $\mathbf{P}$ $(k \times A)$ and $\mathbf{Q}$ $(m \times A)$ are loading matrices that show how the latent variables are related to the original $\mathbf{X}$ and $\mathbf{Y}$ variables. The dimension A of the latent variable space is often quite small, and it is determined by cross validation or some other procedure (Jackson 1991; Wold 1978).

Latent variable models assume that the data spaces $(\mathbf{X}, \mathbf{Y})$ are effectively of very low dimension (i.e., nonfull rank) and are observed with error. The dimension of the problem is reduced by these models through a projection of the high-dimensional $\mathbf{X}$ and $\mathbf{Y}$ spaces onto the low-dimensional latent variable space $\mathbf{T}$, which contains most of the important information. By working in this low-dimensional space of the latent variables $(t_1, t_2, \ldots, t_A)$, the problems of process analysis, monitoring, and optimization are greatly simplified. There are several latent variable methods. Principal component analysis (PCA) models only a single space ($\mathbf{X}$ or $\mathbf{Y}$) by finding the latent variables that explain the maximum variance. Principal components can then be used in regression [principal component regression (PCR)]. Projection to latent structures or partial least squares (PLS) maximizes the covariance of $\mathbf{X}$ and $\mathbf{Y}$ (i.e., the variance of $\mathbf{X}$ and $\mathbf{Y}$ explained, plus correlation between $\mathbf{X}$ and $\mathbf{Y}$). Reduced rank regression (RRR) maximizes the variance of $\mathbf{Y}$ and the correlation between $\mathbf{X}$ and $\mathbf{Y}$. Canonical variate analysis (CVA), or canonical correlation regression (CCR), maximizes only the correlation between $\mathbf{X}$ and $\mathbf{Y}$. A discussion of these latent variable models can be found elsewhere (Burnham et al. 1996).

Comments for Applications for Batch Processes. The above models are applicable to two-dimensional arrays $\mathbf{X}$ and $\mathbf{Y}$. Summary batch data $\mathbf{X}$ described in Table 4.2 can be analyzed with latent variable methods such as PCA, and when using quality, $\mathbf{Y}$, with PLS, between $\mathbf{X}$ and $\mathbf{Y}$. PCA may also be applied to $\mathbf{Z}$ or $\mathbf{Y}$ matrices of Table 4.2. The choice of method depends on the objectives of the problem; however, all of them lead to a great reduction in the dimension of the problem. Some of them (PCR and PLS) model the variation in the $\mathbf{X}$ space as well as in the $\mathbf{Y}$ space. This point is crucial in most of the applications related to process understanding, process monitoring, and process control (mid-course correction) that are discussed in the following sections, as well as for the problem of treating missing data. The properties of PCA and PLS are discussed briefly in the following sections as well as their use for historical data analysis, troubleshooting, and SPC.

4.3.1 PRINCIPAL COMPONENT ANALYSIS

For a sample of mean centered and scaled measurements with n observations on k variables, $\mathbf{X}$, the principal components (PCs) are derived as linear combinations

$t_i = \mathbf{X}\mathbf{p}_i$ in such a way that, subject to $|\mathbf{p}_i| = 1$, the first PC has the maximum variance, the second PC has the next greatest variance and is subject to the condition that it is uncorrelated with (orthogonal to) the first PC, etc. Up to k PCs are similarly defined. The sample principal component loading vectors $\mathbf{p}_i$ are the eigenvectors of the covariance matrix of $\mathbf{X}$ [in practice, for mean centered data, the covariance matrix is estimated by $(n-1)^{-1} \mathbf{X}^T\mathbf{X}$]. The corresponding eigenvalues give the variance of the PCs [i.e., var $(\mathbf{t}_i) = \lambda_i$]. In practice, one rarely needs to compute all k eigenvectors because most of the predictable variability in the data is captured in the first few PCs. By retaining only the first A PCs, the $\mathbf{X}$ matrix is approximated by Equation 4.1.

4.3.2   PARTIAL LEAST SQUARES

PLS can extract latent variables that explain the high variation in the process data, $\mathbf{X}$, which is most predictive of the product quality data, $\mathbf{Y}$. In the most common version of PLS, the first PLS latent variable $\mathbf{t}_1 = \mathbf{X}\mathbf{w}_1$ is the linear combination of the x variables that maximizes the covariance between $\mathbf{t}_1$ and the $\mathbf{Y}$ space. The first PLS weight vector $\mathbf{w}_1$ is the first eigenvector of the sample covariance matrix $\mathbf{X}^T\mathbf{Y}\mathbf{Y}^T\mathbf{X}$. Once the scores for the first component have been computed, the columns of $\mathbf{X}$ are regressed on $\mathbf{t}_1$ to give a regression vector, $\mathbf{p}_1 = \mathbf{X}\mathbf{t}_1/\mathbf{t}_1^T\mathbf{t}_1$; the $\mathbf{X}$ matrix is then deflated (the $\hat{\mathbf{X}}$ values predicted by the model formed by $\mathbf{p}_1$, $\mathbf{t}_1$, and $\mathbf{w}_1$ are subtracted from the original $\mathbf{X}$ values) to give residuals $\mathbf{X}_2 = \mathbf{X} - \mathbf{t}_1\mathbf{p}_1^T$. $\mathbf{Q}$ are the loadings in the $\mathbf{Y}$ space. In one of the PLS algorithms, $\mathbf{q}_1$ is obtained by regressing $\mathbf{t}_1$ on $\mathbf{Y}$, and then $\mathbf{Y}$ is deflated $\mathbf{Y}_2 = \mathbf{Y} - \mathbf{t}_1\mathbf{q}_1^T$. The second latent variable is then computed from the residuals as $\mathbf{t}_2 = \mathbf{X}_2\mathbf{w}_2$, where $\mathbf{w}_2$ is the first eigenvector of $\mathbf{X}_2^T\mathbf{Y}_2\mathbf{Y}_2^T\mathbf{X}_2$, and so on. The new latent vectors or scores $(\mathbf{t}_1, \mathbf{t}_2, \ldots)$ and the weight vectors $(\mathbf{w}_1, \mathbf{w}_2, \ldots)$ are orthogonal. The final models for $\mathbf{X}$ and $\mathbf{Y}$ are given by Equations 4.1 and 4.2.

4.3.3   USING LATENT VARIABLE METHODS FOR HISTORICAL
DATA ANALYSIS AND TROUBLESHOOTING

Latent variable methods provide excellent tools for data exploration; periods of unusual or abnormal process behavior are easily identified and possible causes for such behavior can be diagnosed. The scores and loadings calculated by PCA and PLS and the weights by PLS can be utilized for this purpose. By plotting the latent variables $(t_1, t_2, \ldots, t_A)$ against each other, the behavior of the original data set [be it raw material ($\mathbf{Z}$), process ($\mathbf{X}$), or quality data ($\mathbf{Y}$)] can be observed on the projection space. By examining the behavior in the projection spaces, one can observe regions of stable operation, sudden changes, or slow process drifts. Outlier and cluster detection becomes also easy, for both the process and the quality space. An interpretation of the process movements in this reduced space can be found by examining the loading vectors $(\mathbf{p}_1, \mathbf{p}_2, \ldots, \mathbf{p}_A)$ in PCA or the weights $(\mathbf{w}_1, \mathbf{w}_2, \ldots, \mathbf{w}_A)$ in the case of PLS, and the contribution plots.

Contribution Plots to Scores. A variable contribution plot indicates how each variable involved in the calculation of that score contributes to it. For example, for process data $\mathbf{X}$, the contribution of each variable of the original data set to the score of component q is given by

$$c_j = p_{q,j}(x_j - \bar{x}_j) \text{ for PCA} \tag{4.3}$$

and

$$c_j = w_{q,j}(x_j - \bar{x}_j) \text{ for PLS between } \mathbf{X} \text{ and } \mathbf{Y},$$

where c_j is the contribution of the jth variable at the given observation; $p_{q,j}$ is the loading and $w_{q,j}$ is the weight of this variable to the score of the principal component q; and $\bar{x}_j$ is its mean value (which is zero for mean centered data).

Comments on the Use for Batch Processes. For a PCA analysis on $\mathbf{X}$ ($n \times k$) or a PLS analysis on $\mathbf{X}$ and $\mathbf{Y}$, each point on a t_1 vs. t_2 plot is the summary of measurements on k variables. For a PCA analysis on $\mathbf{Z}$ ($n \times r$) or $\mathbf{Y}$ ($n \times m$), each point on a t_1 vs. t_2 plot is the summary of measurements on r and m variables, respectively. A PLS can be performed also between $\mathbf{Z}$ and $\mathbf{Y}$ to see any direct effects of raw material variability on the product assuming the process remains constant.

Figure 4.1 gives an example of the t_1 vs. t_2 plot obtained after a PLS between raw material characteristics $\mathbf{Z}$ and yield of a process. The raw material was sourced from different suppliers. It was revealed that the raw material characteristics of a specific supplier project on side B of the graph (black triangles), which is a separate area from the other suppliers that project on A. Moreover, the scatter in the properties in the area B is higher. All the materials produce acceptable quality and pass specifications; however, the one supplier whose material is at the lower part of the specifications tends to have an effect on yield. One can investigate which characteristics (variables) of the raw material are responsible for the change in the yield by using the

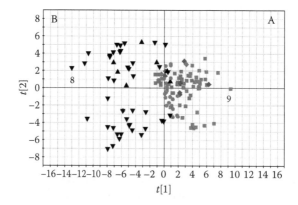

FIGURE 4.1 Projection of raw material characteristics; material from supplier that projects in area B (black triangles) tends to give lower yield than that of other suppliers that project in area A (gray circles, boxes, diamonds). Inverted black triangles are material from one supplier only; black triangles are material from that supplier mixed with others.

contribution plots to investigate which variables are important for the change in the values of the scores corresponding to different suppliers.

The contribution of variable j to the difference of the score values between two observations (say, 8 and 9) for component q is calculated as

$$p_{jq} \times (x_{j,9} - x_{j,8}) \text{ for PCA} \qquad (4.4)$$

and

$$w_{jq} \times (x_{j,9} - x_{j,8}) \text{ for PLS between } \mathbf{X} \text{ and } \mathbf{Y},$$

where p_{jq} is the loading of variable j on component q, and w_{jq} is the weight of variable j on component q (Kourti 2005a).

4.3.4   USING LATENT VARIABLE METHODS FOR STATISTICAL PROCESS CONTROL

From routine operations, we can establish acceptable limits of good process behavior. On a t_1 vs. t_2 plane, such limits will take the form of an ellipse. When the process is in statistical control, the points will be within the ellipse. If there is a problem in the process, the points will plot out of the ellipse.

To monitor the process in real time, however, it would have become cumbersome to have to plot all combinations of PCs (even if we had four components, we would need six charts). A statistic (Hotelling's T^2) can be calculated, and the overall variability of the main events of the system can be monitored with a single chart. The Hotelling's T^2 for scores is calculated as

$$T_A^2 = \sum_{i=1}^{A} \frac{t_i^2}{\lambda_i} = \sum_{i=1}^{A} \frac{t_i^2}{s_{t_i}^2}, \qquad (4.5)$$

where $s_{t_i}^2$ is the estimated variance of the corresponding latent variable t_i. This chart, calculated by using the first A important PCs, essentially checks if a new observation vector of measurements on k process variables projects on the hyperplane within the limits determined by the reference data.

As mentioned above, the A PCs explain the main variability of the system. The variability that cannot be explained forms the residuals [squared prediction error (SPE)]. In some software packages, the term distance to the model (DModX) is used. This residual variability is also monitored, and a control limit for typical operation is being established. By monitoring the residuals, we test that the unexplained disturbances of the system remain similar to the ones observed when we derived the model. For example, a model derived with data collected in the summer may not be valid in the winter when different disturbances affect the system (cooling water temperatures differ, valves may reach limits in capacity of providing heating agent, etc.). It is, therefore, important to check the validity of the model by checking the type of disturbances affecting the system. When the residual variability is out of limit, it is usually an indication that a new set of disturbances have entered the system; it is necessary to identify the reason for the deviation, and it may become necessary to change the model.

SPE_X is calculated as

$$SPE_X = \sum_{i=1}^{k} \left(x_{new,i} - \hat{x}_{new,i} \right)^2, \tag{4.6}$$

where $\hat{x}_{new}$ is computed from the reference PLS or PCA model. Notice that SPE_X is the sum over the squared elements of a row in matrix E in Equation 4.1. This latter plot will detect the occurrence of any new events that cause the process to move away from the hyperplane defined by the refence model.

The above nomenclature applies if the scores were determined from a PCA on the X matrix or a PLS between X and Y. Obviously it can be easily modified for a PCA on Y or Z, or a PLS between Z and Y. It should be emphasized that the models built for process monitoring model only common-cause variation and not causal variation. The philosophy applied in developing multivariate SPC procedures, based on projection methods, is the same as that used for the univariate SPC charts. An appropriate reference set is chosen, which defines the typical operating conditions for a particular process. Future values are compared against this set. A PCA or PLS model is built based on data collected from periods of plant operation at typical operating conditions when performance was good. Periods containing variations due to special events are omitted at this stage. The choice and quality of this reference set is critical to the successful application of the procedure.

The main concepts behind the development and use of these multivariate SPC charts based on latent variables were laid out in early 1990s (Kourti and MacGegor 1995). The calculation of limits for the control charts (Hotelling's T^2 and SPE_X) is discussed in Kourti and MacGegor (1995) and Kourti (2009).

These two charts (T^2 and SPE) are two complementary indices (Figure 4.2); together they give a picture of the wellness of the system under investigation at a glance. As long as the points are within their respective limits, everything is in order. Once a point is detected out of limit, then *contribution plots* can be utilized to give us a list of all the variables that mainly contribute to the out-of-limit point and, hence, allow us to diagnose the problem immediately. Contribution plots can be derived for out-of-limit points in both charts. The reader should be aware that the contribution plots together with process knowledge will help diagnose the problem. The contribution plots may not point directly to the source of the problem but to the variables that respond to the problem. This would be the case if the source of the problem is not a measured variable. For example if a pipe has been plugged and the liquid does not flow freely, one will observe that the pump requires higher effort to push the same amount of liquid through the pipe.

Contributions to SPE. When an out-of-control situation is detected on the SPE plot, the contribution of each variable of the original data set is simply given by $\left(x_{new,j} - \hat{x}_{new,j} \right)^2$. Variables with high contributions are investigated.

Contributions to Hotelling's T^2. Contributions to an out-of-limits value in the Hotelling's T^2 chart are obtained as follows: a bar plot of the normalized scores $(t_i/s_{t_i})^2$ is plotted and scores with high normalized values are further investigated by calculating the variable contributions to the scores as shown in Equation 4.3.

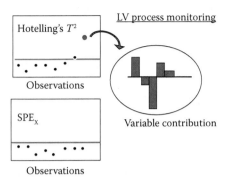

FIGURE 4.2 Latent variable (LV) process monitoring. Two indices are monitored; when either is out of limits, contribution plots show the variable combinations responsible for the deviation.

Variables on this plot that appear to have the largest contributions to it, but also the same sign as the score, should be investigated (contributions of the opposite sign will only make the score smaller). When there are K scores with high values, an "overall average contribution" per variable is calculated, over all the K scores (Kourti 2005a).

Utilizing contribution plots, when an abnormal situation is detected, the source of the problem can be diagnosed such that corrective action is taken. Some actions can be taken immediately, in real time. Others may require interventions to the process. One such example of an abnormal situation appeared in a reactor, in which the reactor temperature should be controlled during an exothermic reaction to 50°C. On a very hot day, the charts indicated abnormalities. Contribution plots pointed to a break in the correlation of cooling water flow and reactor temperature. It turned out that although the cooling water valve was fully open, it could not cope with the demand, as the cooling water was warmer. The valve had to be resized. Multivariate SPC (MSPC) pointed to a problem that had to be corrected. Therefore, the contribution plots are very important tools in understanding factors influencing the process during production and help in an "on-going process understanding" philosophy.

Comments on the Use for Batch Process. Latent variable control charts can be constructed to monitor either a group of response variables **Y** (e.g., product quality variables) or a group of predictor variables **X** (process variables, in the form of batch process summary data such as min, max, and slopes of trajectories). Moreover, multivariate charts can be constructed to assess the consistency of the multivariate quality of raw materials, **Z**, as well as to test the final product **Y** for consistent quality. If there is spectral analysis on some of the materials, then multiblock concepts, discussed later, can be used. A very important advantage of latent variables is that they can be used to monitor predictor variables taking into account their effect on the response variables. A model is built to relate **X** and **Y** using available historical or specially collected data. Monitoring charts are then constructed for future values of **X**. This approach means that the process performance can be monitored even at times when the product quality measurements, **Y**, are not available.

4.4 LATENT VARIABLE MODELING OF BATCH PROCESS TRAJECTORY DATA

As mentioned earlier, batch process summary data forming two-dimensional arrays can be analyzed with the PCA and PLS methodology described above. However, with summary data, detailed information on the trajectories is lost. Such information may be necessary for process monitoring and also for creating models for real-time control purposes (mid-course correction). Including trajectory information introduces a complex data structure as outlined below, and it requires appropriate methodology for modeling. The variable trajectories measured over the duration of the batch are nonlinear with respect to time and form a multivariate time series with dynamic nature. The product quality (properties of the product y_{it}) measured at the end of the batch at time t is a function of the process conditions at time t, but also a function of the process conditions prevailing several lags before, and, in most cases, a function of the conditions prevailing during the entire batch. In modeling, when we deal with dynamic multivariate time series data, to relate input $\mathbf{X}$ to output $\mathbf{Y}$ and to capture the dependence of the final product on events that took place during different time intervals, the $\mathbf{X}$ matrix is expanded to include values of the x variables at several lags (MacGregor et al. 1991). For a batch process where J process variables are measured at K time intervals, or K AONs, for each one of I batches, then if the number of lags is the same for all the x variables, the expanded matrix would be $\mathbf{X}$ ($I \times JK$). The data could also be folded and represented by a three-dimensional data array $\underline{\mathbf{X}}$ ($I \times J \times K$) (Figure 4.3).

Very often, batches do not have the same time duration. Different runs of the same batch process may take different times. Although there are a significant number of batch processes that work with very reproducible cycles, we may occasionally have problems even in these processes. An alignment of the batch trajectories is required before the analysis. This is discussed in detail in a section later. For the discussion here, we assume that the batches have been aligned; therefore, we use time intervals or AONs interchangeably.

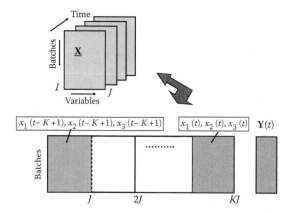

FIGURE 4.3 Quality $\mathbf{Y}(t)$ is a function of variables x_1, x_2, and x_3 at K lags. The slices at different times can be folded and show the appearance of a cube.

In a batch process, we may have the following situations:

1. The measurements may not be equally spaced in time for all the variables (i.e., some variables may be measured every several minutes and others every few seconds).
2. For a part of the batch process that is crucial to monitor, we may have (or may choose to include in the model) more frequent measurements than the rest of the batch (i.e., data obtained every minute during the beginning of the process and every 5 min later in the process).
3. There are variable measurements that may not be available for the entire batch. For example, charge of some materials may happen at short impulses and may be recorded only during the charging period (measure cumulative flow during the charging period); or variables may be measured in the system only for a short duration or infrequently (a sample taken infrequently during the batch or a probe taking measurements infrequently). Other variables may be measured for some duration until they reach values below the limit of detection. It is not correct to place zeroes to the values of these variables when there are no measurements because a correlation structure is forced, which is fictitious. For obvious reasons, the data should not be treated as missing for those periods either. Therefore, in the general case, the batch process may not be able to be represented by a full cube, but rather by a cube with empty columns in the vertical direction. In Figure 4.4, where a series of data sets is shown, $\underline{\mathbf{X}}_1$ and $\underline{\mathbf{X}}_2$ represent batch processes where variables are not measured for all time intervals, whereas $\underline{\mathbf{X}}_3$ is a complete cube (all variables are measured for the same time intervals) similar to that of Figure 4.3.

4.4.1 Modeling Batch Data: Implications of Unfolding

There are several approaches for modeling three-way data derived from batch processes. (Three-way data may be produced by other systems, such as imaging; here we restrict ourselves in methods for batch unit operations.) The choice of the method depends on the use of the model (i.e., prediction of final quality, monitoring, and process control) and the types of the data sets available. Critical discussions on batch process modeling procedures for robust process monitoring, fault detection, and control can be found in selected publications (Nomikos 1995; Westerhuis et al. 1999; Kourti 2003a,b).

In various approaches, the three-way data matrix $\underline{\mathbf{X}}$ ($I \times J \times K$) is unfolded to a two-way array first, and then PCA, or PLS, is applied to this unfolded array. Each

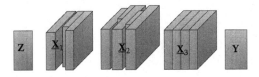

FIGURE 4.4 Representation of raw materials ($\mathbf{Z}$), and quality ($\mathbf{Y}$) variables, and three batch unit operations ($\underline{\mathbf{X}}_1$, $\underline{\mathbf{X}}_2$, and $\underline{\mathbf{X}}_3$). In $\underline{\mathbf{X}}_3$, all the variables are measured for the same time intervals for the duration of the batch; in the other two batch operations, some variables are not measured for the full duration.

of the different rearrangements of the three-way data matrix $\underline{X}$ into a large two-dimensional matrix followed by a PCA on this matrix corresponds to looking at a different type of variability. The following unfolded two-dimensional matrices are most in use for batch processes.

4.4.1.1 Batchwise Unfolding

A method presented by Nomikos and MacGregor (1994, 1995a,b) is termed in the literature "batchwise unfolding." The method unfolds the three-dimensional structure into a two-dimensional array, A $(I \times KJ)$. In this new array, different time slices are arranged next to each other; variables observed at a given time interval are grouped in one time slice; the number of variables in each time slice may vary (Figure 4.3). This arrangement is best for real-time monitoring and process control for mid-course correction. Another way of unfolding results to matrix B $(I \times JK)$; the two matrices are equal with just the columns rearranged. Rearranging the data by matrix B gives a good interpretation to the loadings for process understanding, as discussed later (Figure 4.6). Batchwise unfolding (A or B) can account for variables present and/or measured for a fraction of the duration of the batch and, therefore, is capable of modeling the incomplete cube structure. The number of variables considered for each time interval can vary. Notice that, in this arrangement, one batch is represented by one row of data in the unfolded $\underline{X}$ matrix and the corresponding row in the Y matrix. When L quality data are available per batch, they can be arranged in a Y $(I \times L)$ matrix, and a PLS between A (or B) and Y is straightforward. When information related to raw materials and other preprocessing that took place before the batch (possibly in another vessel) is available, then the data may take a multiblock representation, discussed later (Figure 4.4). Again, one line of data in the preprocessing information matrix Z corresponds to one batch. Modeling with this unfolding takes into account simultaneously both the auto and cross correlation of all the variables. This is because the batch representation by batchwise unfolding allows for analyzing the variability among the batches by summarizing the information in the data with respect to both variables and their time variation. With this particular representation, by subtracting the mean of each column before performing the multiway PCA/PLS (MPCA/MPLS), the average trajectory of each variable is subtracted, and we look at the deviations from the average trajectory. This way a nonlinear problem is also converted to one that can be tackled with linear methods such as PCA/PLS. Finally, the method is capable of modeling three-way structures generated when formulating the control problem of batch processes using latent variables, which is discussed later.

4.4.1.2 Variablewise Unfolding

This results to matrix C $(IK \times J)$ or D $(KI \times J)$—the two matrices are equal with just the rows rearranged (Ündey et al. 2003; Wold et al. 1998). This way of unfolding was discussed in Nomikos (1995): "the only other meaningful unfolding of X is to arrange its horizontal slices, corresponding to each batch, one below the other into a two dimensional X $(IK \times J)$ where the first K rows are the measurements from the first batch in the database. A PCA performed on this unfolded matrix is a study of the dynamic behavior of the process about the overall mean value for each variable.

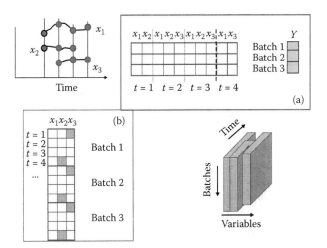

FIGURE 4.5 Example of batchwise (a) and variablewise (b) unfolding, when variable x_3 is not measured at $t = 1$ and variable x_2 is not measured at $t = 4$.

Although this variation might be of interest in some situations, it is not the type of variation of interest in SPC of batch processes."

An example of the implications of unfolding is shown in Figure 4.5, where batchwise and variablewise unfolding are shown for a set of three batches with three variables, when variable x_3 is not measured at $t = 1$ and variable x_2 is not measured at $t = 4$. Batch wise unfolding shown in (a) of Figure 4.5 can describe the process with all the available measurements inserted at the appropriate times. But in the variable wise unfolding shown in (b) of Figure 4.5, notice that when variables are measured or present for a fraction of the duration of the run, this two-way matrix has empty spaces. Therefore, this forces the user to set the values of variables that are present only for a fraction of the length of the batch run equal to zero. They cannot be treated as missing data because they consistently appear at the same location. In another alternative, one could break the batches in phases (each phase with a different number of variables) and to model each phase separately (i.e., one phase for $t = 1$ and variables x_1 and x_2, a second phase for $t = 2,3$ and variables x_1, x_2 and x_3 and a third phase for $t = 4$ and variables x_1, and x_3).

The implications of unfolding variablewise are discussed by Westerhuis et al. (1999), Chen et al. (2002), and Kourti (2003a,b). Several authors mention as an advantage of the variablewise unfolding against the batchwise unfolding the fact that for on-line batch monitoring, one would not have to deal with missing data and forecast the remainder of the trajectory at real-time applications. However, recent studies (García-Muñoz et al. 2004) have shown that excellent results can be obtained with batchwise unfolding and the proper missing data methods.

4.4.2 Analysis of Batch Trajectory Historical Databases

With batchwise unfolding, after performing PCA or PLS utilizing the two-way unfolded matrix **A** or **B**, one obtains scores per batch. Therefore, one can plot scores against each other, as described earlier for two-way data arrays. The loadings of the unfolded matrix present special interest for process understanding.

Figure 4.6 shows the type of information that can be extracted when modeling batch data by batchwise unfolding. Here the loadings of seven variables are plotted for all the time intervals, for the first principal component. First, the loadings for the first variable (reactor temperature) for the entire duration of the batch are plotted, followed by the loadings of the second variable (flow of component A), and so on, for the rest of the variables. These are the loadings obtained from the unfolded matrix **B**. (To avoid confusion, these loadings are the same loadings as those obtained if one uses matrix **A**; they are simply rearranged such that all the time measurements for a variable appear together.) With this setup, the loadings obtained from PCA give practically the history of the process. In this particular case, we are studying a semi-batch process where seven variable trajectories per batch are monitored. The reactor is jacketed, the reaction of component A is exothermic, and cooling water through the jacket is used to remove the heat. The variables plotted are reactor temperature, flow of component A, temperature difference of cooling water in the jacket, the difference of the average temperatures between jacket and reactor, the cooling water flow, the heat removal from reactor, and the agitation speed. Such a figure helps identify some important features of the process. In this particular case, we can notice that the heat removal is positively correlated to the driving force (the temperature difference between reactor and jacket), the jacket water flow, and the cooling water temperature difference. High heat removal corresponds to increased flow of component A.

The loadings obtained with this unfolding give a detailed picture of the auto and cross correlations of the variables for the duration of the batch and, indeed, a detailed picture of the batch operation and the heat balance. If a PLS model is built, by plotting the PLS weights, we can get indications of time–variable combinations responsible for affecting specific product properties (**Y**). This detailed picture, obtained by the loadings due to utilizing the batchwise unfolding of trajectories has

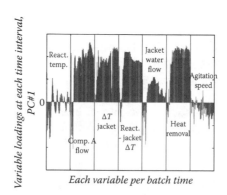

FIGURE 4.6   Variable loadings per batch duration for the first principal component. (After Kourti, T., *IEEE Control Syst. Mag.*, 22(5), 10–25, © 2002 IEEE.)

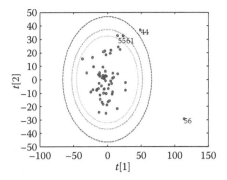

FIGURE 4.7 Score plots, t_1 vs. t_2, for 61 completed batches. The ellipses are the 90%, 95%, and 99% joint confidence regions. (After Kourti, T., *IEEE Control Syst. Mag.*, 22(5), 10–25, © 2002 IEEE.)

been extremely valuable in analyzing numerous batch industrial processes, and it is a feature of multivariate batch analysis highly appreciated by industry.

Scores can be projected on t_1 vs. t_2 space as shown in Figure 4.7, and unusual batches can be detected.

Comments for Bio Applications. The fact that the batchwise approach gives rich information about the process is discussed by other authors. Albert and Kinley (2001) have reported that both unfolding approaches were implemented in their application. However, they mention that batchwise models (which they call the batch-to-batch models) "are discussed in greater detail" in their work, "owing to their superior information content and suitability for on-line control charting." For instance, "loading plots produced by the batch-to-batch model enables the visualization of the complex dynamic correlation structure between variables throughout the batch duration."

4.5 MULTIVARIATE BATCH STATISTICAL PROCESS CONTROL

The analysis of batch process performance and the development of control charts for process monitoring of future batches by utilizing historical data usually involves two stages:

A. Analysis of past batch performance by utilizing data collected from past batches and determination of the good (nominal) operation
B. Checking future batches against good operation

Each one of these stages involves several steps.

Stage A: Historical Data Analysis/Check for Observability of Faults

1. Selecting data from past batches both good and bad.
2. Trajectory alignment.
3. Centering and scaling.
4. Modeling historical data.

5. Checking for observability, that is, making sure that "bad batches" and "faulty operation" as designated by the operators and company personnel can be detected from the collected process measurements. This is an important step to decide if representative data are being collected and how the model will be used for monitoring.
6. Diagnosing the reasons for bad batches and faulty operation in the past using contribution plots.
7. Making proper changes in the process. It is often the case that this initial historical data analysis will point to some problems in the process. Nomikos (1996) reports that reoccurring problems due to a specific fault in the design were detected from historical data analysis. At this stage, if there is room for improving the process, changes in the process operation may be made.
8. Historical data for the next step must be collected after the changes have been made. If no changes are made, the existing historical data can be used in the next stage.

Stage B: Check Future Batches against Good Operation

9. Selecting the good batches from the historical data.
10. Building the model.
11. Determining the control charts to be used for monitoring and calculating the limits for these charts.
12. Making sure that these charts are capable of detecting problems for past batches that produced bad quality product or had unusual operation.
13. Checking future batches against good operation.
    13a. On-line monitoring
    13b. Checking a batch at the end of the run. Some companies choose not to do on-line monitoring but only check the batch behavior right at the end of the run (post-analysis) and release the product based on this information. They are still several hours ahead of the lab analysis.
14. For an out-of-limits batch, using contribution plots, determine the combination of process variables responsible for the deviation.

Figure 4.8 shows the behavior of an unusual batch (batch 56 of Figure 4.7) plotted on a t_1 scores plot. It should be emphasized here that in batch monitoring, the score is plotted against time, and the score value at the end of the run (i.e., at time = 250) is the value that appears in the plot of Figure 4.7. The score value at different time intervals in the duration of the batch is an estimate of the final value (Nomikos and MacGregor 1994, 1995a; Kourti 2003a, 2005a; García-Muñoz et al. 2004).

It must be clarified that by on-line monitoring in this paper, we mean on-line process monitoring, where process operating variables are recorded together with data from on-line analyzers. This leads to multivariate SPC. The term on-line monitoring is also used in the literature in several fields to refer to the use of on-line analyzers to measure one or more properties in the process. Using on-line measurements from on-line analyzers only is equivalent to multivariate statistical quality control

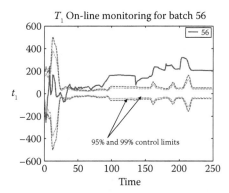

FIGURE 4.8 Value of the score for the first principal component is plotted against time for the duration of batch #56, against normal operation limits. (After Kourti, T., *IEEE Control Syst. Mag.*, 22(5), 10–25, © 2002 IEEE.)

(MSQC). If the quality (property) we monitor goes out of limits, there is a warning that one should investigate the process; but it is difficult to isolate the process problem only from MSQC. On the contrary, when the process variables are monitored alone or simultaneously with on-line quality, the diagnosis and fault isolation becomes significantly easier.

When setting the control charts, the user should keep in mind that by definition, in SPC, we check that the deviation from the target is within certain limits. Therefore, we need to subtract from a known target and check the deviations of the process variables against limits. The known target in the case of batch processes are the average variable trajectories (calculated from a set of training data and from batches corresponding to good operation) or the desired trajectories derived by optimization procedures (Duchesne and MacGregor 2000).

Comments for Bio Applications. A large percentage of the recent industrial applications reported in the literature are using the variablewise unfolding for process monitoring due to the availability of commercially available software that is based on this approach. If the reader is interested in applications utilizing batchwise unfolding, theoretical principles can be found in Nomikos and MacGregor (1994, 1995b) and industrial applications in Albert and Kinley (2001), Zhang and Lennox (2004), and Ferreira et al. (2007).

4.6 BATCH TRAJECTORY SYNCHRONIZATION/ALIGNMENT

Very frequently, batches have variable time duration. In the simplest case, two batches have different lengths, but the trajectories overlap in the common time part. If this is the case, the problem is very easily solved. Provided there are enough long batches, a model can be derived using the information from the long batches while the absent part of the trajectory for the shorter batches is treated as missing data. In the general case, however, the trajectories of the variables have different shapes for most of the common time (in this case, even batches with equal time duration need alignment). This occurs because the progress of the process in the batch (be it reaction, distillation, or fermentation) is a complicated function of several phenomena and not simply a function of time. For example, all other parameters being the same, a batch with a given monomer/initiator/temperature combination will achieve 100% conversion faster if there are no impurities in the system, and the variable trajectories will have different shapes than when impurities are present. Because impurities cannot be measured, the length of the batch and the shape of the trajectory cannot be predicted ahead. Therefore, it is very often the case that the batch trajectories have to be expressed against some other variables (or combination of variables) in order to be aligned. Of course, if the batch length only is a function of some other factor that is known *a priori* (recipe, grade, etc.), then the total length can be predicted, and the time can be scaled appropriately (time divided by expected length). Here, we emphasize the length of the trajectory only because if the covariance structure of the variables is also affected, this linear scaling is not appropriate.

One approach to align batches suggested in the initial work by Nomikos (1995) was to replace time by another measured variable that progresses monotonically in time and has the same starting and ending value for each batch, and to perform the analysis and on-line monitoring relative to its progress. This variable is being termed indicator variable. Such a variable could be an on-line measure of conversion in a chemical reactor or a measure of lance position in injection molding. In fact, any variable that changes monotonically during the batch can be used as an indicator variable, either directly measured as, for example, the cumulative monomer fed (Kourti et al. 1996) or calculated from other measured variables based on process knowledge, such as the extent of reaction (Neogi and Schlags 1998) or another calculated quantity in fermentations (Jorgensen et al. 2004).

However, in some situations, the indicator variable approach is not adequate. Methodologies developed in the speech recognition literature where they encounter similar problems were explored for the alignment of the variable trajectories. Several approaches that utilize dynamic time warping (DTW) methods from that literature and combine it with latent variable methods have been very successful in aligning batches of different durations and allowing for the analysis and diagnosis of operating problems (Kassidas et al. 1998). An evolving or real-time version of the DTW methodology has also been developed to allow for the on-line monitoring of batch processes having these difficult variable duration characteristics (e.g., fermentations and other biological batch processes). After the alignment, the new $\mathbf{X}$ matrix contains the aligned variable trajectories. It was also suggested (Kassidas et al. 1998) to include the total time as an extra variable in the $\mathbf{Z}$ space that usually contains auxiliary information. Taylor (1998) suggested that using the cumulative warped information as an extra variable in the *new aligned* $\mathbf{X}$ space makes DTW much more powerful for fault detection and also easier to use for on-line monitoring. (This is because, for on-line real-time monitoring, the total time of the batch is not known in advance to be included in $\mathbf{Z}$, but the cumulative warp at the current moment is known and treated as any other variable in $\mathbf{X}$.) The same is true when using the cumulative time deviation from average time as an extra variable in the new aligned $\mathbf{X}$ for the indicator variable method. This was applied by Westerhuis et al. (1999) and

García-Muñoz et al. (2008). Ündey et al. (2003) addressed the problem of disconti-nuity in process variable measurements due to operation switching (or moving to a different phase) that causes problems in alignment and modeling. Derivative DTW has also been proposed to avoid singularity points and reduce the bias of alignment results toward the reference trajectory compared with DTW (Zhang and Edgar 2008) and applied also for batch and chromatographic peak alignment (Bork 2011).

When using either DTW or the indicator variable, the raw data are being manipu-lated before being used for the latent variable analysis. The issues related to this manipulation are discussed in Kourti (2003a). A notable publication on real time synchronization of batch trajectories using DTW has appeared recently (González-Martínez et al. 2011).

4.7 CENTERING AND SCALING THE DATA

Based on the definition of multivariate SPC, there should be no doubt that we should check for deviations of the variable trajectory of the current batch from its corre-sponding average trajectory obtained from a set of good batches running at typical operating conditions. That means that the average trajectory (from typical operation) is subtracted from the data, and that we should deal with deviations from the mean at each given time interval. This corresponds to mean centering the data of the two-way batchwise unfolded matrix. This way of mean centering effectively subtracts the trajectory, thus converting a nonlinear problem to one that can be tackled with linear methods such as PCA and PLS.

We should stress here that the above discussion is for modeling that will be used for batch process monitoring; the requirements are different if PCA or PLS is used to model data collected from designed experiments. In that case, several recipes/process conditions are tried in a set of designed experiments and the product quality is obtained for these different conditions. For these cases, chances are that the trajec-tories of the same variable obtained with different recipes are significantly different from each other. If we wanted to relate recipe, trajectory, and product quality, we could still express the trajectories as a function of a monotonic variable (percentage of added reactant A, where A is the main reactant). In this case, the shape of the trajectory does matter as discussed in Duchesne and MacGregor (2000) because we are trying to find the optimal combination of recipe and trajectory to obtain the desired quality product. However, when we are into the monitoring situation, we are in production and, therefore, we have already selected the optimal recipe and the shape of the trajectory, and we wish to keep repeating it, to produce the product with the desired specifications. During production, what prevents us from repeating the exact desired trajectories are disturbances (impurities, problems in raw materi-als, and inability to cool reactor in a very hot day) or unit problems (faulty sensors, polymer plugged sensors, and plugged pipes). These are the problems that we try to detect and isolate by MSPC.

Scaling will also define the problem we try to solve. For a set of dynamic data batchwise unfolded, the two-way array is scaled to unit variance by dividing each column by its standard deviation. Nomikos (1995) discussed this scaling and men-tioned that, "The variables in each column of the unfolded $\underline{\mathbf{X}}$, after they are mean centered, are also scaled to unit variance by dividing by their standard deviation, in order to handle differences in the measurement units between variables and to give equal weight to each variable at each time interval." However, if one wishes to give greater or less weight to any particular variable or to any particular period of time in the batch, these weights are easily changed. Another way of scaling is to scale each variable at each time interval by its overall (throughout the batch duration) standard deviation. Nomikos (1995) reported that "the result from such scaling is that periods with more variability with respect to other periods of the same measurement variable will be weighted more and will have a greater influence in the MPCA model. But if the variability in a particular period is very large, this will result in the rest of this variable's history being ignored in the MPCA model!"

It is the author's experience that scaling per variable means that periods of high noise will be weighted more and periods under tight control will get a small weight. The reader should consider, however, that variables that are under tight control are the ones that matter the most; when a variable is kept under tight control for a given period in a process, this variable is very important in defining the quality (or meeting other constraints in the process—environmental and safety). Hence, small fluctuations are not acceptable for such a variable. It is, therefore, of paramount importance to detect even small deviations for that variable at that crucial period. However, if the variable is given a small weight in the model for that period, this may not be feasible. The effect of centering and scaling is discussed in detail in Kourti (2003a).

4.8   MULTISTAGE OPERATIONS—MULTIBLOCK ANALYSIS

There are multiple steps in pharmaceutical or biopharmaceutical manufacturing, and each step may involve multiple unit operations. One can build a model for the full process that will take into account the interactions between units and their relative importance to the final product quality by weighting them differently. This is the approach of multiblock PLS (MB-PLS).

In the MB-PLS approach, large sets of process variables ($\mathbf{X}$) are broken into meaningful blocks, with each block usually corresponding to a process unit or a section of a unit. MB-PLS is not simply a PLS between each $\mathbf{X}$ block and $\mathbf{Y}$. The blocks are weighted in such a way that their combination is most predictive of $\mathbf{Y}$. Several algorithms have been reported for multiblock modeling, and for a good review, it is suggested that the reader consult the article by Westerhuis et al. (1998).

Multivariate monitoring charts for important subsections of the process, as well as for the entire process, can then be constructed, and contribution plots are used for fault diagnosis as before. In a multiblock analysis of a batch process, for example, one could have the combination of three blocks ($\mathbf{Z}$, $\underline{\mathbf{X}}$, and $\mathbf{Y}$). Block $\mathbf{Z}$ could include information available on recipes, preprocessing times, hold times, as well as information of the shifts (which operator was in charge) or the vessels used (i.e., which reactor was utilized); $\underline{\mathbf{X}}$ would include process variable trajectories; and $\mathbf{Y}$ would be quality variables. Analysis of this type of data could even point to different ways that the operators operate the units and therefore relate product quality improvements to specific; operator, or in other cases could different behavior of vessels, and identify faulty vessels, etc. The reader is referred to the work of García-Muñoz et al. (2003, 2008) for detailed examples where the multiblock analysis is utilized in batch processes for troubleshooting and for determining the batch operating policies to achieve specific product quality while minimizing the duration of the batch run.

Several alternative ways to perform multiblock appear in commercial software. One approach that is being frequently used to deal with a data structure of several blocks involves two stages: PCA is performed for each one of the Z and X blocks, and then the scores and/or residuals derived from these initial models are related to Y with a PLS. In an alternative version, PLS is performed between Z and Y, and X and Y, and the resulting scores are related to Y. The users should exercise caution because these approaches may fail to take into account combinations of variables from different blocks that are most predictive of Y. For example, in situations where process parameters in X are modified to account for variability of raw material properties in Z (i.e., when X settings are calculated as a feed-forward control to deviations of Z), a PLS between Z and Y will show that Z is not predictive of Y variability; similarly a PLS between X and Y will show that X is not predictive of Y; an MB-PLS of $[Z, X]$ and Y will identify the correct model. Finally, MB-PLS handles missing data in a very effective way.

In some multistage operations, the path of the product through the various process units can be traced easily, and eventually, one can relate a specific lot number to several process stages (via an MB-PLS). In such cases, the process conditions of these units can be used to predict the quality of the product. There are situations, however, where a product (or the composition of the effluent stream of a process) is a result of a multistage operation, but its path cannot be traced clearly due to mixing of streams from several parallel units in one vessel and then splitting to a number of other vessels. A discussion on monitoring difficult multistage operations can be found in Kourti (2003b). In those cases, the best alternative to achieve consistent operation is to monitor each unit, separately, by a PCA model. By assuring a consistent operation per unit, one hopes for a consistent product. Once an unusual event is detected in one unit, one may decide not to mix the product further or investigate lab quality before proceeding to the next stage.

4.9 CONSIDERATIONS IN EMPIRICAL BATCH MODELING

4.9.1 OBSERVABILITY

Is the fault observable from the measurements? For an empirical, data-driven monitoring model, where the objective is to detect faults when they occur, the model should be tested with known faults to determine the "observability" of these faults. In common with all on-line monitoring methods, these multivariate SPC methods can only detect "observable" events, that is, events that influence at least one or more of the measured variables. No monitoring procedure can detect events that do not affect any of the measured variables. This is analogous to the requirement of "observability" in state estimation of mechanistic models. Some events that lead to quality problems may pass undetected if they have no impact on the measured variables. The only way of improving this situation is to add new measurements (sensors) that are responsive to these events.

Therefore, when analyzing past data from a set that contained batches that produced good product and bad product, and batches that had questionable operation, we should be able to detect on the projection space (by plotting the scores of MPCA/MPLS against each other), or in the residuals, clusters corresponding to the bad and faulty batches separated from the cluster of the good operation. If this is not possible, it may mean that the measured process variables do not contain the signature of the fault(s) and more representative variables should be collected.

It is, therefore, crucial to test for observability of faults before proceeding to the on-line application of the method. Furthermore, once the model based on the good batches is built, one should test the control charts that will be used for process monitoring, with known faults. Suppose that the SPE chart and the Hotelling's T^2 chart at A components will be used to monitor the process. Then, if the data from a batch that was known to produce a bad quality product pass through the model, the SPE chart, or the Hotelling's T^2 chart, or both should signal the problem. If they do not, then the specific problem cannot be seen with this model (i.e., this set of charts). Sometimes the problem can be corrected by rebuilding the model after weighing some process variables (at certain process periods) with a different weight, or introducing new calculated variables based on process knowledge.

4.9.2 CHOICE OF MPCA OR MPLS FOR BATCH MONITORING AND ANALYSIS

Very frequently when building a model for monitoring (i.e., utilizing good batches only), MPLS shows that only a very small percentage in the variability of Y is explained. This may sometimes be due to high measurement noise, but in general, it is expected that when a batch process runs well, the product variability is not high.

It is the author's experience that for batch process monitoring, MPCA is preferable. By monitoring process variables with MPCA, any deviation from normal operation will be detected, even if this deviation is not related to product quality. For example, a pending equipment failure (such as a faulty valve) will alarm in PCA monitoring. This event may not alarm in a PLS monitoring scheme (because it does not affect quality immediately, the variable might have been excluded from a PLS scheme, or given low weight). Only after the equipment fails then the product quality will be affected. In general, incipient faults are detectable by PCA before affecting product quality.

Sometimes monitoring process data with MPCA is the only alternative. This happens when quality data are not available at certain stages of a process. Sometimes product quality is only determined by the performance of the product later, during another process. Monitoring the batch process variables would detect abnormal situations and would provide an early indication of poor subsequent performance. In these cases, the process data may contain more information about events with special causes that may affect the product structure and, thus, its performance in different applications.

4.9.3   Utilizing Theoretical and Process-Related Knowledge

Common sense implies that empirical methods become more powerful when combined with process-related knowledge and knowledge from theoretical models. Utilizing such knowledge will help determine which variables to include, calculate new variables (transform raw data), and decide on the frequency of data sampling. It is, therefore, always important that process champions work closely with the persons that will perform the statistical analysis unless the analysts have such process knowledge themselves. Knowledge of the specific process as well as engineering knowledge of the unit operations, basic mass, and energy balances, and frequently past experience and common sense can be utilized to convert process variables to new meaningful variables, and also to weigh the variables appropriately for modeling. By carefully constructing new variables from the data, nonlinear relations may be converted to linear ones.

From the early applications in multivariate analysis, researchers have followed the approach of converting some measured variables to new meaningful variables. Energy balances were used in batch process modeling (Nomikos and MacGregor 1994, 1995a); the extent of the reaction (Neogi and Schlags 1998) or the cumulative amount of reactants was added (Kourti et al. 1996) as indicator variables to align batch data. Quantities such as the difference of the outlet and inlet temperature of the cooling water (ΔT_{jacket}) or ΔT_{jacket} multiplied by the cooling water flow are routinely used to get a variable proportional to the heat transferred between jacket and reactor. This way, variability caused by seasonal fluctuations of the cooling water inlet temperature is avoided.

Process-specific information (such as operator shifts, evaporation time, and holding time) has already been used for batch processes (Kourti et al. 1995) and transitions (Duchesne et al. 2002). Specific process knowledge will determine sampling interval in batch analysis, where it is not necessary to have the data equally spaced. Data can be collected more frequently during crucial phases (known from theoretical models and experience) and less frequently in other stages.

Should we include both controlled variables and controller outputs? When we try to build a PLS model between the product quality and the process conditions, we need the controlled variables (i.e., prevailing process conditions). For example, if the temperature in the vessel affects the product quality, we need know the temperature at which the process took place. Chances are that this temperature will not fluctuate much, as it is probably controlled, as a critical process parameter. However, if we try to monitor special events in the process, then we should also include the controller output (e.g., percentage of open cooling water valve and consequently heat flow between jacket and reactor). This information, in combination with other process variables, will give an indication of disturbances.

When it comes to how we are going to weigh the variables, we should think of what we are trying to model and what these variables represent. The weight of a variable depends on the objective of the model prediction, or monitoring, or both. Variables that are perfectly controlled during a batch process (i.e., not too much variability during the process) do not add to predictions of variability of the product, and sometimes people give them low weight or tend to exclude them from the model.

This is acceptable if the model will be used for prediction only. However, it is not a good idea if the model will also be used for monitoring; if there are problems during the process with these variables, these problems will be very difficult to detect if these variables were excluded.

4.10   PROCESS CONTROL TO ACHIEVE DESIRED PRODUCT QUALITY

The term "control" currently appears in the biopharmaceutical literature to describe a variety of concepts such as, end-point determination, feedback control, SPC, or simply monitoring. Process control refers to a system of measurements and actions within a process intended to ensure that the output of the process conforms to pertinent specifications.

Here the terms related to process control are used as follows:

1. Feedback control, to indicate that the corrective action is taken on the process based on information from the process output
2. Feed-forward control, to indicate that the process conditions are adjusted based on measured deviations of the input to the process

4.10.1   FEED-FORWARD ESTIMATION OF PROCESS CONDITIONS

The concept of adjusting the process conditions of a unit based on measured disturbances (feed-forward control) is a concept well known to the process systems engineering community for several decades. The methodology is also used in multistep (multiunit) processes where the process conditions of a unit are adjusted based on information of the intermediate quality achieved by the previous unit (or based on raw material information). An example of a feed-forward control scheme in the pharmaceutical industry, where multivariate analysis was involved, is described by Westerhuis et al. (1997). The authors related crushing strength, disintegration time, and ejection force of the tablets with process variables from both the wet granulation and tableting steps and the composition variables of the powder mixture. They also included physical properties of the intermediate granules. The granule properties may differ from batch to batch due to uncontrolled sources such as humidity and temperature. This model is then used for each new granulation batch. A feed-forward control scheme was devised that can adjust the variables of the tableting step of the process based on the intermediate properties to achieve desirable final properties of the tablets.

To the author's knowledge, there are several unpublished examples in the chemical and other industries where information on the raw data $\mathbf{Z}$ is used to determine the process conditions $\mathbf{X}$ or $\underline{\mathbf{X}}$ to achieve the desired quality $\mathbf{Y}$, utilizing projection methods. Sometimes such information from $\mathbf{Z}$ may simply be used to determine the length of the run, whereas in other cases, it may be a multivariate sophisticated scheme that determines a multivariate combination of trajectories for the manipulated variables. To achieve this, historical databases can be used to develop multiblock models $\mathbf{Z}$, $\underline{\mathbf{X}}$ (or $\mathbf{X}$), and $\mathbf{Y}$.

Kourti (2011) presented a feed-forward scheme utilizing multivariate projection space for a pharmaceutical product. That example illustrates a feed-forward control scheme for unit $N + 1$ based on input information on the "state-of-the-intermediate product" from unit N. The settings for unit $N + 1$ are calculated and adjusted such that the target value for quality $\mathbf{Y}$ is met. A multivariate model was built (from batch data) to relate product quality to the process parameters of unit $N + 1$ and the "state-of-the-intermediate product" from unit N. The state-of-the-intermediate product is a multivariate projection of all the raw materials and the process parameters up to unit N. From this model, a quantitative understanding was developed showing how process parameters in $N + 1$ and the state-of-the-intermediate product from N interact to affect quality.

4.10.2 END-POINT DETERMINATION

There have been reports in the literature where real-time analyzers are used for "end-point detection" or "end-point control." In most of these situations, a desired target concentration is sought as, for example, the percentage of moisture in drying operations.

An example is described by Findlay et al. (2003), where NIR spectroscopy is used to determine granulation end point. The moisture content and particle size determined by the NIR monitor correlate well with off-line moisture content and particle size measurements. Given a known formulation, with predefined parameters for peak moisture content, final moisture content, and final granule size, the NIR monitoring system can be used to control a fluidized bed granulation by determining when binder addition should be stopped and when drying of the granules is complete.

Latent variable methodology allows for taking into consideration the process signatures in a multivariate way for end-point detection problems. Marjanovic et al. (2006) describe a preliminary investigation into the development of a real-time monitoring system for a batch process. The process shares many similarities with other batch processes in that cycle times can vary considerably, instrumentation is limited, and inefficient laboratory assays are required to determine the end point of each batch. The aim of the work conducted in this study was to develop a data-based system able to accurately identify the end point of the batch. This information can then be used to reduce the overall cycle time of the process. Novel approaches based on multivariate statistical techniques are shown to provide a soft sensor able to estimate the product quality throughout the batch and a prediction model able to provide a long-term estimate of the likely cycle time. This system has been implemented on-line, and initial results indicate that it offers the potential to reduce operating costs.

4.10.3 MULTIVARIATE MANIPULATION OF PROCESS VARIABLES

Control of batch product quality requires the simultaneous on-line adjustment of several manipulated variable trajectories such as temperature, material feed rates, etc. Traditional approaches based on detailed theoretical models are based either on nonlinear differential geometric control or on-line optimization. Many of the schemes suggested in the literature require substantial model knowledge or are computationally intensive and, therefore, difficult to implement in practice. Empirical modeling offers the advantage of easy model building.

Lately, latent variable methods have found their way to control of batch product quality and have been applied in industrial problems.

Zhang and Lennox (2004) utilized latent variable methodology for soft sensor development that could be used to provide fault detection and isolation capabilities, and it can be integrated within a standard model predictive control framework to regulate the growth of biomass within a fermenter. This model predictive controller is shown to provide its own monitoring capabilities that can be used to identify faults within the process and also within the controller itself. Finally, it is demonstrated that the performance of the controller can be maintained in the presence of fault conditions within the process.

Work has also been reported for complicated control problems where adjustments are required for the full manipulated variable trajectories (Flores-Cerrilo and MacGregor 2004). Control through complete trajectory manipulation using empirical models is possible by controlling the process in the reduce space (scores) of a latent variable model rather than in the real space of the manipulated variables. Model inversion and trajectory reconstruction are achieved by exploiting the correlation structure in the manipulated variable trajectories. Novel multivariate empirical model predictive control strategy for trajectory tracking and disturbance rejection for batch processes, based on dynamic PCA models of the batch processes, has been presented. The method presented by Nomikos and MacGregor (1994, 1995a,b) is capable of modeling three-way structures generated when formulating the control problem of batch processes using latent variables.

4.10.4 SETTING RAW MATERIAL MULTIVARIATE SPECIFICATIONS AS A MEANS TO CONTROL QUALITY

Duchesne and MacGregor (2004) presented a methodology for establishing multivariate specification regions on raw/incoming materials or components. The thought process here is that if the process remains fixed, we should control the incoming material variability. PLS is used to extract information from databases and to relate the properties of the raw materials supplied to the plant and the process variables at the plant to the quality measures of the product exiting the plant. The specification regions are multivariate in nature and are defined in the latent variable space of the PLS model. The authors emphasize that although it is usually assumed that the raw material quality can be assessed in a univariate manner by setting specification limits on each variable separately, this is valid only when the raw material properties of interest are independent from one another. However, most of the times, the properties of products are highly correlated. To develop models to address the problem, MB-PLS is used for Z, X, and Y. Z contains measurements on N lots of raw material data from the past; X contains the steady-state processing conditions used to process each one of the N lots; and Y contains final product quality for these N lots. The methodology could be easily extended to batch process $\underline{X}$.

It should become one of the priorities in industries to express the raw material orders as a multivariate request to the supplier.

4.11   USING LATENT VARIABLE METHODS FOR OPTIMIZATION

4.11.1   EXPLOITING DATABASES FOR CAUSAL INFORMATION

Recently, there has been a lot of interest in exploiting historical databases to derive empirical models (using tools such as neural networks regression or PLS) and use them for process optimization. The idea is to use already available data rather than collecting new data through a design of experiments. The problem is that for process optimization, causal information must be extracted from the data so that a change in the operating variables can be made that will lead to a better quality product or higher productivity and profit. However, databases obtained from routine operation contain mostly noncausal information. Inconsistent data, range of variables limited by control, noncausal relations, spurious relations due to feedback control, and dynamic relations are some of the problems the user will face using such happenstance data. These are discussed in detail in the section "Hazards of fitting regression equations to happenstance data" in Box et al. (1978), where the advantage of experimental designs as a means of obtaining causal information is emphasized. In fact, in a humorous way, the authors warn the young scientists that they need a strong character to resist the suggestion of their boss to use data from past plant operation every time they suggest performing designed experiments to collect data.

In spite of this, several authors have proposed approaches to optimization and control based on interpolating historical bases. However, in all these cases, their success was based on making strong assumptions that allowed the database to be reorganized and causal information to be extracted. One approach was referred to as "similarity optimization," which combined multivariate statistical methods for reconstructing unmeasured disturbances and nearest neighbor methods for finding similar conditions with better performance. However, it too was shown to fail for many of the same reasons. In general, it was concluded that one can only optimize the process if there exist manipulated variables that change independently of the disturbances and if disturbances are piecewise constant, a situation that would be rare in historical process operations.

The reader should, therefore, exercise caution of how historical databases are used when it comes to retrieving causal information. However, databases obtained from routine operation are a great source of data for building monitoring schemes.

4.11.2   PRODUCT DESIGN

Given the reservations about the use of historical databases, one area where some success has been achieved is in identifying a range of process operating conditions for a new grade of product with a desired set of quality properties and in matching two different production plants to produce the same grade of product. If fundamental models of the process exist, then these problems are easily handled as constrained optimization problems. If not, optimization procedures based on response surface methodology can be used. However, even before one performs experiments, there exists information within the historical database on past operating conditions for a range of existing product grades (García-Muñoz et al. 2006).

In this case, the historical data used are selected from different grades and, therefore, contain information on variables for several levels of past operation (i.e., there is intentional variation in them and are not happenstance data). The key element in this empirical model approach is the use of latent variable models that both reduce the space of X and Y to a lower dimensional orthogonal set of latent variables and provide a model for X as well as Y. This is essential in providing solutions that are consistent with past operating policies. In this sense, PCR and PLS are acceptable approaches, whereas MLR, neural networks, and RRR are not.

The major limitation of this approach is that one is restricted to finding solutions within the space and bounds of the process space X defined by previously produced grades. There may indeed be equivalent or better conditions in other regions where the process has never been operated before and, hence, where no data exist. Fundamental models or more experimentation would be needed if one hopes to find such novel conditions.

A very good discussion on these issues can be found in García-Muñoz et al. (2008). The authors illustrate a methodology with an industrial application where the batch trajectories are designed to satisfy certain customer requirements in the final product quality properties while using the minimal amount of time for the batch run. The cumulative time, or used time, is added as an extra variable trajectory after the alignment of the batches.

4.12 SITE TRANSFER AND SCALE-UP

Product transfer to different sites and scale-up fall into the same class of problems: one needs to estimate the process operating conditions of plant B to produce the same product that is currently produced in plant A.

Attempts have been made to solve such problems with latent variable methods, utilizing historical data from both locations from transferring other products.

The following are the main points to keep in mind when addressing such a problem.

1. The quality properties of the product should always be checked within a multivariate context because univariate charts may be deceiving. The multivariate quality space for both sites should be the same. Correct product transfer cannot be achieved by comparing end-point quality on univariate charts from the two sites (or from pilot scale and manufacturing). The product quality has to be mapped from site to site in a multivariate way (the products in both sites have to project on the same multivariate space).
2. The end-point quality may not be sufficient to characterize a product. The path to end product is important. Whenever full mechanistic models exist, these models describe the phenomena that are important for the process and therefore determine this path. When changing sites, the full mechanistic model will describe the desired path in the new site taking into account

size, mass and energy balances, and/or other phenomena related to the process. When mechanistic models do not exist, this mapping of the "desired process paths" or "process signatures" has to happen with empirical data.

A methodology has been developed for product transfer and scale-up based on latent variables (García-Muñoz et al. 2005). The methodology utilizes databases with information on previous products and their corresponding process conditions from both sites. The two sites may differ in equipment, number of process variables, locations of sensors, and history of products produced.

ACKNOWLEDGMENTS

The author would like to acknowledge Gordon Muirhead and Bernadette Doyle, GlaxoSmithKline, for their continuous mentoring and support.

REFERENCES

Albert, S. and R.D. Kinley. 2001. Multivariate statistical monitoring of batch processes: An industrial case study of fermentation supervision. *Trends Biotechnol.* 19: 53–62.

Bork, Ch. 2011. Chromatographic peak alignment using derivative dynamic time warping. Presented at IFPAC-2011, Baltimore, MD, January 20, 2011.

Box, G.E.P., W.G. Hunter, and J.S. Hunter. 1978. Statistics for experimenters. *An Introduction to Design, Data Analysis and Model Building.* New York: Wiley & Sons. Wiley Series in Probability and Mathematical Statistics.

Burnham, A.J., R. Viveros, and J.F. MacGregor. 1996. Frameworks for latent variable multivariate regression. *J. Chemom.* 10: 31–45.

Chen, Y., T. Kourti, and J.F. MacGregor. 2002. Analysis and monitoring of batch processes using projection methods—An evaluation of alternative approaches. CAC 2002, Seattle, September 23–27, 2002, AIChE Annual Meeting, Indianapolis, IN.

Cinar, A., S.J. Parulekar, C. Undey, and G. Birol. 2003. *Batch Fermentation: Modeling, Monitoring and Control.* New York: CRC Press.

Duchesne, C. and J.F. MacGregor. 2000. Multivariate analysis and optimization of process variable trajectories for batch processes. *Chemom. Intell. Lab. Syst.* 51: 125–137.

Duchesne, C., T. Kourti, and J.F. MacGregor. 2002. Multivariate SPC for start-ups, and grade transitions. *AIChE J.* 48 (12): 2890–2901.

Duchesne, C. and J.F. MacGregor. 2004. Establishing multivariate specification regions for incoming materials. *J. Qual. Technol.* 36: 78–94.

Ferreira, A.P., J.A. Lopes, and J.C. Menezes. 2007. Study of the application of multiway multivariate techniques to model data from an industrial fermentation process. *Anal. Chim. Acta* 595 (1–2): 120–127.

Findlay, P., K. Morris, and D. Kildsig. 2003. PAT in fluid bed granulation. Presented at AIChE Annual Meeting, San Francisco, 2003.

Flores-Cerrillo, J. and J.F. MacGregor. 2004. Control of batch product quality by trajectory manipulation using latent variable models. *J. Process Control* 14: 539–553.

Ganguly, J., S. Yoon, L. Obando, and J.P. Higgins. 2011. Multivariate data analysis in biopharmaceuticals. In *Process Analytical Technology Applied in Biopharmaceutical Process Development and Manufacturing*, edited by C. Undey, D. Low, J.C. Menezes, and M. Koch (in print).

García-Muñoz, S. 2004. Batch process improvement using latent variable methods. PhD Thesis, McMaster University, Hamilton, Ontario, Canada.

García-Muñoz, S., T. Kourti, J.F. MacGregor, F. Apruzzece, and M. Champagne. 2006. Optimization of batch operating policies. Part I. Handling multiple solutions. *Ind. Eng. Chem. Res.* 45: 7856–7866.

García-Muñoz, S., T. Kourti, J.F. MacGregor, A.G. Mateos, and G. Murphy. 2003. Troubleshooting of an industrial batch process using multivariate methods. *Ind. Eng. Chem. Res.* 42: 3592–3601.

García-Muñoz, S., J.F. MacGregor, and T. Kourti. 2004. Model predictive monitoring for batch processes with multivariate methods. *Ind. Eng. Chem. Res.* 43: 5929–5941.

García-Muñoz, S., J.F. MacGregor, and T. Kourti. 2005. Product transfer between sites using joint Y_PLS. *Chemom. Intell. Lab. Syst.* 79: 101–114.

García-Muñoz, S., J.F. MacGregor, D. Neogi, B.E. Latshaw, and S. Mehta. 2008. Optimization of batch operating policies. Part II. Incorporating process constraints and industrial applications. *Ind. Eng. Chem. Res.* 47: 4202–4208.

González-Martínez J.M., A. Ferrer, and J.A. Westerhuis. 2011. Real time synchronization of batch trajectories for on-line multivariate statistical process control using dynamic time warping. *Chemom. Intell Lab. Syst.* 105: 195–206.

Jackson, J.E. 1991. *A User's Guide to Principal Components.* New York: Wiley.

Jorgensen, J.P., J.G. Pedersen, E.P. Jensen, and K. Esbensen. 2004. On-line batch fermentation process monitoring—Introducing biological process time. *J. Chemom.* 18: 1–11.

Kassidas, A., J.F. MacGregor, and P.A. Taylor. 1998. Synchronization of batch trajectories using dynamic time warping. *AIChE J.* 44: 864–875.

Kourti, T. 2002. Process analysis and abnormal situation detection: From theory to practice. *IEEE Control Syst. Mag.* 22 (5): 10–25.

Kourti, T. 2003a. Multivariate dynamic data modeling for analysis and statistical process control of batch processes, start-ups and grade transitions. *J. Chemom.* 17: 93–109.

Kourti, T. 2003b. Abnormal situation detection, three way data and projection methods—Robust data archiving and modeling for industrial applications. *Annu. Rev. Control* 27 (2): 131–138.

Kourti, T. 2004. Process analytical technology and multivariate statistical process control. Wellness index of process and product—Part 1. *PAT—J. Process Anal. Technol.* 1 (1): 13–19.

Kourti, T. 2005a. Application of latent variable methods to process control and multivariate statistical process control in industry. *Int. J. Adapt. Control Signal Process.* 19: 213–246.

Kourti, T. 2005b. Symposium report on abnormal situation detection and projection methods. *Chemom. Intell. Lab. Syst.* 76: 215–220.

Kourti, T. 2009. Multivariate statistical process control and process control, using latent variables. In *Comprehensive Chemometrics*, edited by S. Brown, R. Tauler, and R. Walczak, Volume 4, pp. 21–54. Oxford: Elsevier.

Kourti, T. 2011. Pharmaceutical manufacturing: The role of multivariate analysis in design space, control strategy, process understanding, troubleshooting, and optimization. In *Chemical Engineering in the Pharmaceutical Industry: R&D to Manufacturing*, edited by D.J. Am Ende, pp. 853–887. New York: Wiley.

Kourti, T., J. Lee, and J.F. MacGregor. 1996. Experiences with industrial applications of projection methods for multivariate statistical process control. *Comput. Chem. Eng.* 20 (Suppl.): S745–S750.

Kourti, T. and J.F. MacGregor. 1995. Process analysis, monitoring and diagnosis using multivariate projection methods—A tutorial. *Chemom. Intell. Lab. Syst.* 28: 3–21.

Kourti, T., P. Nomikos, and J.F. MacGregor. 1995. Analysis, monitoring and fault diagnosis of batch processes using multiblock and multiway PLS. *J. Process Control.* 5 (4): 277–284.

Kundu, S., V. Bhatnagar, N. Pathak, and C. Undey. 2011. Chemical engineering principles in biologics: Unique challenges and applications. In *Chemical Engineering in the Pharmaceutical Industry: R&D to Manufacturing*, edited by D.J. Am Ende, pp. 29–55. New York: Wiley.

Larson, T.M., J. Davis, H. Lam, and J. Cacia. 2003. Use of process data to assess chromato-graphic performance in production-scale protein purification columns. *Biotechnol. Prog.* 19: 485–492.

Lopes, J.A., P.F. Costa, T.P. Alves, and J.C. Menezes. 2004. Chemometrics in bioprocess engi-neering: Process analytical technology (PAT) applications. *Chemom. Intell. Lab. Syst.* 74: 269–275.

Lennox, B., G.A. Montague, H.G. Hiden, G. Kornfeld, and P.R. Goulding. 2001. Process mon-itoring of an industrial fed-batch fermentation. *Biotechnol. Bioeng.* 74 (2): 125–135.

MacGregor, J.F., T. Kourti, and J.V. Kresta. 1991. Multivariate identification: A study of sev-eral methods. International Symposium on Advanced Control of Chemical Processes ADCHEM '91 (an IFAC Symposium), Toulouse, France, October 14–16. Proceedings, edited by K. Najim and J.P. Babary, pp. 369–375.

Marjanovic, O., B. Lennox, D. Sandoz, K. Smith, and M. Crofts. 2006. Real-time monitor-ing of an industrial batch process. Chemical Process Control. Presented at CPC7, Lake Louise, Alberta, Canada, January 8–13, 2006.

Menezes, J.C., A.P. Ferreira, L.O. Rodrigues, L.P. Brás, and T.P. Alves. 2009. Chemometrics role within the PAT context: Examples from primary pharmaceutical manufacturing. In *Comprehensive Chemometrics*, edited by S. Brown, R. Tauler, and R. Walczak, Volume 4, pp. 313–355. Oxford: Elsevier.

Neogi, D. and C.E. Schlags. 1998. Multivariate statistical analysis of an emulsion batch pro-cess. *Ind. Eng. Chem. Res.* 37: 3971–3979.

Nomikos, P. 1995. Statistical process control of batch processes. PhD Thesis, McMaster University, Hamilton, ON, Canada.

Nomikos, P. and J.F. MacGregor. 1994. Monitoring of batch processes using multi-way prin-cipal component analysis. *AIChE J.* 40 (8): 1361–1375.

Nomikos, P. and J.F. MacGregor. 1995a. Multivariate SPC charts for monitoring batch pro-cesses. *Technometrics* 37 (1): 41–59.

Nomikos, P. and J.F. MacGregor. 1995b. Multiway partial least squares in monitoring batch processes. *Chemom. Intell. Lab. Syst.* 30: 97–108.

Nomikos, P. 1996. Detection and diagnosis of abnormal batch operations based on multiway principal component analysis. *ISA Transactions* 35: 259–267.

Taylor, P.A. 1998. Computing and Software Department, McMaster University, Hamilton, Ontario, Canada (May 1998), personal communication.

Ündey, C., S. Ertunç, and A. Çinar. 2003. Online batch/fed-batch process performance moni-toring, quality prediction, and variable-contribution analysis for diagnosis. *Ind. Eng. Chem. Res.* 42 (20): 4645–4658.

Ündey, C., S. Ertunc, T. Mistretta, and B. Looze. 2010. Applied advanced process analytics in biopharmaceutical manufacturing: Challenges and prospects in real-time monitoring and control. *J. Process Control* 20 (9): 1009–1018.

Ündey, C., E. Tatara, and A. Çinar. 2004. Intelligent real-time performance monitoring and quality prediction for batch/fed-batch cultivations. *J. Biotechnol.* 108 (1): 19, 61–77.

Westerhuis, J.A., P.M.J. Coenegracht, and F.L. Coenraad. 1997. Multivariate modeling of the tablet manufacturing process with wet granulation for tablet optimization and in-process control. *Int. J. Pharm.* 156: 109–117.

Westerhuis, J.A., A. Kassidas, T. Kourti, P.A. Taylor, and J.F. MacGregor. 1999. On-line syn-chronization of the trajectories of process variables for monitoring batch processes with varying duration. Presented at the 6th Scandinavian Symposium on Chemometrics, Porsgrunn, Norway, Aug. 15–19, 1999.

Westerhuis, J.A., T. Kourti, and J.F. MacGregor. 1998. Analysis of multiblock and hierarchical PCA and PLS models. *J. Chemom.* 12: 301–321.

Westerhuis, J.A., T. Kourti, and J.F. MacGregor. 1999. Comparing alternative approaches for multivariate statistical analysis of batch process data. *J. Chemom.* 13: 397–413.

Wold, S. 1978. Cross-validatory estimation of the number of components in factor and principal components model. *Technometrics* 20 (4): 397–405.

Wold, S., N. Kettaneh, H. Fridén, and A. Holmberg. 1998. Modeling and diagnostics of batch processes and analogous kinetic experiments. *Chemom. Intell. Lab. Syst.* 44: 331–340.

Zhang, Y. and T.F. Edgar. 2008. A robust dynamic time warping algorithm for batch trajectory synchronization. American Control Conference, June 11–13, 2008, pp. 2864–2869, Seattle, WA (ISSN: 0743-1619; Print ISBN: 978-1-4244-2078-0).

Zhang, H. and B. Lennox. 2004. Integrated condition monitoring and control of fed-batch fermentation processes. *J. Process Control* 14: 41–50.

5 Multivariate Data Analysis in Biopharmaceuticals

Joydeep Ganguly, Louis Obando,
John P. Higgins, and Seongkyu Yoon

CONTENTS

5.1 INTRODUCTION

Data analysis has long been recognized as an important source of competitive advantage in the biopharmaceutical industry. One can classify data analysis into three categories: overview of data, prediction, and classification. Multivariate data analysis (MVDA) tackles these three categories with a chemometric approach. The flexibility of multivariate methods has made them useful for the analysis and modeling of complicated and cumbersome data. These methods are increasingly used in a wide range of applications in biopharmaceutical development and manufacturing. These typically include process monitoring, early fault detection, quality control, and final product quality prediction (i.e., deriving relationships between process parameters and product quality attributes). Typical business benefits that can be achieved with

MVDA include achieving enhanced process consistency, prevention of batch discards, identifying process improvement opportunities, and identifying root cause explanations in investigations. The ICH Q8, Q9, and Q10 guidelines that form the foundation for quality by design (QbD) in the pharmaceutical industry clearly define the expectations on the use of MVDA in the context of process analytical technology (PAT) and QbD.

Multivariate calibration, which models the relationship between a multitude of signals measured on various samples and the concentrations of constituents of these samples, is another application of MVDA. With the recent interest in developing a new paradigm for process validation that requires continuous trending of large datasets (FDA 2011), MVDA has become a critical tool—not just in the development of a design space under the QbD framework but also in the subsequent characterization and continuous monitoring of manufacturing data. Additionally, there has been wide acceptance of these techniques by the regulatory agencies and increasing expectations of manufacturers to use these tools to understand and monitor processes.

Application of MVDA generally falls into two broad categories: intrabatch (real time) and interbatch (retrospective, off-line). Real-time MVDA is useful for immediate fault detection and isolation in manufacturing processes as well as achieving real-time process control. Retrospective (off-line) applications include the use of MVDA for the rapid analysis of completed batches for consistency, for support of investigations to determine root cause, and for solving complex problems involving multiple inputs and outputs. MVDA can be applied to a wide variety of problems, but the most value is realized when the raw data contain information that provides insight into the product's properties. For example, MVDA has been shown to be extremely useful for fermentation processes. This is because analysis of the off-gases, consumption rate of nutrients, and metabolite levels together provide information about the state of the cells. If after using all the data, no correlation can be found, this is a strong indication that no univariate correlations will be found.

5.2   MULTIVARIATE DATA MODELING FUNDAMENTALS

5.2.1   Multivariate Modeling Basics

Multivariate statistical methods for the analysis, monitoring, and diagnosis of process operating performance have received increasing attention. Application of MVDA, which utilizes not only the product quality data (as traditional approaches have done) but also the available process data, can be achieved using multivariate projection methods [principal component analysis* (PCA) and partial least squares (PLS)†]. These methods are becoming increasingly accepted and utilized by the

---

* PCA is the technique used for finding a transformation that transforms an original set of correlated variables to a new set of uncorrelated variables, called principal components. The components are obtained in order of decreasing importance, the aim being to reduce the dimensionality of the data. The analysis is used to uncover approximate linear dependencies among variables.
† PLS is a generalization of PCA where a projection model is developed predicting Y from X via scores of X. It can also be seen as a generalized multiple regression method that can cope with multiple collinear X and Y variables.

biopharmaceutical industry. We outline basic facts for multivariate data modeling and discuss a few practical considerations.

PCA is the technique used for finding a transformation that transforms an original set of PCA. It is a multivariate statistical method where a dataset containing many variables is reduced to a few variables called scores (t). The components or t-scores contain information about the variation of each variable in the dataset and the correlation of each variable to every other variable in the dataset. As such, t-scores describe the variation and correlation structure of each observation or batch in the dataset compared with other observations or batches in the dataset. The PCA plot is a plot of one component (t-score) against another component, usually t_1 vs. t_2. The PCA plot is essentially a distribution that shows how the variation and correlation structure compare for all of the observations or batches in the dataset.

For batch modeling, there are two types of modeling methods: observation model vs. batch-level model. The observation model provides batch evolution information over maturity variable progress. The average batch trajectory is calculated with all good batches in the training dataset. This allows process monitoring over time (or maturity variable). The batch mode allows monitoring of batch-to-batch variation. The full discussion between the two types of modeling is well described in the literature (Wold et al. 2009). MVDA of batch processes in real time is accomplished using the observation-level mode in which each sampling during the production of the batch is treated as an independent observation. It is not possible to perform real-time MVDA at the batch level because the batch must be complete to be able to generate the vector string describing the entire batch. Some software tools perform the observation-level analysis followed by batch-level MVDA when the batch is complete. A graphical representation of the two approaches is provided in Figure 5.1.

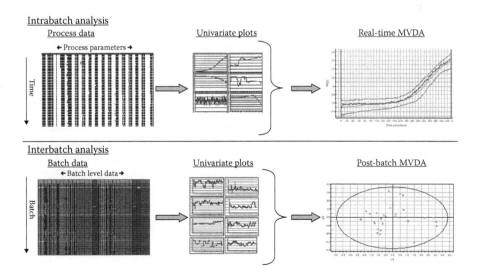

FIGURE 5.1 Overview of datasets and relationship to MVDA approaches.

5.2.2 Training vs. Testing Datasets

Datasets utilized for multivariate model building consist of two types: a training dataset is required for model training (or calibration) and a testing dataset for model testing to check the model robustness (validation). The training dataset must be representative of the process to be modeled. It should include most common causes of variations that would occur during normal operation of the processes and would be the basis for multivariate application. If the model is for process monitoring, the dataset must include batches with which type I* and II† errors can be tested. If the model is for prediction, the dataset must contain similar batches to the training set so one can verify the predictability.

5.2.3 Data Verification and Preprocessing

Before the dataset is used for model building, one should perform verification. The following is a minimum checklist to ensure correct modeling.

1. Relevant process measurements are included and each instrument has appropriate resolution.
2. Data collection frequency is properly chosen depending on the objective and process time constants.
3. The dataset is a representative of normal operation.
4. The test set contains a suitable operation period.

Analogous to univariate statistical process control (SPC), samples or batches are selected as the training set and are used to calculate statistical limits and an average. Future batches are then compared to those limits to assess whether they are different from historical performance. In the process of sample selection, there may be outlier batches that are excluded because they inflate the ±3σ limits and reduce the sensitivity of the tool. The exclusion of such batches does not imply that those batches were of unacceptable quality because those batches were tested and released based on approved release criteria. Samples should be selected that are representative of the process capability and span a range of product and process variation. Inclusion of outlier batches will produce insensitive models, and restrictions to a small number of batches will produce oversensitive models.

Sufficient numbers of batches are required to build multivariate models that sufficiently capture the process variation and from which reasonable and useful limits can be calculated for comparison of new batches. Just as with univariate SPC, there is no fixed number of batches required to build a multivariate model, but technically, at least two batches are needed for variance calculations. The confidence in the output of a model increases as the number of batches in the model increases; however, MVDA models built with as few as four batches can be useful as long as it is clearly noted that

---

* *Type I error*, also known as an "error of the first kind" or a "false positive," is referred to as the error of rejecting a null hypothesis when it is in fact true.
† *Type II error*, also known as an "error of the second kind" or a "false negative," is referred to as the error of failing to reject a null hypothesis when it is in fact not true.

the model is still in the training phase. An example of this might be an MVDA model constructed with four batches from scale-up and process validation campaigns. It is not uncommon that these are the only full-scale batches prior to transfer to manufacturing, and a preliminary MVDA could be constructed using the available batches to accompany the product into production. The model could then be updated with the first-production batches to add robustness to the model. This is seldom an issue with established products because numerous batches are typically available.

Once the dataset is verified, the next step is to preprocess the datasets for multivariate modeling. Usually, mean-centering and unit-variance scaling are commonly used. Depending on modeling objectives, different types of scaling methods can be used. When a model is initially fitted, default scaling of the work set is based on the standard deviation of the parameters in the work set. As a result, even a slight variation in the data for variables with small standard variations can drastically affect the multivariate overview plots. However, model sensitivity and robustness can be improved by adjusting factors of the variables used in the model. Reducing the weight of noncritical variables (by changing the modifier) increases the predictive power and reliability of the overview plots. Process expertise, particularly knowledge of the importance of different parameters for a specific unit operation, can be used for this purpose.

5.2.4 Model Building

Once the dataset is verified, the next step is to build the model. Depending on the applications, different types of models are used. Table 5.1 summarizes the model and data types depending on the applications.

TABLE 5.1
Model and Data Types for MVDA Application Types

Application	Model Type	Data Type
Process monitoring of a continuous unit operation	PCA and PLS	Continuous operation data
Process monitoring of a batch unit operation	Batch modeling using multiway PCA/PLS	Batch operation data, initial condition, end condition (lab data)
Raw material assessment	PCA/PLS	Raw material property data (certificate of analysis, additional characterization; e.g., NIR spectroscopic data)
Yield prediction, product quality assessment	PLS	Quality measurement with corresponding input data
Raw material identification	PCA, PLS-DA[a]	Characterization data of raw material library
Scale-up/down assessment	PCA/PLS/batch modeling	Small-scale, pilot, and at-scale operation data

[a] PLS-DA, partial least squares discriminant analysis.

Model building is an iterative procedure. After the model is fitted, one needs to check if applications can be performed satisfactorily with the resulting model. In the case of a regression model, the variable importance in projection plots of the first and the total components should be checked after the model is first built. The average variable importance for the model is always 1.0, so increasing the weight of one parameter decreases the importance of all other parameters, such that the variable importance average remains 1.0. It is the responsibility of the scientist building the MVDA model to determine the appropriate number of principal components to retain in a model; this is also widely referred to as the *rank* of the model. Selection of model rank is a critical step in the construction of any MVDA model. An expert in MVDA should be consulted during the development of multivariate models, model updates, and general troubleshooting.

5.2.5   ROBUSTNESS TESTING

The purpose of robustness testing is to check model effectiveness in distinguishing between normal and abnormal process behavior, eliminating the possibility of both type I and type II errors (falsely concluding that a batch is out of control or falsely concluding that a batch is in control, respectively). The model objective is to explain and summarize all the systematic variation in the variable (X). Through robustness testing, one can test batches with known perturbations and batches processed normally to check the model's ability to distinguish a "normal" batch from an "abnormal" batch and eliminate the possibility of both type I and type II errors. Batches used for robustness testing are not included in the model dataset.

5.2.6   MAINTAINING MODELS

The principles for updating univariate SPC charts generally apply to updating multivariate models. Multivariate models are updated by changes to the observations or to the variables. The training set can be changed by the addition or subtraction of samples (observations) or by modifications to selection and weights of the variables in the model. Model updates are required in the following situations.

1. *Process modifications that result in what is defined as normal operating condition(s).* For example, a piping modification lowers pressure readings that result in "alarms" given the normal operating conditions have shifted, and as such, the MVDA model should be updated to incorporate the new acceptable levels of pressure. A given variable could be disabled in the model until sufficient batches are available for the model update.
2. *Expansion of the operating space.* For example, batches are produced with higher product weights that exceed the $+3\sigma$ limit. If process experts agree that the higher product weight is not a deviation and will become the new norm and is reflective of suitable batch quality, the model should be updated to reflect the positive change in mean.
3. *When multivariate models are constructed with a limited number of batches, as deemed by the MVDA expert.* For example, process development

activities typically produce few batches. Production batches should be added to the model as they become available to build robustness and accuracy into the model.

4. *There are changes to the modeled variables.* The exclusion of variables or the inclusion of new variables will require recreating the model and can result in new limits, confidence intervals, and trajectories.

5. *Temporary deactivation of variables.* In situations where a parameter is temporarily unavailable, the parameter should be deactivated in the model to prevent false alarms due to that parameter. For example, if a sensor is being replaced, the data may be temporarily unavailable or inaccurate. In such situations, the parameter should be deactivated in the model until the data stream is deemed reliable.

5.2.7 DATA VISUALIZATION AND MULTIVARIATE CHARTS

The output of multivariate models is typically visualized using five multivariate parameters: scores, Hotelling's T^2 (will be referred to as T2 hereafter) statistic, batch maturity, distance to the model of X-space (DModX) statistics, and the contribution plot. Each is discussed in further detail below.

1. *Scores.* The multivariate representation of the samples is plotted to give a trajectory of the batch in multivariate space.

2. *T2.* The absolute distance of a sample from the center of the modeled space but within the space defined by the principal components of the model. The T2 statistic operates on the modeled portion of the data and summarizes the scores to detect if an observation is statistically within the modeled score space. A sample with a high T2 either has a very high or very low score in at least one principal component. Samples with low values of the T2 are very average and close to the center of the modeled space. It is sensitive to deviations of individual variables from the mean.

3. *Batch maturity.* A prediction of age or maturity of a batch process based on the raw data. Batch processes tend to have a characteristic profile that changes with time: a distinguishing feature from steady-state process that is not expected to change. For example, cell fermentation processes have a characteristic oxygen update rate, carbon dioxide evolution rate, and a respiratory quotient that changes as the batch ages. The batch maturity metric regresses the raw data against time using the training set using PLS regression. This PLS model is then used to predict the age or maturity of new batches (intrabatch and interbatch). It is useful to determine if the batch is maturing faster or slower than expected because it could indicate an abnormal condition within the process.

4. *DModX statistics.* A measure of a sample's distance to the modeled space that operates on the unmodeled or residual part of the data. The statistic known as the DModX, F-distance, or Q-residual in different software packages is an analysis of the residuals (unmodeled data). A new sample that is unlike the training set will be poorly modeled, resulting in high residuals

and, therefore, a high DModX. The DModX is ideal for detecting differences in the interactions between variables and is complimentary to the T2. It is very common to see deviations in the T2 with no corresponding deviation in the DModX. For example, if the product tank temperature were to exceed the +3σ limit, it would be immediately detected by the T2 but might not exceed the DModX limit because its interactions with other variables are preserved; that is, the increase in product tank temperature was accompanied by an increase in the parameters positively correlated to it (e.g., jacket temperature) and a decrease in parameters negatively correlated to it (e.g., dissolved oxygen). However, if the increase in product tank temperature was accompanied by an increase in agitation rate resulting in a new interaction between agitator and tank temperature, the T2 would detect the increases in the individual variables and the DModX would detect the broken relationship between tank temperature and agitator speed. Samples with low values of DModX have residuals and variable interactions that are consistent with the training set.

5. *Contribution plots.* The link between the multivariate statistics and the actual variables is accomplished through the use of contribution plots. For example, a sample with a high score can be interrogated to determine which variable or group of variables have contributed to the high score. The same is true for the other multivariate metrics of T2, DModX, and batch maturity. The contribution plot is extremely useful for drilling down to identify the variable or variables that are causing or indicative of a process deviation. Additionally, univariate run charts of the actual data (not a multivariate statistic) with the average and ±3σ limits can be generated directly from the contribution plot to further confirm that a variable is beyond its expected range or predefined limit.

These parameters form the basis of MVDA for both interbatch and intrabatch applications. They can be calculated and plotted in real time in a multivariate process monitoring scenario or in a postbatch analysis manner.

5.2.8 DOCUMENTATION OF MODELING

Best practice dictates that the following model content should be documented to ensure that models can be reproduced by others.

1. Batches of which operation period that are used to create the model
2. Number of observations per batch
3. The parameters that are part of the dataset (including calculated variables)
4. Any adjustments made to the models (weighting, scaling, exclusion of outliers or batches)
5. Sampling/execution interval and maturity variable
6. Phases and phase identifiers (if applicable)
7. Model summary (model type, number of components, R^2, and Q^2)

5.3 MVDA APPLICATIONS IN DEVELOPMENT AND MANUFACTURING SCALES

5.3.1 IN-PROCESS MONITORING AND CONTROL USING PCA

Achieving process consistency through adherence to operational limits using in-process monitoring and control is essential where the critical quality attributes (CQAs) of the product are not readily measurable. This may be due to limitations of sampling or detectability (e.g., viral clearance or microbial contamination), or when intermediates and products cannot be highly characterized and well-defined quality attributes cannot be identified. Any process measurements supporting the overall process control strategy need to be closely monitored. Typically there are a significant number of process parameters that need to be monitored to perform in-process control. Both critical process parameters (CPPs) as well as non-CPPs need to be assessed to achieve an increased level of process understanding. Process monitoring using PCA can often meet this monitoring need. The following figure is an example of a PCA plot for a cell culture process. The plot is usually used to determine if batches subsequent to those used to build the model are conforming to the defined process signature. Parameters that were included in the development of this signature include dissolver oxygen, pH, temperature, oxygen sparge, weight of base tanks, viable cell density, and pressure. In addition, calculated parameters (such as area under the viable cell density curve) were also included in the model. In the attached PCA plot (Figure 5.2), the new batches processed recently (black squares) plot inside the T2 ellipsoid, confirming that they are consistent with historical batches at the 99% confidence interval.

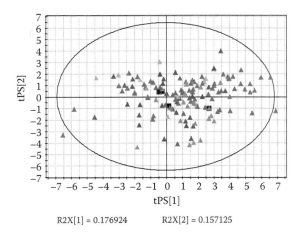

FIGURE 5.2   PCA model for cell culture bioreactor.

5.3.2 MULTIVARIATE STATISTICAL PROCESS CONTROL: BATCH MONITORING USING MULTIWAY PCA AND PLS

Biopharmaceutical processes typically consist of a series of batch unit operations: upstream processes such as cell culture and fermentation as well as downstream purification processes such as harvesting, filtration, and chromatography. Due to the batch process characteristics that include a large number of variables that are often highly correlated, classical SPC is not an optimal approach to process monitoring. Multivariate SPC (MSPC) based on multiway PCA and PLS provides an optimal framework for monitoring batch unit operations. These approaches have proved their usefulness in batch process monitoring and have been widely studied and extensively applied in industry. They are the most powerful and often the only approach available for monitoring batch processes when there is no deterministic model available. The success of these techniques can be attributed to the robust mathematical algorithms and their ability to handle nonlinearity and missing data.

To run an MSPC system in real time, one needs access to a real-time data historian or repository system, sensor data, as well as software that supports real-time process monitoring based on the available data. MSPC software applies a multivariate model to on-line real-time data from the data historian or distributed control system to give the user a real-time indication of current batch progress against a predefined model. Relevant process parameters of key unit operations can be trended within a multivariate framework using an MSPC system. The batch process monitoring can be accomplished by two modes: observation mode that allows process monitoring over time (intrabatch) and batch mode that allows batch-to-batch monitoring. The observation model monitors parameter evolution over time, and the batch mode model depicts batch-to-batch variation in multivariate space. Figure 5.3 demonstrates how on-line process monitoring can be conducted. It shows the average multivariate trajectory of all historical batches in the training dataset with the associated control limits that are typically ±3σ. The current batch that is evolving

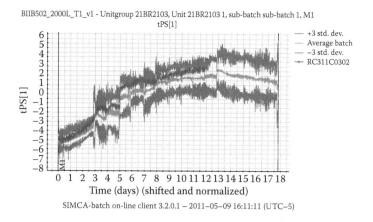

FIGURE 5.3 Observation mode monitoring of cell culture process.

with time can be seen in the plot. The observation-level PLS model was built for a batch process using historical data.

MSPC on-line production models can be built for most upstream and downstream product process and phases as well as cleaning processes such as clean in place and steam in place. During batch manufacturing campaign when the model is active, a link will be established into the continuous data around the given process step.

5.3.3 RESPONDING TO SIGNALS

Where multivariate control charts are used to monitor process health on multiple dimensions, the power of multivariate statistics should be utilized to the fullest as an early warning system of process upsets, drift, sensor failures, or recipe deviations. Excursions beyond any of the multivariate limits do not in themselves indicate that a true process deviation has occurred. Rather, it signals that a variable or group of variables are either at the extremes of the modeled space or outside the modeled space defined by the training batches. It is important to realize that the multivariate limits are solely determined by the batches in the training set and the relative weighting of variables. Moreover, the multivariate limits typically define a subspace of the licensed space (from a regulatory context) for a particular product. Therefore, a true process deviation can only be determined by referring back to the univariate charts and comparing the process parameter's value to prespecified limits or ranges (such as those that exist for CQAs, CPPs, or non-CPPs).

Upon detection of out-of-trend on any multivariate control chart, the appropriate response is to use contribution plots to determine which variable or group of variables are causing the deviation, followed by the use of univariate control charts to determine if the variable has exceeded a predetermined control limit. If the deviating variable is either a CQA, CPP, or non-CPP and exceeds any established limit or range, then consultation with the appropriate process and quality experts follows to determine the appropriate course of action. The response to out-of-trend events for CPPs or CQAs is typically addressed through the existing manufacturer's quality system. The logical approach to using MVDA tools for real-time process monitoring is outlined in Figure 5.4.

5.3.4 SCALE-UP VERIFICATION AND DESIGN SPACE APPLICATION

Besides using MVDA for process monitoring, one can also use it for scale-up verification. Transfer of a certain product from one production site to another or scaling up a product from bench scale to pilot plant and finally to a production plant can be an expensive and difficult problem. For example, if differences between sites were well understood, one might be able to minimize the effort around process transfer and validation. Ideally, if validating or qualifying one site, all sites and trains would automatically be validated. However, the challenge is how to efficiently address the comparability between scales and sites. It is not always clear whether there is sufficient data to prove comparability. Existing methods may make suboptimal use of available datasets from bench and pilot scales for clinical campaigns. These existing datasets may also be deficient in information for effective scale-up. Therefore,

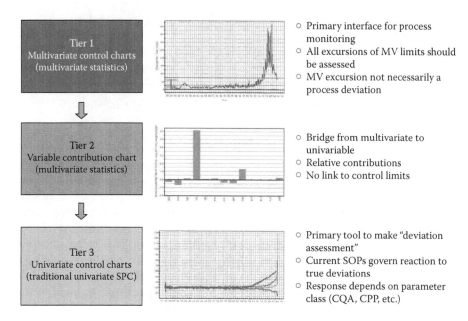

FIGURE 5.4 Logical approach to response to signals using a real-time MVDA.

designed experiments may need to be performed to enhance the data available at any particular scale (bench, pilot, or manufacturing) or to minimize the number of new experiments if scaling-up a new product.

If datasets based on operating conditions from the existing plants at each scale are available, powerful empirical latent variable modeling and model inversion approaches can be used to obtain preliminary solutions based on the existing historical databases. An alternative and generally improved latent variable modeling and inversion approach to the scale-up and product transfer problem was proposed later by García-Muñoz et al. (2005) and demonstrated on a batch process in the pulp and paper industry. These approaches are based on multivariate statistical methodology (in particular, multiblock PLS methods). Related multivariate technology has been developed by the same research group for the rapid development of new products (Muteki and MacGregor 2006, 2007).

The same methodology can also be applied for the product-transfer problem. However, this method has not yet been accepted as a normal practice in the biopharmaceutical industry. Instead, a simpler method has been used as a best practice. PCA is used to compare bench, pilot, and manufacturing data. Figure 5.5a shows a PCA model fitted with the bench-scale model to historical manufacturing data. The PCA model is derived from two historical manufacturing datasets. The multivariate statistical model suggests that data from bench-scale (light gray) and manufacturing (dark gray) are statistically similar. Figure 5.5b shows the observation-mode model. The predicted score was derived from the manufacturing dataset. The bench-scale model fits well to the batch model. Off-line observation mode–multivariate score plots for the manufacturing bioreactor operation shows batch-scale data compared to

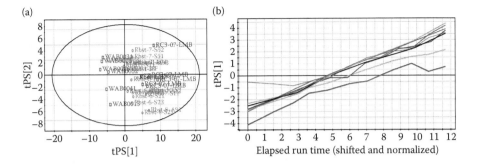

FIGURE 5.5 (a) PCA model fitted with the bench-scale model to historical manufacturing data; (b) batch-observation level model.

the average (lighter-gray trajectory in the middle) and ±3σ based on manufacturing. Lower cell viability in production is the main contributor to the difference between the bench-scale batches and the model.

The correlation structure of the data observed at each scale and the relationships between these correlation structures need to be maintained to ensure effective scale-up. Such correlation structures can be obtained by applying PLS models for two or more different scales, and the resulting multiblock models can be used in an optimization framework to find optimal conditions to operate the scaled-up process. Along with this new multivariate analysis technology, a new approach to design of experiments was also developed to allow for a minimal number of experiments that will optimally augment the existing databases (Muteki and MacGregor 2007) or lead to a minimal number of experiments in scaling up a new process (Muteki and MacGregor 2007). These analysis and design technologies can be applied to address scalability and product transferability problems.

5.3.5 RAW MATERIAL ANALYSIS

A significant source of variation for biopharmaceutical products emanates from raw materials used during the biomanufacturing such as growth media, buffers, inocula, and resins. Lot-to-lot variation in the raw material is known to impact protein product quality attributes and protein therapeutics manufacturing characteristics. To minimize the raw material variations and their impact to drug substance quality attributes, various quality tests are typically applied. For example, a growth test has been used to evaluate cell culture growth rate affected by an animal-based raw material. As a common practice, biopharmaceutical manufacturers try to minimize lot-to-lot variations to improve product consistency. A screening method based on multivariate method can be applied to tackle this difficult problem. Product quality prediction is also possible based on PLS projection methods to estimate expected product quality attributes when a specific raw material lot is used. This tool can be used to make the best possible selection from raw material lots available.

Most raw materials are provided with a certificate of analysis that confirms quality of the raw material lots and includes various physical and biochemical test results established by the supplier. The screening tool can use the parameters reported in the certificate of analysis, and the consistency of new raw material lots is compared with historical data through application of the PLS model. Product quality attributes are estimated and compared to the product quality specification. If the consistency check and the product quality predictions are satisfactory, the raw material can be released for production use.

In the following section, we consider one example with one ingredient of cell culture growth medium. The screening tool consists of two components: a consistency check module and a product quality attribute prediction module. The consistency check module allows for comparison of all attributes of the new raw material lot against those of the raw material lots previously used at a same manufacturing facility. The consistency check uses PCA. The product quality prediction model uses PLS regression based on correlation of historical raw material and product quality data from previous manufacturing records.

The dataset used for model building consists of multiple raw material lots from two vendors used. There are 14 lots provided by A vendor and 18 lots from B. The certificate of analysis includes 27 parameters. A multivariate statistical method was used for building the consistency check model with the historical data. Not all parameters appear consistent between the vendors. To account for the difference of the datasets provided from different vendors, different models may be required.

A PCA model built with 32 raw material lots was used for assessment of all raw material lots. As shown in Figure 5.6a, there are two clusters with four outlier lots (AB1, AB2, C1, and C2). These four lots were successfully used in production, but they were not clustering with the other lots from the same vendors. The clustering by vendor is due to the differences in the lot components. Figure 5.6b shows the component contributions to the differences. It can be seen that the B vendor lots have higher concentrations of certain components. There are smaller differences in other components. The clustering of data in the PCA model and consistent differences in certain parameters suggest that the datasets from the two different vendors cannot be combined for further assessment unless their differences are completely explained and understood. Otherwise, the consistency check will not be able to identify outlier lots from either vendor.

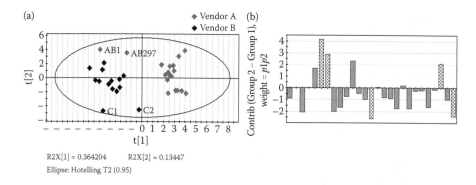

R2X[1] = 0.364204   R2X[2] = 0.13447
Ellipse: Hotelling T2 (0.95)

FIGURE 5.6   (a) PCA score plot with all lots; (b) component contributions to the differences.

Raw materials have been a source of process and product quality variation, and there have been substantial efforts in the industry to compensate for the raw material variation with control of process parameters to produce consistent yield and product quality. The raw material screening system presented can contribute to reliable and consistent operation of the protein manufacturing process by seeking to limit lot-to-lot variation in raw materials. By assessing the raw material characteristics before procurement or use in manufacture, a consistent raw material lot, which is predicted to show consistent repeatable product quality, can be chosen. If lots can be identified that will perform closer to historical means of process performance and product quality, the raw material screening tools will reduce variation in the manufacturing process.

5.4 CONCLUSION

The goals of applying PAT tools to manufacturing processes are to provide an opportunity for real-time control of the process, continuous improvement, and reduction of variability resulting in a greater assurance of product quality. An optimal strategy for application of appropriate tools is needed to successfully meet these goals. A key principle to ensure success of a PAT application is to employ the simplest technology possible to achieve the stated goals. In the case of biopharmaceutical processes, typically a significant quantity of raw material, process parameters, and product quality data is already collected. Application of MVDA tools can take advantage of these data that are already being collected to generate increased process understanding that leads to the stated business and product quality benefits already outlined. In particular, upstream bulk manufacturing processes do not lend themselves to the application of PAT analyzer-based techniques that provide for direct measurement of process performance parameters such as protein expression levels or cell viability. Hence, empirical statistical models built using PCA and PLS provide an optimum approach to achieving a successful PAT implementation strategy in biopharmaceuticals.

Multivariate process trending will be a critical piece of the new process validation paradigm proposed by the FDA. Many tools and techniques, some statistical and other more qualitative, can be used to detect variation, characterize it, and determine the root cause. It is recommended to continue monitoring and sampling at the level established during the process qualification stage until sufficient data are available to generate significant variability estimates. Timely assessment of defect complaints, out-of-specification findings, process deviation reports, process yield variations, batch records, incoming raw material records, and adverse event reports are critical components of the continued monitoring.

REFERENCES

FDA. 2011. Guidance for industry. Process validation: General principles and practices. http://www.fda.gov/downloads/Drugs/GuidanceComplianceRegulatoryInformation/Guidances/UCM070336.pdf (accessed Feb. 2011).

García-Muñoz, S., T. Kourti, and J.F. MacGregor. 2005. Product transfer between sites using joint-Y PLS. *Chemom. Intell. Lab. Syst.* 79: 101–114.

Muteki, K. and J.F. MacGregor. 2006. Multi-block PLS modeling for L-shaped data structures, with applications to mixture modeling. *Chemom. Intell. Lab. Syst.* 85: 186–194.

Muteki, K. and J.F. MacGregor. 2007. Sequential design of mixture experiments for the development of new products. *J. Chemom.* 21: 496–505.

Wold, S., N. Kettaneh-Wold, J.F. MacGregor, and K.G. Dunn. 2009. Batch process modeling and MSPC. In *Comprehensive Chemometrics*, edited by S. Brown, R. Tauler, and R. Walczak, Volume 2, pp. 163–197. Oxford: Elsevier.

6 Process Analytical Technology Advances and Applications in Recombinant Protein Cell Culture Processes

Ana Veronica Carvalhal and Victor M. Saucedo

CONTENTS

6.1 PROCESS ANALYTICAL TECHNOLOGY AND RECOMBINANT PROTEIN PRODUCTION PROCESSES

The Food and Drug Administration (FDA) recognition that process optimization in industry may be delayed or never happen to minimize regulatory risks is an excellent first step toward application of emerging technologies by the industry. Nevertheless, the success of this strategy will only be possible if the FDA and other regulatory agencies place an equal amount of effort in ensuring communication, openness, and the right support to the industry.

One of the key process analytical technology (PAT) concepts is that it is not just a method of monitoring but also a way of *controlling* the process while the process is ongoing (herein mentioned as in-process control) to achieve the target product quality (U.S. FDA 2004; ICH 2009). This implies moving from an expected fixed process (the "black box" approach) to a flexible or adaptable process (Figure 6.1). To achieve in-process control, true *process understanding* must be achieved, and the required new and reliable tools need to be developed. *PAT tools* already exist and are being applied to industrial processes (Section 6.3), but again, more effort needs to be placed on developing innovative tools. Implementation challenges of PAT tools are described in Section 6.2. Another perspective of PAT is the possibility of process change (or process control) based on the life cycle of a product with the perspective of *continuous improvement*.

Biotechnology companies have been investing resources on process understanding to ensure that manufacturing processes achieve the target critical quality attributes (CQAs), operate in a large operational design space [quality by design (QbD) strategy], and apply technologies that enable an adaptable process (PAT strategy). Further reading on the definition of those concepts may be found in Chapters 4 and 7 that deal with different aspects of multivariate analysis and techniques applied to biopharmaceutical processes/problems.

Current recombinant protein production processes present the advantage of being robust and well established in manufacturing facilities or contract manufacturing

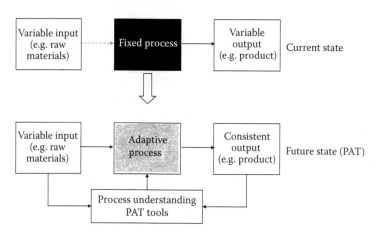

FIGURE 6.1   PAT concept [concept based on Lam and Kiss (2007)].

organizations. It is also recognized by both industry and regulatory agencies that there are many factors that impact product quality and process attributes that are still not well understood. Nevertheless, the industry continues to gain process understanding that has increased significantly in the past decades. Improving process productivity is critical from a cost of goods and manufacturing capability perspective, but biologics product quality and its direct correlation with efficacy and safety is the critical aspect in recombinant protein production processes. As an example, Figure 6.2 shows the unexpected significant variability of an antibody protein property across multiple manufacturing scale production runs and different licensed manufacturing sites. Although the data presented in Figure 6.2 are not related to an attribute critical for the activity of the molecule, they are a concern in terms of assessing manufacturing consistency and target range comparability. In the FDA's PAT guidance to industry, it is recognized that "When a quality problem arises in current processes it is increasingly difficult to identify the root cause" (U.S. FDA 2004). This difficulty arises due to the inherent complexity of a biologic system, the variability of raw materials, and production at different manufacturing sites, among others.

Figure 6.3 outlines the main steps in antibody production, as it will be the main focus of this chapter. The production cells [mammalian, such as Chinese hamster ovary (CHO) cells, or microbial cells, such as *Escherichia coli*] are typically frozen in a cell bank that contains several hundred vials (i.e., Master and Working Cell Banks) that are kept for many years in optimized conditions. At the time of production, one or more vials are thawed and maintained in selective conditions at the seed train level. Selective conditions are maintained typically by adding or removing specific components from the culture media and ensuring that the cells maintain their capacity to produce the protein of interest. After the seed train stage, cells are grown in consecutive bioreactors of increasing scale (called the inoculum train stage) to scale up the cell biomass to sufficient levels to allow the inoculation of the final large-scale production vessel.

Typically, current antibody cell culture production processes are performed in manufacturing as fed-batch processes where specific nutrients (e.g., sugars, amino acids, and vitamins, among others) are added to the culture medium to maintain the

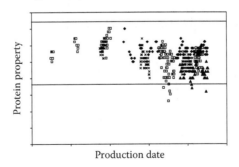

FIGURE 6.2 Unexpected variability example of a protein property for a manufactured antibody (different symbols indicate different manufacturing sites; lines represent expected range). This protein property does not impact product quality, but it represents the variability that can occur in biotechnology processes.

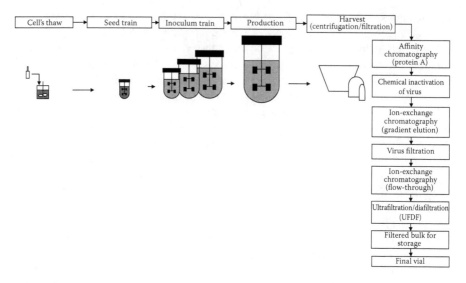

FIGURE 6.3 Flow diagram for typical CHO-derived antibody manufacturing production process.

viability of the culture and the production rate of the product of interest. Bioreactors are vessels containing agitators for mixing and control systems for maintaining the desired level of dissolved oxygen (by sparging air or pure oxygen), pH (by adding a base or an acid), and temperature. Usually, antibodies are secreted by mammalian cells, and before the purification process, it is necessary to separate the cells from the media containing the protein of interest by centrifugation/filtration units (harvest process). Table 6.1 indicates the main factors taken into consideration during cell culture process development. The antibody is then purified by sequential chromatographic steps including affinity and ion-exchange chromatography (IEC). The protein properties must be assessed postaffinity chromatography because a certain level of purity is necessary to characterize the antibody using techniques such as size-exclusion chromatography (quantification of antibody aggregation levels), capillary electrophoresis (glycosylation characterization), or IEC (quantification of antibody relative charged fractions). Recombinant protein CQAs may impact its target product profile, that is, product potency, specific receptor binding affinity if applicable, serum half-life, immunogenicity, or clearance. In recombinant protein production processes, product quality is typically determined primarily by the selected cell clone and fermentation process conditions. The purification steps tend to only slightly affect product quality attributes with the possibility of impacting the charge variant distribution and decreasing the level of aggregates, but not typically impacting glycosylation of antibodies produced in CHO cells.

Purification will significantly remove process-related impurities such as DNA, host cell proteins, and different compounds used during the process. Formulation is also critical in maintaining the drug product quality. At this stage, the antibody product quality may be, for example, impacted by protein oxidation and/or aggregation.

TABLE 6.1

Main Factors Considered during Cell Culture Process Development

Process Stage	Parameter or Process Attribute
Clone selection	Selection of cells that allow recombinant protein production with the correct product quality attributes
Thawing of cells	Viability
	Specific cell growth rate
Seed train, inoculum train, production culture	Seed density
	Temperature
	pH
	Dissolved gases (e.g., oxygen, CO_2)
	Media concentration/composition
	Batch feed volume: concentration/composition and addition timing
	Metabolic by-product accumulation
	Osmolality
	Culture duration
	Accumulated biomass
	Viability
	Recombinant protein volumetric and specific productivity
Harvest	Cell lysis

In the production of recombinant proteins, typical examples of CQAs are related to product variants, product and process impurities, stability, and product integrity. Acceptable ranges for CQAs will need to be met in manufacturing by properly selecting and controlling critical process parameters (CPPs); otherwise, the product would likely be discarded. The difficulty of measuring CQAs in real time makes it difficult to act on the process ("adaptable" PAT process) based on these attributes. QbD and PAT initiatives aim to develop processes that may be operated in broader conditions targeting solely the correct product quality specifications. CQAs will remain the same independent of the process strategy, whereas PAT may introduce new CPPs that current technologies are incapable of monitoring, controlling, or understanding.

Current processes offer a degree of robustness; nevertheless, the coefficient of variation for the titer and/or protein properties may vary up to 20% at the end of cell culture production. Furthermore, in the evaluation of commercial recombinant protein production, loss of manufacturing runs greatly impacts manufacturing costs, and it is, therefore, very important to understand the main root causes of run loss. Run loss can occur with cell culture production contamination events and equipment deterioration, failure, or improper setup. PAT may further increase process robustness by identifying additional factors affecting the process that are not currently understood, considered, or controlled.

It is well accepted that PAT applications are still in their early beginning, where tools are being developed and their robustness assessed. In the future, the risk of using PAT tools will be similar to that of the current pH and temperature controls in cell culture processes. An increase in future PAT tool usage will alter production processes and create the need to develop the necessary control strategies during manufacturing. The "adaptable" PAT process will also bring challenges to industry regarding how to define the processes in manufacturing, authoring manufacturing batch records, training personnel, and recording deviations from the process.

Global implementation of PAT will not happen overnight. It will be a slow process, step by step, and only in specific processes. Instances where PAT may show promise include processes that have long production windows, have low efficiency, generate high levels of waste, or have high cost of production (Hinz 2006). PAT should also be considered for new products where this type of approach could provide more data to help operate the process. In the future, despite the lower risk of implementing PAT, cost may limit its application in small companies or low-volume products.

6.2   IMPLEMENTATION CHALLENGES OF PAT TOOLS IN RECOMBINANT PROTEIN PRODUCTION CELL CULTURE PROCESSES

The monoclonal antibody industry has the opportunity to simplify and accelerate the development and manufacturing of new products by increasing process understanding. Process knowledge should contribute in at least one of three forms: (1) identify the relationships among important quality attributes and process parameters, (2) identify and explain sources of variability, or (3) reduce process variability. The general tools identified by FDA to increase process knowledge are (U.S. FDA 2004)

1. Multivariable tools for process design, data acquisition, and analysis
2. Process analyzers
3. Process control tools
4. Continuous improvement and knowledge management tools

Multiple publications describe different PAT tools in pharmaceutical and biotechnological industries (Willis 2004; Ganguly and Vogel 2006). However, references do not discuss how PAT tools are developed and implemented. This chapter discusses the development and implementation challenges of PAT tools in upstream processes of the monoclonal antibody industry.

6.2.1   ON-LINE SENSORS IN BIOREACTORS

The capability to improve a process depends on the information available. A process can potentially be better characterized when a larger number of independent variables are measured and analyzed. Good measurements have to be specific and selective to quantify unique variables, such as a protein sensor that could correctly quantify only the protein of interest regardless of other proteins, analytes, or process conditions. Finally, the measurements are most valuable when they provide real-time information about the process, thus allowing real-time response. Additional benefits are achieved when the measurements are obtained *in situ* and with no manual intervention. Several on-line sensors are used in bioprocessing. The principles, limitations, and implementation challenges for on-line measurements in the upstream monoclonal antibody processes are discussed next.

6.2.1.1   On-Line pH Measurement

The importance of culture pH on mammalian cell culture performance has been demonstrated in numerous studies (Schmid et al. 1990; Ozturk and Palsson 1991; Zhangi et al. 1999; Xie et al. 2002). The addition of base or acid to maintain the culture pH at a predefined set point based on an on-line pH sensor signal is an example of a PAT application (in-process control). A decrease in product yield has been reported in large-scale manufacturing due to drifting pH sensor signals (Evans and Larson 2006). Theories dealing with different sensor and process drifts have been proposed (Illingworth 1981; Langheinrich and Nienow 1999). No recommendations have been given to solve the pH drift observed by Evans and Larson (2006). Therefore, pH sensor drift needs to be understood and mitigated to reduce variability during scale-up. The objective of this section is to show that a better understanding of pH sensor technology could lead to reduced process variability.

A typical pH sensor consists of an electrochemical cell between a measuring electrode and an outer reference electrode. The outer reference electrode potential stays constant when the properties of the process solution change (Bard and Faulkner 1980), and the signal reported is the difference between the two. The reference electrode has to be in contact with the process liquid, and this exposure can result in impurities contaminating the reference electrode and the liquid junction, thereby changing the reference potential (Illingworth 1981; Diamond et al. 1994). Glass pH sensors have proven to be robust and selective against redox potential variations of the media and interference by other components.

A frequent assumption in bioreactor operations is that pH sensor signals are stable. However, experience in development and manufacturing does not always support this assumption, often evidenced by the common practice of adjusting the on-line sensor during a culture to match an off-line value. The errors between on-line pH measurements using two glass sensors and off-line measurements (BGA 400, Nova Biomedical, Waltham, MA) in a 2-L bioreactor running a CHO cell culture are shown in Figure 6.4. The difference between the on-line and off-line pH increases with time. The actual process pH (off-line) decreases with time as the control system maintains a pH set point based on a drifting sensor signal. This phenomenon is representative of trends sometimes seen in development and manufacturing. This behavior is not favorable because large errors can negatively impact the recombinant protein production.

Glass pH sensors convert the sensor output potential into a pH reading using the Nernst equation. The Nernst equation uses the sensor offset, or the asymmetric potential, V_o, which reflects the difference between the outer and inner surface potentials of the pH sensing glass membrane (Bard and Faulkner 1980). The glass membrane potential of sensors, V_o, regardless of the reference electrolytes, goes through two phases during cell culture processes (Figure 6.5): a positive shift after the autoclaving process and a slow negative drift during the cell culture process. The drift observed during the cell culture process is a slow recovery process as shown in phase

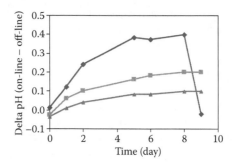

FIGURE 6.4 Drift of on-line measurements with respect to off-line during a cell culture experiment. (▲) On-line pH, metal oxide gel; (■) on-line pH, inorganic silica gel; (◆) standard on-line pH. The standard on-line pH measurement is corrected on day 9. The on-line pH sensors overestimate the pH, as measured by the off-line measurements.

II of Figure 6.5. This phenomenon occurs in bioreactors where the pH sensors are not submerged in liquid during autoclaving. During autoclaving, the glass membrane outer surface is exposed to high-pressure steam, whereas the inner surface of the pH sensing glass is constantly exposed to the inner reference electrolyte. These different conditions result in an unequal impact on the pH sensing glass (Saucedo et al. 2011).

Different approaches can be explored to minimize on-line pH drift in cell culture processes. One approach might be to optimize the glass membrane chemistry to reduce the impact of autoclaving at high temperatures. Another approach is to change operating procedures to protect the glass membrane from saturated steam during autoclaving or steam-in-place (SIP). Even though pH glass sensors operate based on the same theoretical principles, subtle differences in design and operating procedures can have a significant impact on cell culture performance. Understanding how the sensor design interacts with the process can achieve a desirable process

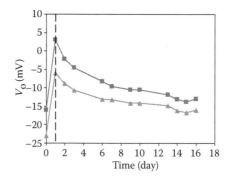

FIGURE 6.5 Time profile of the glass membrane voltage for two pH prototypes during a cell culture experiment: (■) metal oxide gel; (▲) inorganic silica gel. The increase in phase I is due to autoclaving. The glass membrane recovers in phase II (after dashed line) when in contact with the media.

performance. This knowledge allows practitioners to modify procedures and accomplish optimal pH control.

6.2.1.2 On-Line Near-Infrared for Analyte Measurement

Most cell culture process monitoring requires manual sampling operations, sample preparation, and a wide array of off-line measuring technologies to understand the process performance. Off-line measurements offer limited understanding and performance because only a few sporadic measurements are available. On-line monitoring of important analytes such as glucose, lactate, cell density, glutamine, ammonia, and other amino acids is limited by sterilization and long-term sensor stability problems of several technologies (Rhiel et al. 2002a).

Near-infrared (NIR) spectroscopy is a technology that could provide real-time data on the most important analytes in a cell culture process. Chemical information from an *in situ* NIR sensor is obtained by analyzing a spectrum of NIR radiation transmitted through the sample of interest. Theoretically, all covalent bonds of C–H, N–H, O–H, and S–H stretch, bend, and contribute to the NIR spectra (Scarff et al. 2006). Ionic analytes such as pH or metals do not affect NIR spectra directly, but it has been shown that they can be measured indirectly with NIR spectroscopy (Mattes et al. 2009). An NIR sensor can monitor several components, eliminating the need for multiple sensors simultaneously. Spectroscopic measurement is noninvasive and does not depend on the exposure of the process to chemical reagents, and NIR probes can be steam sterilized. Hence, concerns about sterilization and long-term sensor stability are irrelevant here.

NIR spectroscopy has been well understood and used in industrial applications since the early 1980s. The applications have been generally qualitative, where the main purpose has been to detect moisture levels in solid samples and powder blending (El-Hagrasy et al. 2001; Gupta et al. 2005) and raw material identification (Gonzalez and Pous 1995; Blanco et al. 1996). Only recently has NIR spectroscopy matured enough to suggest potential applications in industrial cell culture bioprocesses (Henriques et al. 2010). The main problem hindering its use in bioreactors is the fact that water has large absorption bands located in the heart of the NIR range at 1400 nm and between 1800 and 2200 nm. The overwhelming absorption of water in the NIR spectra, until recently, made extraction of information from spectra very difficult. Improvements in fiber optics, electronics, optics, sensor design, and computational power have increased the signal-to-noise ratio and the resolving power in NIR spectrophotometers, increasing the likelihood of success in aqueous environments (Chung et al. 1995; McShane and Cote 1998; Rhiel et al. 2002a; Tamburini et al. 2003; Luypaert et al. 2007).

The analytical ability of NIR spectroscopy to measure off-line millimolar concentrations of various chemical analytes in fermentations, cell cultures, and natural and simulated matrices from clinical chemistry was fully established by the end of the 1990s, but required a significant amount of mathematical pretreatment and data analysis (Amrhein et al. 1999; Riley and Crider 2000). Quantitative NIR spectroscopy applications require calibration using measurements from secondary methods, process scans (training set), and a mathematical model that correlates the process scans and the process values obtained from the secondary measurements. Successful multivariate calibration models must have all of the matrix and analyte variation in the training set to make accurate measurements or predictions on subsequent samples (Rhiel et al. 2002a). Relationships among the many analytes in cell culture processes are well known, and the implications of cellular metabolism on model calibration have been thoroughly examined (Haaland and Thomas 1988; Rhiel et al. 2002b). Spiking samples with known, random concentrations of material can improve calibration procedures, eliminating the concentration correlations within a set of samples. Researchers have built successful calibration models for glucose, lactate, glutamine, and ammonia (Arnold et al. 2003), but there have been no published studies that show the use of these models to predict similar or different cell processes robustly. Other studies successfully built *in situ* models to predict product titer in real time, although significant issues with prediction noise and offset still remain (Triadaphillou et al. 2007).

Several challenges must be resolved before widespread implementation of NIR spectroscopy in cell culture bioreactors, including improved repeatability between runs, model robustness, and reduced sensor-to-sensor and instrument-to-instrument variability. Building NIR spectroscopy calibrations requires significant resources, and maintaining this technology may represent a significant organizational challenge (Menezes et al. 2010). The challenges and recommendations to build a robust calibration model are presented next.

Figure 6.6 shows that NIR spectroscopy measures glucose in real time, with a potential in glucose concentration control. The model in this application is robust because it predicts glucose in a cell culture bioreactor where the cell line and media compositions are different from those used in the calibration (Saucedo et al. 2009). Most of the NIR spectroscopy results published do not specify the procedure required to achieve successful repeatability in cell culture bioreactors. Slight differences in the bending and torsion of the fiber-optic cables affect the absorption spectra even when the sensor remains locked in place inside the bioreactors. One practice to overcome this limitation is having the fiber optic always attached to the sensor and secured to the sensor during all runs. This practice is possible only for

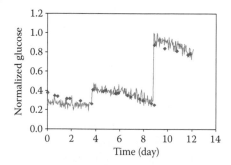

FIGURE 6.6 NIR spectroscopy prediction of glucose in a cell culture bioreactor using the "hybrid" calibration model built from a process with a different cell line. Robustness is also achieved by performing a reference procedure before each run; therefore, models are valid from run-to-run: (solid line, –) NIR predictions; (♦) off-line measurements.

bioreactors with SIP. For smaller bioreactors requiring autoclaving, the sensors and cables need to be moved, invalidating any previously developed calibration models. Then, random optic variations occur to the spectra, and the calibration models are no longer valid. This problem is not significant in applications where the NIR sensitivity is well established such as moisture content determination or where the sensitivity to the measuring component is significantly higher than fiber optic changes. However, it does present a challenge for successful implementation of NIR spectroscopy in cell culture bioreactors. One solution that has been implemented is an adjustment to a reference before each experiment. Such an adjustment consists of modifying a scan of air taken by the reflectance sensor at a constant temperature so that the difference with an internal reference is practically zero for all frequencies. Such an adjustment removes the variability caused by procedures that require moving the sensor from run-to-run (Figure 6.6).

NIR spectroscopy has the potential to monitor several cell culture components in real time when models are built from actual cell culture experiments (McShane and Cote 1998; Arnold et al. 2003; Navratil et al. 2005; Rodrigues et al. 2008).

In general, robust models are those capable of accurately predicting future experiments. The NIR calibration becomes more robust as more spectra are collected from several representative experiments. This approach demonstrates that NIR spectroscopy can be a reliable tool for real-time monitoring of bioreactors as long as the NIR operates within the measuring range used in calibration. This method is robust for run-to-run and is valid for batch processes with minimum changes in operating conditions (Rodrigues et al. 2008). Calibration models are not universal and are not meant to work across different cell culture processes. Each new cell line, medium composition, and process requires a new NIR spectroscopy calibration model. Hence, it is necessary to develop calibration procedures that build robust models that can work in a variety of conditions.

A "hybrid" NIR spectroscopy calibration procedure for cell culture bioreactors, with the potential to monitor analytes across different mammalian culture media, cell lines, and conditions, can also be developed (Saucedo et al. 2009). It is "hybrid" because it combines the method of spiking the analyte of interest in a controlled cell culture bioreactor where other important variables are held constant. The "hybrid" calibration performance to measure glucose where the cell line and medium composition are different from the calibration is shown in Figure 6.6. A proper design of experiments can generate spectra covering a wide range of conditions and collect data to build a robust model in a couple of days, compared to months required to collect data from actual cell culture experiments. Design of experiments is very important to build reliable NIR spectroscopy models efficiently; however, this aspect of implementing NIR spectroscopy sensors is very often ignored.

6.2.1.3 On-Line Cell Density Measurement

Biomass is an important variable to quantify (Table 6.1). Traditional off-line methods such as hemocytometer, cell counters, and agar plates are based on cell number and are laborious and time consuming. New methods use optical principles. Cell density methods use a single frequency light source and are based on the light scattered, not the light absorbed by cells. The wavelengths used for cell density measurements, normally between 550 and 650 nm, interact mainly with the cells and not with the dissolved components in the media.

Commercially available on-line cell density sensors generally fall within two categories. The first category, optical density, uses an optical-based detection technique. The second type uses a capacitance or conductance measurement. Optically based sensors typically use transmittance for low cell density applications. In this configuration, the sample is located between the detector and the emitter. At high cell densities, where only a small fraction of the emitted light can reach a detector, sensors use backscatter or reflectance, in which the emitter and the detector are on the same side.

In a capacitance or conductance measurement, electromagnetic waves interact with matter and change its electrostatic properties. This phenomenon is called dielectric spectroscopy and is the principle for capacitance sensors. Dielectric or capacitance-based sensors are known to have a good linear response in viscous fermentations, and their performance may vary significantly across cell lines and cell morphology; hence, it is recommended to use spectra at different frequencies (Cannizzaro et al. 2003). It is also reported that dielectric sensors predict accurately during the growth phase of CHO cell cultures. However, the accuracy decreases in later phases when cellular properties such as size may change significantly (Opel et al. 2010). Off-line techniques are still popular in process development applications because existing technologies cannot robustly measure cell density, size, and viabilities. However, they are more accepted in manufacturing, especially in *E. coli* processes, because conditions are more repeatable from run-to-run.

6.2.1.4 On-Line Dissolved Oxygen Measurement

Dissolved oxygen is an important control parameter in cell culture bioreactor operations (Table 6.1). Air or oxygen-enriched air is supplied to bioreactors to meet cell oxygen demand. Cells utilize oxygen for energy generation and cell growth. Unlike pH that needs to be tightly controlled, dissolved oxygen (dO_2) can be maintained within wider operating conditions without affecting the cell growth rate or product quality (Lin et al. 1993). dO_2 is also used routinely to measure the oxygen uptake rate (OUR), a useful operating parameter for microbial and CHO cultures. Some processes use OUR as a control variable (Ramirez and Mutharasan 1990; Kovács et al. 2007).

Dissolved oxygen is commonly measured with two main types of sensors: galvanometric and polarographic (electrolytic). In either type of sensor, a change in oxygen concentration (partial pressure) changes the sensor signal according to an electrochemical reaction on the cathode surface (Ik-Hwan et al. 1993).

Dissolved oxygen sensors using these principles have been operating in bioreactors of all scales for decades, though they have limitations. For example, servicing polarographic membranes is costly in large operations; the membranes are susceptible to malfunctions, resulting in operational losses. More recently, the performance of bioreactors running CHO cell culture has been linked to noisy dO_2 measurements. dO_2 control can shift the process into nonoptimal operation when dO_2 measurements are noisy (Chung et al. 2003). Therefore, there is a motivation to find new dO_2 technologies. Optical dO_2 sensors have been gaining popularity for applications that require miniaturization. They are cheap, are reliable, and have a fast response required for real-time control (Hanson et al. 2007; Naciri et al. 2008).

Optical sensors based on light quenching when oxygen interacts with a fluorescent dye are becoming the preferred commercial alternative techniques to polarographic sensors (Hanson et al. 2007; Naciri et al. 2008). This technology meets sterility requirements through comparability with autoclaving and gamma irradiation. It can also be used in state-of-the-art processes such as high-throughput, mini- and microreactors and disposable bioreactors. They also have a fast response time to implement the control algorithms required in cell culture bioreactors. Reported time response for optical dissolved oxygen sensors is $t_{90} < 15$ s (Polestar Technologies, Needham Heights, MA) compared to $t_{95} < 30\text{--}60$ s for common polarographic sensors. Microelectromechanical systems can miniaturize dO_2 sensors and are economical. They have applications in different areas such as microscale water samples and microbial granules (Lee et al. 2007), but references with applications in bioprocesses are not known.

6.2.1.5 On-Line Glucose Measurement

Glucose is the main carbon source in cell culture processes and fermentations (see Sections 6.3.1 and 6.3.2). Cells metabolize glucose for growth and product yield. Hence, it is of great importance to monitor and control glucose. Glucose oxidase-based sensors are the most commonly used method for measuring glucose. This approach is successful in off-line commercial equipment. However, there is an active research area in changing the enzyme immobilization and sample presentation to work in on-line and *in situ* applications (Zhu et al. 2002; Wen et al. 1997; Phelps et al. 1995). Glucose oxidase converts glucose to gluconic acid and hydrogen peroxidase, which is then measured amperometrically. Some disadvantages of amperometric enzymatic sensors are loss of activity at high temperatures such as sterilization and membrane fouling by protein and cellular products (Xu et al. 2002). Glucose control applications based on enzymatic sensors have been reported for several years (Zhou et al. 1995; Cruz et al. 2000; Luan et al. 2005). In some experiments, the samples are automatically obtained from the bioreactor and injected to a glucose analyzer (Kong et al. 1998; Xu et al. 2002) via an autosampler system as described in Section 6.3.1. Although these applications report satisfactory implementation results, these methods depend on mechanical sampling devices that reduce reliability and add complexity to the operation.

An *in situ* sensor regenerated with fresh glucose oxidase in a cellulose matrix has also been developed (Phelps et al. 1995). The regeneration and recalibration functions are fully automated, but this technology is not commercially available. Glucose oxidase has also been immobilized in an *in situ* sensor and sterilized with UV with promising results (Scully et al. 2007). Other technologies have the potential to overcome the limitations of glucose oxidase. A glucose sensor based on cationic *bis*-boronic acid appended to benzyl viologens and the anionic fluorescent dye, 8-hydroxypyrene-1,3,6-tridulfonic acid trisodium salt (HPTS), has been shown to selectively recognize up to 12 saccharides (Schiller et al. 2007). The concentration of glucose in solution interacts with the complex formed by HPTS and varies with the detected fluorescence intensity. This concept inherits the limitations of

HPTS-based sensing technology, which suffers from drift due to photobleaching (Schulman et al. 1995; Bultzingslowen et al. 2002; Kermis et al. 2008). The technology can potentially measure glucose in cell culture bioreactors, but it cannot be steam sterilized; hence, it is more appropriate for gamma radiation-sterilized disposable bioreactors.

Surface-enhanced Raman spectroscopy can add specificity and improve the glucose detection signal even in *in vivo* applications (Stuart et al. 2006). More recently, optical coherence tomography (OCT) has been utilized to monitor glucose in medical applications. Hydrogel particles were functionalized with glucose-specific lectin affecting the media scattering coefficient and detected by OCT (Ballerstadt et al. 2007). Most glucose technologies are developed for monitoring diabetics. For this reason, OCT technology does not meet many requirements necessary for cell culture bioreactors, such as sterilization. Nevertheless, many requirements such as time response, limit of detection, selectivity, pH, and temperature ranges are similar to cell culture bioreactors.

6.2.2 AUTOMATION AND ROBOTICS

Biotechnology companies need to stay competitive by responding quickly to development requirements. Cell culture process development needs to screen different clones and produce material for toxicological studies at higher rates. One way to respond to this need and achieve process understanding using multivariate tools is to miniaturize cell culture systems and run many of these systems in parallel. New high-throughput systems are based on microreactors (Chambers 2005; Davies et al. 2005). Microreactor volumes can range from 600 µl to 300 ml. As this approach conceptually has the advantage of running many bioreactor experiments simultaneously, automating these systems is quite challenging.

Dissolved oxygen and pH sensors are the most common on-line sensors in minireactors, which have operating volumes in the order of liters. However, these sensors represent an area of development to be implemented in microreactors. The most suitable sensors for this application are based on optical principles. Fluorescent pH sensors suitable for microreactors have been reported (Hanson et al. 2007; Kermis et al. 2008). These pH sensors are based on the fluorescent dye HPTS. Although these sensors cannot be autoclaved or steam sterilized, they can be gamma radiated. Even though optical pH and dissolved oxygen sensors are small and can fit microreactors, it is necessary to modify the microbioreactors to interface with the hardware. This is a new technology emerging in commercial platforms (Chen et al. 2009).

Another challenge in both micro- and minibioreactors is the intensive labor required for sampling and sample and data handling. Sampling tools have been developed in the last years (Harms et al. 2002). Solutions for miniscale bioreactors are available. Recently, an automated sampling, analysis, and data management system for mammalian cell culture monitoring has been reported (Derfus et al. 2010) that has the potential to reduce labor and without sacrificing the quality of data collected. Sampling solutions for microbioreactors are also evolving with a variety of new technologies such as lab-on-a-chip (Legmann et al. 2009).

6.2.3 KNOWLEDGE MANAGEMENT TOOLS: MODELING

Modeling can be applied in different areas of cell culture: (1) equipment design, (2) process scale-up and scale-down, (3) process optimization, and (4) real-time applications. Computational fluid dynamics (CFD) has been used for bioreactor design and scale-up and scale-down. CFD allows studying the effects of critical parameters and the validation of experimental studies. Mixing performance and various equipment configurations in different scales have been improved using CFD (Bezzo et al. 2003; Davidson et al. 2003; Paul et al. 2004; Heath and Kiss 2007). Normally, CFD is a computationally intensive technique and is not used for real-time applications. Computer-based process simulation is popular for process design and process optimization. It uses mathematical algorithms to solve the mass and energy balances to minimize wasted resources, achieving the product in the least amount of processing steps and the most economical way (Zhou and Titchener-Hooker 1999). Simulation tools are most useful when designing a new plant, but have a smaller impact once a plant has been built.

Real-time models are the most important type of applications from the PAT point of view. These models use real-time data to improve the performance of an ongoing process. The process can be tailored in real time to, for example, maximize titer or minimize process variability to ensure consistent product quality. The metabolic flux method is a modeling approach that has real-time capabilities and has inherent physiological process knowledge (Vallino and Stephanopoulos 1993; Varma and Palsson 1994; Henry et al. 2007; Goudar et al. 2006). This method uses linear programming to solve a set of mass and kinetic equations to determine the metabolic fluxes of components and metabolites generated or consumed in a cell. This approach has shown potential for use in process optimization, but many kinetic or yield parameters and assumptions are required to solve the models. The sensitivity analysis of metabolic fluxes and the implementation of real-time models for control or optimization need accurate and reliable real-time intracellular and extracellular metabolite sensors.

Data-driven models are also used for real-time applications. These models create relationships between the inputs and outputs of systems, regardless of the physical, chemical, or biochemical principles governing the systems. They are easier to build and capture a wide variety of multivariable and nonlinear interactions. Among the most common in industry are the partial least squares (PLS) models (Wise and Gallagher 1996; Wold et al. 1989). However, these models are not useful for extrapolation, do not bring more insights into the process, and require large data sets to build. Standard PLS models can be used for some types of nonlinear processes, and variations on PLS have been developed to capture more nonlinear behaviors (Baffi et al. 1991; Qin and McAvoy 1992). PLS models have been used for on-line prediction of product yield and process monitoring (Mandenius 2000; Larson et al. 2002; Lopes et al. 2004; Kirdar et al. 2008). Although the literature shows the feasibility and benefits of this technology, in practice, industry needs to put more effort into data infrastructure to implement real-time prediction models. Data need to be available, consistent, and properly networked across the enterprise to generate useful results. Moreover, real-time models have to be implemented in easy-to-use interfaces, and the organization has to be in place to support them continuously.

Although real-time models can be used for real-time applications, their benefits depend on how they are used. The information of the models can also be used for advanced process control (Proell and Karim 1994). Ultimately, the information from real-time models can be used by decision engines to continuously improve the process. This is known as real-time optimization (Saucedo and Karim 1997).

6.2.4 FUTURE OF PAT TOOLS

Most of the on-line and *in situ* sensors used in the development and manufacturing of monoclonal antibody processes have been used in other industries for many years, for example, pH and dO_2 probes. They have been evolving to meet the biotechnology process needs. However, as the biotechnology industry differentiates itself from other industries, it requires its own sensing technologies to meet its needs. On-line sensors with high selectivity, a wide range of detection, robustness, capability to meet sterility requirements, and ease of use will facilitate development, increase process understanding, and ultimately be used for in-process control as a PAT application. Sensors that measure physical and intracellular cell properties will also play a key role in new developments.

Many exciting emerging technologies are showing the proof of concept in the academic environment. Some of the most promising technologies include optics, nanotechnology, new chemistries, and new data analysis tools (Esbensen et al. 2004; Stuart et al. 2006; Armani et al. 2007; Larsson et al. 2007; Potyryailo and Morris 2007; Schiller et al. 2007). Although numerous investigations use these techniques, implementation in a commercial setting will require significant effort and collaboration between academia and industry. In the meantime, measurements will continue to be at-line in the absence of robust on-line sensors. A short-term goal is bringing benefits coupling at-line sensors and instruments with automated sampling systems. These systems can generate more data and with more quality than off-line systems.

The fields of modeling and experimental design are more advanced and mature than the on-line sensors. Many mathematical tools are waiting for the on-line sensors to situate themselves in the trust of industry. In the meantime, sensor and automation industries can do two things to prepare the path for more knowledge management tools. First, publish more proof of concept work creating awareness of modeling importance. Second, software and automation industries must have the initiative and vision to develop easy-to-use tools, whereas biotechnology industries need to plan future data needs and requirements. It is understandable why it may take a long time to implement new PATs when novel on-line sensors are not available. However, PAT applications already exist in the biotechnology industry.

6.3 APPLICATIONS OF PAT TOOLS IN RECOMBINANT PROTEIN PRODUCTION CELL CULTURE PROCESSES

PAT is not a new strategy for the biotechnology industry or cell culture processes (Zhou et al. 1995; Åkesson et al. 1999, 2001; Cruz et al. 2000; Johnston et al. 2002; Luan et al. 2005; Zawada and Swartz 2005; Hamilton et al. 2009; Mun et al. 2010). Research and process development teams have long invested resources and expertise in process understanding. Nevertheless, PAT, as a unique strategy where the final process is a process that adapts in real time, has not been applied frequently. This may be due to concerns with the challenges of implementing postmarket process changes or costly and complex new technologies. The following sections describe three interesting PAT applications in cell culture processes for the production of recombinant proteins.

6.3.1 Controlled Nutrient Continuous Feeding: Applications in CHO Cell Culture Processes

The use of mammalian cells has been the preferred choice as the host cell for the production of recombinant proteins, particularly large, complex, and glycosylated proteins used for human and animal administration. One of the main challenges in producing proteins using mammalian cells relates to the metabolism of cells *in vitro*, where by-products such as lactate and ammonia as a result of sugar and amino acids metabolism, respectively, are accumulated in cultures.

Glucose is taken up by the cell and converted into energy via glycolysis (Figure 6.7): 5%–8% of the glucose will feed the ribose-P-pentose shunt (essential for nucleotide synthesis), whereas the rest is converted into pyruvate. Nevertheless, *in vitro*, the majority of the formed pyruvate is converted to lactate and alanine, thus not significantly feeding the tricarboxylic cycle or Krebs cycle resulting in an inefficient glucose to lactate metabolic yield coefficient, YLact/Glc, of 1.5–2. Excessive lactate production can inhibit cell growth by decreasing the pH of the culture in the absence of pH control and by increasing the osmolality of the medium when pH is controlled (base addition). Glutamine has long been shown to be an amino acid easily metabolized by cells and feeding the Krebs cycle (Neermann and Wagner 1996). Additionally, CHO cells may efficiently metabolize other ammonia-generating amino

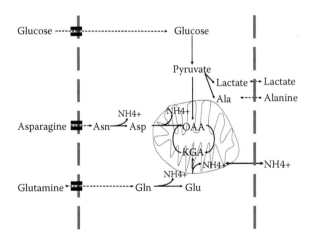

FIGURE 6.7 General schematic of glucose, asparagine, and glutamine mammalian cell metabolism (Asn, asparagine; Asp, aspartate; Gln, glutamine; Glu, glutamate; KGA, α-ketoglutarate; OAA, oxaloacetate).

acids (e.g., asparagine, aspartate, and glutamate) via the Krebs cycle. For each consumed glutamine or asparagine, two ammonia molecules are produced (Figure 6.7). Several possible explanations have been outlined for the potential negative effects of high levels of ammonia in animal cell cultures, including (1) disturbance of electrochemical gradients (Glacken et al. 1988); (2) inhibition of enzymatic reactions, e.g., the conversion of glutamate to α-ketoglutarate via glutamate dehydrogenase (Glacken et al. 1988); (3) intracellular pH changes that may lead to disturbance of proton gradients (Glacken et al. 1988; McQueen and Bailey 1990a,b, 1991); (4) inhibition of endocytosis and exocytosis (Docherty and Snider 1991); and (5) increased demand for maintenance energy (Martinelle and Häggström 1993). Decreasing potentially detrimental by-product production (lactate and ammonia) will, in many cases, benefit culture performance. Lowering lactate production in mammalian cell culture processes lowers the base addition and osmolality that can lead to improved process robustness. Decreasing ammonia production may improve cell growth (cell line dependent) and may impact the glycosylation profile of the recombinant protein being produced (Glacken et al. 1988; Gawlitzek et al. 2000; Chena and Harcumb 2006). Ammonia concentration is known to decrease $\beta(1,4)$-galactosyltransferase and $\alpha(2,3)$-sialyltransferase activity or expression (Gawlitzek et al. 2000; Chena and Harcumb 2006), impacting the distribution of glycoforms in the recombinant protein. Both process and genetic approaches to decrease lactate and ammonia levels in mammalian cell cultures have been pursued, with only partial success (Table 6.2).

TABLE 6.2
Strategies for Lowering Lactate and Ammonia Levels during Mammalian Cell Culture[a]

Strategy	Reference
Reducing temperature below 37°C	Reuveny et al. 1986; Sureshkumar and Mutharasan 1991; Chuppa et al. 1997
Reducing pH below 7.15 reduces glucose specific consumption	Miller et al. 1988
Overexpression of glutamine synthetase or pyruvate carboxylase	Bebbington et al. 1992; Irani et al. 1999
Downregulation of glucose transporter, enolase enzyme (glycolysis pathway) (by antisense RNA)	Paredes et al. 1999
Replacing glucose (galactose, fructose, mannose) or glutamine (glutamine dipeptides, glutamate)	Glacken et al. 1988; Roth et al. 1988; Minamoto et al. 1991; Duval et al. 1992; Butler and Christie 1994; Altamirano et al. 2004, 2006
Reducing glucose/glutamine extracellular concentration at culture start	Glacken et al. 1988; Miller et al. 1989; Zeng and Deckwer 1995; Zhou et al. 1997
Reducing glucose/glutamine extracellular concentration during the culture run (nutrient-controlled feeding)	Zhou et al. 1995; Cruz et al. 2000; Luan et al. 2005

[a] In bold; PAT applications are highlighted.

One of the process approaches was based on the observation that the lower the initial glucose or glutamine concentration in the culture media, the lower the specific consumption of these nutrients (Cruz et al. 2000). Maintaining glucose and glutamine at a low level during the cell culture process has been explored by several authors in the literature (Zhou et al. 1995; Chen and Forman 1999; Cruz et al. 2000), showing the need to further improve the strategies to consistently maintain the concentration of these nutrients at desired low levels. As indicated by Miller et al. (1989), when glucose is increased above 5.5 mM in the culture medium, a rapid transient increase in glucose-specific consumption and lactate production is observed. Strategies for maintaining glucose or glutamine at low or constant levels have been reported in the literature based on the following: (1) a constant feeding rate (based on previously obtained data) (Ljunggren and Häggström 1994); (2) the direct measurement of the metabolite (Cruz et al. 2000); or (3) via indirect measurements that correlate with specific nutrient consumption, such as OUR, base consumption, or ATP (5' adenosine-triphosphate) levels. Independent of the strategy, it is very important that each nutrient being controlled is maintained at concentration ranges that do not lead to a metabolic shift.

None of the available literature to date explores why the correlation of lower nutrient cell specific consumption with lower extracellular levels of these nutrients is observed (Glacken et al. 1988; Miller et al. 1989; Zeng and Deckwer 1995). One of the critical aspects that might explain such correlation is related with the underlying transport mechanisms of these nutrients and their specific K_m values (Ljunggren and Häggström 1994). K_m is an association constant representing the substrate concentration at which the velocity of substrate uptake is half maximal. Typically, it is mentioned that nutrients such as glucose and glutamine are efficiently metabolized by the cells, which means that cells are able to quickly take up these nutrients from the extracellular environment and present the necessary enzymatic machinery that allows their conversion through metabolic pathways. Excessive nutrient supplementation to the culture media (at concentrations above K_m) will unnecessarily increase their uptake velocity, which may lead to inefficient metabolism. Controlled nutrient feeding for maintaining key metabolic nutrients below the K_m will decrease and significantly impact the specific consumption of these nutrients. The challenge in the application of controlled nutrient feeding strategies is that nutrient K_m values may vary depending on the clone (type and expression level of transporters) and culture media (presence of media components that could inhibit or enhance the nutrient transport into the cell) being used. Glucose uptake is mediated by facilitated diffusion in most mammalian cells. The transporter exists in five isoforms with different kinetic properties ranging from a K_m of 1–20 mM (Elsas and Longo 1992; Brown 2000; Wood and Trayhurn 2003). Glutamine can be taken up by all three major transport systems for amino acids: the A, ASC, and L systems (Hyde et al. 2003). Another factor to be taken into consideration is the interrelation between sugar and amino acid metabolism, more specifically studied regarding glucose and glutamine metabolism. Both provide the carbon skeletons for cellular molecules, and both are catabolized for energy generation. They are considered as partially substitutable and partially complementary substrates (Hu and Himes 1989).

The approach by Ana Veronica Carvalhal at Genentech described here focused on optimizing a strategy where the target key metabolites would be maintained at constant low levels below their respective K_m values, based on daily metabolite and cell biomass measurements and adjustment of feed rate and feed concentration. This constant process adjustment is a PAT example application in cell culture processes. Furthermore, there is the benefit of applying this strategy for multiple nutrients key in cell metabolism, protein expression, and product quality.

Shown here is an example of two operational strategies in maintaining extracellular glucose concentration at a low and constant level as compared with a standard fed-batch process where glucose is maintained at high concentrations (between 3 and 14 g/L). A commercially available autosampler system connected to a glucose analyzer (BioProfiler, Nova Biomedical) has been successfully used where, every 4 h, a sample was automatically removed from the bioreactor and the glucose concentration was analyzed. With this information, the system allowed a feedback control system by turning on a pump to deliver a concentrated glucose solution to the bioreactor if glucose was below a predefined value. The rate of addition could be optimized by manipulating the pump flow (manually) and the pump on-time (software). The other used strategy was based on daily adjustment of the feed flow rate and glucose feed concentration based on specific glucose consumption and specific growth rate. In this strategy, the settings are based on the previous two data points. This last control strategy offered more flexibility for the controlled feeding of several nutrients, for which there is currently no autosampling capability.

Figure 6.8 shows the glucose concentration during a glucose-controlled experiment using both strategies described above. Figure 6.9 shows the impact of glucose-controlled feeding on total residual glucose and lactate accumulation during a production run for different cell lines. With the glucose-controlled feeding, it was possible to reduce the specific glucose consumption rate twofold.

The controlled glucose feeding strategy conferred two main advantages regarding process robustness and product characteristics. Process robustness was achieved by maintaining osmolality at low levels (300–350 mOsm) due to the lower concentrations of both glucose and lactate, and by the reduced need for base addition to counteract the acidic character of lactate. Additionally, lower glucose consumption results in a decrease in a galactosylation-related attribute (Figure 6.10), which is suggestive of an innovative glycosylation control strategy.

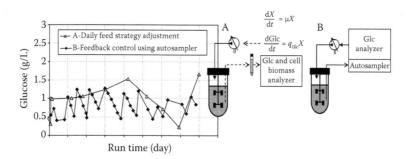

FIGURE 6.8 Glucose concentration in controlled glucose feed bioreaction: feedback control using autosampler (closed symbols; diagram B) and feed strategy based on daily (off-line) measurements (open symbols; diagram A). (Courtesy of Carvalhal, A.V., Genentech.)

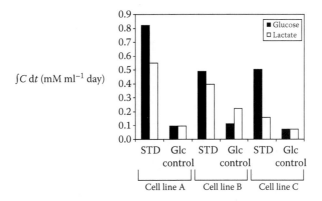

FIGURE 6.9   Glucose and lactate accumulated concentration along a production run for three different CHO cell lines (STD: standard fed-batch process—high glucose concentration; Glc control: process with controlled glucose addition—low glucose concentration). (Courtesy of Carvalhal, A.V., Genentech.)

This approach was also applied to amino acids that directly impact the production of ammonia, such as glutamine, asparagine, aspartate, and glutamate. By controlling one or a combination of these nutrients at a low and constant level, it was possible to maintain them at a specified level during the production run, decrease their specific consumption rate, and impact ammonia production. Critical to these strategies was the concentration at which the nutrient was controlled, which may vary with clone and media used. As an example, maintaining asparagine at two different concentrations showed a lower specific consumption of this nutrient when it was controlled below a critical concentration (Figure 6.11).

Controlled nutrient feeding is a PAT application in mammalian cell culture for the production of recombinant proteins. Based on frequent nutrient analysis (either

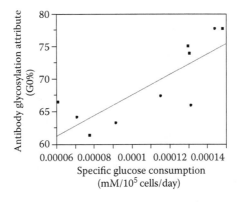

FIGURE 6.10   Lower G0% (protein glycosylation property) is correlated with lower specific glucose consumption rate obtained by PAT-controlled nutrient feeding strategies (statistically significant: $p < 0.05$, bivariate fit: $p < 0.0045$). Different symbols represent different CHO cell lines. (Courtesy of Carvalhal, A.V., Genentech.)

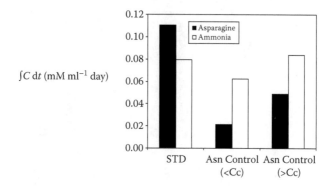

FIGURE 6.11 Asparagine and ammonia accumulated concentration along a production run for three different cases (STD: standard fed-batch process; controlled asparagine feeding above or below critical nutrient concentration, Cc). (Courtesy of Carvalhal, A.V., Genentech.)

on-line or off-line), the controlled nutrient feeding is adjusted based on a mathematical model, pump output, and nutrient feed concentration, resulting in an adaptable process. Other PAT cell culture feeding strategies are being developed such as dynamic feeding where a stoichiometrically balanced batch feed is automatically added to the culture production process at specified intervals (every few hours) based on an on-line signal (e.g., capacitance or OUR) or autosampler capabilities (e.g., glucose or viable cell count) (Li et al. 2010).

6.3.2 CONTROLLED CONTINUOUS GLUCOSE FEEDING: APPLICATION IN *E. COLI* FERMENTATION PROCESS

Although still novel in mammalian cell culture applications, glucose feeding has been applied for many years for the purpose of recombinant protein production using *E. coli*. Examples of products produced in *E. coli* are enzymes (e.g., rennin, amylases, proteases, and cellulases) and therapeutic proteins (e.g., insulin growth hormones and interferons) (Eiteman and Altman 2006). Processes using microbial systems such as *E. coli* have some significant differences from mammalian cell culture processes. *E. coli* cells have a significantly higher growth rate, different nutrient needs, and recombinant protein production pathways (Andersen and Krummen 2002). Nevertheless, *E. coli* and mammalian processes have similar process goals with respect to minimizing cell growth-related inhibitory metabolic by-products. In the case of *E. coli* processes, glucose consumption under high-growth-rate or anaerobic conditions can lead to acetate formation, which is known to be inhibitory to cell growth and product production (Jensen and Carlsen 1990; Andersen et al. 2001; Shiloach and Rinas 2006). Acetate formation may generally be observed in two scenarios: (1) a mixed-acid formation of acetate in the absence of oxygen (anaerobic conditions) and (2) in the presence of oxygen (aerobic conditions) when there is excess glucose supporting very high growth rates (Wolfe 2005). Several process and genetic strategies to control acetate accumulation have been tested, developed, and published in literature (Eiteman and Altman 2006). Table 6.3 shows a selection

TABLE 6.3

Feed Process Strategies to Minimize Acetate Production in Microbial Cell Culture Processes[a]

Strategy	Reference
pH-stat: feeding in response to pH (feeding triggered when pH rises due to carbon source depletion)	Johnston et al. 2002; Kim et al. 2004
Predefined glucose feeding (constant or increased in a gradual, stepwise, linear, or exponential mode)	Lee 1996
DO-stat: feeding in response to DO (feeding triggered with DO raises due to carbon source depletion)	Andersen et al. 2001
Carbon dioxide evolution rate (CER): feeding based on CER (proportional to carbon source consumption rate)	Kovács et al. 2007
Feeding based on cell concentration (feeding will be based on cell growth)	Kim et al. 2004
Feeding based on complex DO testing (use of duration of the DO responses to a glucose pulse to calculate the q_{glc})	Åkesson et al. 1999, 2001; Johnston et al. 2002; Lin et al. 2001; Zawada and Swartz 2005; Hamilton et al. 2009

[a] In bold; PAT applications are highlighted.

of glucose feeding strategies that have been applied to *E. coli* processes. Dynamic glucose feeding strategies are examples of PAT applications because the process is modified (adapted) in real time, based on a variable measured on-line. Preventing acetate formation in a dynamic approach requires the use of glucose feeding algorithms that are responsive to the variability of the process due to metabolic disturbances, limitation of nutrients other than glucose, and protein expression. Because bacteria grow and consume glucose very quickly relative to the timescale for current glucose concentration measurements, a feeding strategy similar to the one indicated above for mammalian cells does not seem to be feasible. In the future, technologies such as NIR spectroscopy may be valuable tools for that purpose. Any attempt at dynamic glucose feeding in an *E. coli* fermentation must then be based on indirect variables that are correlated with cell growth, such as base demand to neutralize acidic species generated with growth, nutrient consumption rates, dissolved oxygen (dO_2 or DO) signals, or carbon dioxide evolution rate (CER) or OUR (see Table 6.3). From these indirect measurements, the most effective and best-studied approaches have been the methods based on dissolved oxygen, a variable that can be measured accurately at high speed by means of a traditional sensor. Figure 6.12 shows the glucose feed rate and dissolved oxygen values for an *E. coli* fermentation under control of a classical DO-stat glucose feeding algorithm.

The complexity of glucose feeding strategies based on DO varies significantly. In its most familiar form, the DO-stat method, the glucose feed rate is manipulated by a proportional–integral–derivative controller to maintain a fixed DO value. The controller function is based on the fact that the DO increases when the substrate supply rate is decreased under glucose-limited conditions. More complex forms of

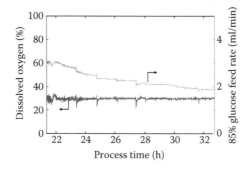

FIGURE 6.12 Example of a variable-controlled glucose feeding rate based on DO measurements in a development of *E. coli* process. (Courtesy of Hamilton, R., Genentech.)

the method make use of a probing feeding strategy where a glucose feed pulse is performed to evaluate qualitatively if glucose is limited or in excess (Åkesson et al. 1999, 2001). This glucose test pulse may be accomplished by briefly increasing the glucose feed rate followed by another brief period of feed rate reduction. If the DO is not affected by this feed pulse, then glucose must be in excess, and the glucose feeding rate is decreased. Oppositely, if DO is perturbed (decreased followed by an increase in DO), glucose is limiting, and glucose feed rate needs to be increased. The magnitude by which glucose feed rate is changed may vary with the method in use (Lin et al. 2001; Zawada and Swartz 2005).

6.3.3 PREVENTING ANTIBODY REDUCTION BY MAINTAINING pO$_2$ CONTROL IN THE HARVESTED CELL CULTURE FLUID

In the production of recombinant proteins at manufacturing scale, protein integrity and purity are assessed, and the results are noted in regulatory submissions. Drug substance (i.e., final formulated process pool containing the protein of interest) structural integrity and purity are typically verified by tryptic peptide map analysis (capillary electrophoresis-sodium dodecyl sulfate method) of the reduced and nonreduced mass and anti-CHO host protein immunoblots. These analyses are also done as a routine procedure during process development for improving process understanding and possibly detecting any protein degradation event. Although antibody degradation by released proteases is negligible, for example, antibody reduction evidence has recently been discussed by several biotechnology companies (Atkinson 2009; Laird et al. 2010; Mun et al. 2010). Protein reduction means the breaking of disulfide bonds established by two cysteine residues in close proximity. Figure 6.13 shows the disulfide bonds established in a typical IgG monoclonal antibody. The structure of IgGs has been thoroughly described (Roitt et al. 1993) as being composed of a Fab region (defining antibody–antigen binding properties) and an Fc region (defining antibody–effector cells/molecules binding properties). IgGs have two identical heavy chains (~50 kDa) and two identical light chains (~25 kDa). Depending on the IgG subclasses (IgG1, IgG2, IgG3, and IgG4) that differ on the number/location of

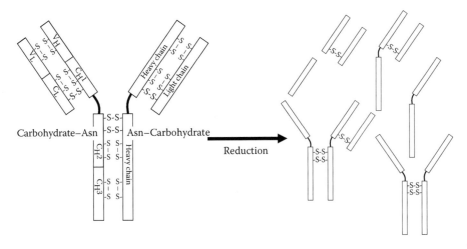

FIGURE 6.13 IgG1 antibody structure and schematic representation of antibody reduction.

interchain disulfide bonds, the length of the hinge region varies. IgG1 seems to be the most common IgG both in relative abundance in human plasma and in its use for recombinant antibody production (Wang et al. 2007). The interchain disulfide bonds are required for the protein stability and antibody function. As described by Kao et al. (2009), recombinant antibody disulfide bond reduction has been seen during the processing of the harvested cell culture fluid (HCCF) after the centrifugation/filtration step for removal of cells and cell debris (Figure 6.3).

Antibody reduction is a major concern because it may significantly impact the process yield. Based on the work from Mike W. Laird at Genentech described here, the antibody reduction seems to be observed after mechanical cell lysis due to the harvest procedure, more specifically at a cell lysis range of 30%–70% depending on the antibody, the cell host, the production system, and production parameters used (Kao et al. 2009, 2010; Trexler-Schmidt et al. 2010). The proposed mechanism for the observed antibody reduction and reduction inhibition methods are discussed in detail by Kao et al. (2009). The reduction events seem to be caused by intracellular factors, identified as thioredoxin or thioredoxin-like enzymes and their associated enzyme pathway intermediates (e.g., NADPH) (Kao et al. 2009, 2010; Trexler-Schmidt et al. 2010), which are released into the HCCF by cell lysis. Interestingly, although the mechanism has not yet been fully elucidated, supplying oxygen to the HCCF fluid maintains an oxidizing environment and avoids antibody reduction (Mun et al. 2010). The use of oxygen sparging (air or pure oxygen) of the HCCF for inhibiting antibody reduction is another application of PAT. The HCCF air sparging rate is controlled as part of the production process, more specifically during the HCCF hold before the first purification step. Depending on the lysis level obtained during the harvest step, the quantity of sparged air needed will vary and can be supplied as necessary to maintain a 30% dO_2 level (Figure 6.14) (Mun et al. 2010).

In essence, supplying oxygen to the HCCF allows the "burning" of the fuel sources released from the cells and, thus, prevents the oxygen level from dropping to levels

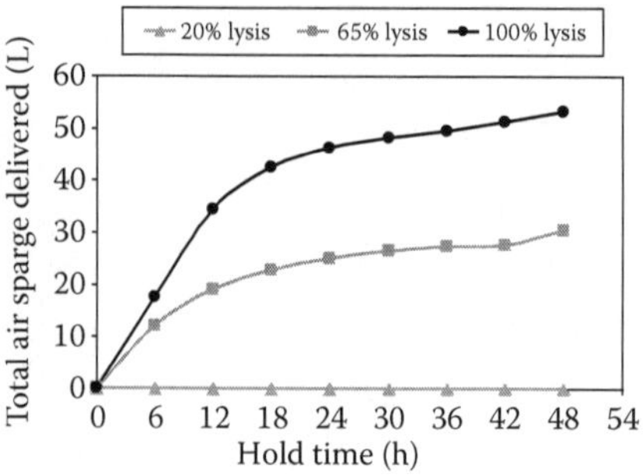

FIGURE 6.14 PAT application for antibody reduction inhibition by air sparging HCCF. Air sparging needs depends on cell lysis levels and will vary to maintain a set point of 30% dO_2 in the HCCF tank. (Courtesy of Mun, M., Genentech.)

low enough to allow the antibody reduction reactions to occur. By this sparging-on-demand approach, the amount of air sparging delivered may vary depending on the performance of each manufacturing run for the same product/process. Because the reducing enzyme levels and released "fuels" levels will vary as a function of the end state of the cell culture process combined with the degree of cell lysis during the harvest operation, this is an example of an "adaptable" process.

6.4 CONCLUSIONS

The PAT concept is not new for industry. However, the regulatory agency's intent to make it an industry-regulatory agreement makes this a very challenging and exciting time for both the pharmaceutical and biotechnology industries. To achieve the desired process understanding and control, the necessary effort has to be placed in developing tools and applying them in either the development or industrial stages. New sensors in biotechnology face big challenges. They must be selective to new chemical molecules, have wide detection limits, and should be robust to work in process environments. With the proper investment and risk assessment, the true advantage of PAT as a strategy enabling consistent product quality will be more real in the near future.

ACKNOWLEDGMENTS

The authors thank Bob Kiss, Mike Laird, Harry Lam, and Tina Larson from Genentech for their invaluable expertise in PAT/QbD strategies; Brad Snedecor, Sid Haskell, and Ryan Hamilton from Genentech for their inspiring work on glucose feeding strategies for *Escherichia coli* processes; Melissa Mun, Stefanie Khoo, Melody Trexler-Schmidt, and Michael Laird from Genentech for their innovative work on preventing antibody reduction by maintaining pO_2 control in the harvested cell culture fluid; Thomas DiRocco from Genentech for his contribution on controlled nutrient feeding strategies for the mammalian cell culture processes; Kara Calhoun for providing data regarding protein property variability at different manufacturing sites; Michael Milligan, Nicholas Lewin-Koh, and Daniel Coleman from Genentech for their contribution on NIR in cell culture processes; and C.D. Feng from Rosemount Analytical for his contribution on pH sensor analysis. The authors would like to especially thank Adeyma Arroyo, Mike Laird, Ryan Hamilton, Melissa Mun, Gayle Derfus, Prateek Gupta, Ashraf Amanullah, Tina Larson, and Bob Kiss for reviewing this chapter and for all their suggestions.

REFERENCES

Åkesson, M., P. Hagander, and J.P. Axelsson. 1999. A probing feeding strategy for *Escherichia coli* cultures. *Biotechnol. Tech.* 13: 523–528.

Åkesson, M., P. Hagander, and J.P. Axelsson. 2001. Avoiding acetate accumulation in *Escherichia coli* cultures using feedback control of glucose feeding. *Biotechnol. Bioeng.* 73 (3): 223–230.

Altamirano, C., A. Illanes, S. Becerra, J.J. Cairo, and F. Godia. 2006. Considerations on the lactate consumption by CHO cells in the presence of galactose. *J. Biotechnol.* 125 (4): 547–556.

Altamirano, C., C. Paredes, A. Illanes, J.J. Cairo, and F. Godia. 2004. Strategies for fed-batch cultivation of t-PA producing CHO cells: Substitution of glucose and glutamine and rational design of culture medium. *J. Biotechnol.* 110: 171–179.

Amrhein, M., B. Srinivasan, D. Bonvin, and M.M. Schumacher. 1999. Calibration of spectral reaction data. *Chemom. Intell. Lab. Syst.* 46: 249–264.

Andersen, D.C. and L. Krummen. 2002. Recombinant protein expression for therapeutic applications. *Curr. Opin. Biotechnol.* 13: 117–123.

Andersen, D.C., J. Swartz, T. Ryll, N. Lin, and B. Snedecor. 2001. Metabolic oscillations in an *E. coli* fermentation. *Biotechnol. Bioeng.* 75: 212–218.

Armani, A.M., R.P. Kulkarni, S.E. Fraser, R.C. Flagan, and K.J. Vahala. 2007. Label-free, single molecules detection with optical microcavities. *Science* 317: 783–787.

Arnold, S.A., J. Crowley, N. Woods, L. Harvey, and B. McNeil. 2003. In-situ near infrared spectroscopy to monitor key analytes in mammalian cell cultivation. *Biotechnol. Bioeng.* 84: 13–19.

Atkinson, E.M. 2009. A manufacturing site journey: Twelve months from chemical synthesis to cell culture capability. ESACT Conference, Dublin, Ireland.

Baffi, G., E.G. Martin, and A.J. Morris. 1991. Non-linear projection to latent structures revisited: The quadratic PLS algorithm. *Comput. Chem. Eng.* 23: 395–411.

Ballerstadt, R., A. Kholodnykh, C. Evans, A. Boretsky, M. Motamedi, A. Gowda, and R. McNichols. 2007. Affinity-based turbidity sensor for glucose monitoring by optical coherence tomography: Toward the development of an implantable sensor. *Anal. Chem.* 79: 6965–6974.

Bard, A.J. and L.R. Faulkner. 1980. *Electrochemical Methods—Fundamentals and Applications*. New York: John Wiley & Sons.

Bebbington, C.R., G. Renner, S. Thomson, D. King, D. Abrams, and G.T. Yarranton. 1992. High-level expression of a recombinant antibody from myeloma cells using a glutamine synthetase gene as an amplifiable selectable marker. *Nat. Biotechnol.* 10 (2): 169–175.

Bezzo, F., S. Macchietto, and C.C. Pantelides. 2003. General hybrid multizonal/CFD approach for bioreactor modeling. *AIChE J.* 49: 2133–2148.

Blanco, M., J. Coello, M. Elaamrani, H. Iturriaga, and S. Maspoch. 1996. Partial least squares regression for the quantitation of pharmaceutical dosages in control analysis. *J. Pharm. Biomed. Anal.* 15: 329–338.

Brown, G.K. 2000. Glucose transporters: Structure, function and consequences of deficiency. *J. Inherited Metab. Dis.* 23: 237–246.

Bultzingslowen, V., A.K. McEvoy, C. McDonagh, B.D. MacCraith, I. Klimant, C. Krause, and O.S. Wolfbeis. 2002. Sol–gel based optical carbon dioxide sensor employing dual luminophore referencing for application in food packaging technology. *Analyst* 127: 1478–1483.

Butler, M. and A. Christie. 1994. Adaptation of mammalian cells to non-ammoniagenic media. *Cytotechnology* 15: 87–94.

Cannizzaro, C., R. Gugerli, I. Marison, and U. Stockar. 2003. On-line biomass monitoring of CHO perfusion culture with scanning dielectric spectroscopy. *Biotechnol. Bioeng.* 84: 597–610.

Chambers, R.S. 2005. High-throughput antibody production. *Curr. Opin. Chem. Biol.* 9: 46–50.

Chen, A., R. Chitta, D. Chang, and A. Amanullah. 2009. Twenty-four well plate miniature bioreactor system as a scale-down model for cell culture process development. *Biotechnol. Bioeng.* 102: 148–160.

Chen, M. and L.W. Forman. 1999. Polypeptide production in animal cell culture. Patent 5,856,179.

Chena, P. and S.W. Harcumb. 2006. Effects of elevated ammonium on glycosylation gene expression in CHO cells. *Metab. Eng.* 8: 123–132.

Chung, H., M.A. Arnold, M. Rhiel, and D. Murhammer. 1995. Simultaneous measurement of glucose and glutamine in aqueous solutions by near-infrared spectroscopy. *Appl. Biochem. Biotechnol.* H50: 109–125.

Chung, D.C., C.C. Chang, and J.A. Groves. 2003. High cell density induces spontaneous bifurcations of dissolved oxygen controllers during CHO cell fermentations. *Biotechnol. Bioeng.* 48: 224–232.

Chuppa, S., Y.-S. Tsai, S. Yoon, S. Shackleford, C. Rozales, R. Bhat, G. Tsay, C. Matanguihan, K. Konstantinov, and D. Naveh. 1997. Fermentor temperature as a tool for control of high-density perfusion cultures of mammalian cells. *Biotechnol. Bioeng.* 55: 328–338.

Cruz, H.J., J.L. Moreira, and M.J.T. Carrondo. 2000. Metabolically optimised BHK cell fed-batch cultures. *J. Biotechnol.* 80: 109–118.

Davidson, K.M., S. Sushil, C. Eggleton, and M.R. Marten. 2003. Using computational fluid dynamics software to estimate circulation time distributions in bioreactors. *Biotechnol. Prog.* 19: 1480–1486.

Davies, A., A. Greene, E. Lullau, and W.M. Abbott. 2005. Optimization and evaluation of a high-throughput mammalian protein expression system. *Protein Expression Purif.* 42: 111–121.

Derfus, G., D. Abramzon, M. Tung, D. Chang, R. Kiss, and A. Amanullah. 2010. Cell culture monitoring via an auto-sampler and an integrated multi-functional off-line analyzer. *AIChE J.* 26: 284–292.

Diamond, D., E. McEnroe, M. McCarrick, and A. Lewenstam. 1994. Evaluation of a new solid-state reference electrode junction material for ion-selective electrodes. *Electroanalysis* 6: 962–971.

Docherty, P.A. and M.D. Snider. 1991. Effects of hypertonic and sodium-free medium on transport of a membrane glycoprotein along the secretory pathway in cultured mammalian cells. *J. Cell. Physiol.* 146: 34–42.

Duval, D., C. Demangel, S. Miossec, and I. Geahel. 1992. Role of metabolic waste products in the control of cell proliferation and antibody production by mouse hybridoma cells. *Hybridoma* 11 (3): 311–322.

Eiteman, M.A. and E. Altman. 2006. Overcoming acetate in *Escherichia coli* recombinant protein fermentations. *Trends Biotechnol.* 24 (11): 530–536.

El-Hagrasy, A.S., H.R. Morris, F. M'Amico, R.A. Lodder, and J.K. Drennen III. 2001. Near-infrared spectroscopy and imaging for the monitoring of powder blend homogeneity. *J. Pharm. Sci.* 90: 1298–1307.

Elsas, L.J. and N. Longo. 1992. Glucose transporters. *Annu. Rev. Med.* 43: 377–393.

Esbensen, K., D. Kirsanov, A. Legin, A. Rudnitskaya, J. Mortensen, J. Pedersen, L. Vognsen, S. Makarychev-Mikhailov, and Y. Vlasov. 2004. Fermentation monitoring using multi-sensor systems; Feasibility study of the electronic tongue. *Anal. Bioanal. Chem.* 378: 391–395.

Evans, H. and T. Larson. 2006. Disparity in on-line and off-line pH measurements of a cell culture process. IFPAC Conference, Baltimore.

Ganguly, J. and G. Vogel. 2006. Process analytical technology (PAT) and scalable automation for bioprocess control and monitoring—A case study. *Pharm. Eng.* 26: 1–9.

Gawlitzek, M., T. Ryll, J. Lofgren, and M.B. Sliwkowski. 2000. Ammonium alters *N*-glycan structures of recombinant TNFR-IgG: Degradative versus biosynthetic mechanisms. *Biotechnol. Bioeng.* 68 (6): 638–646.

Glacken, M.W., E. Adema, and A.J. Sinskey. 1988. Mathematical description of hybridoma culture kinetics: I. Initial metabolic rates. *Biotechnol. Bioeng.* 32: 491–506.

Gonzalez, F. and R. Pous. 1995. Quality control in manufacturing process by near infrared spectroscopy. *J. Pharm. Biomed. Anal.* 13: 419–423.

Goudar, C., R. Biener, C. Zhang, J. Michaels, and K. Konstantinov. 2006. Towards industrial application of quasi real-time metabolic flux analysis for mammalian cell culture. *Adv. Biochem. Eng./Biotechnol.* 101: 99–118.

Gupta, A., G.E. Peck, R.D. Miller, and K.R. Morris. 2005. Real-time near-infrared monitoring of content uniformity, moisture content, compact density, tensile strength, and Young's modulus of roller compacted powder blends. *J. Pharm. Sci.* 94: 1589–1597.

Haaland, D. and E.V. Thomas. 1988. Partial-least squares method for spectral analysis. *Anal. Chem.* 60: 1193–1202.

Hamilton, R., J. Gunson, and M. Laird. 2009. Adaptive glucose feeding schedule for an *E. coli* fermentation process. Poster at AIChE Meeting, Nashville, TN.

Hanson, M.A., G. Xundong, Y. Kostov, K.A. Brorson, A.R. Moreira, and G. Rao. 2007. Comparisons of optical pH and dissolved oxygen sensors with traditional electrochemical sensors during mammalian cell culture. *Biotechnol. Bioeng.* 97: 833–841.

Harms, P., Y. Kostov, and G. Rao. 2002. Bioprocess monitoring. *Curr. Opin. Biotechnol.* 13: 124–127.

Heath, C. and R. Kiss. 2007. Cell culture process development: Advances in process engineering. *Biotechnol. Prog.* 23: 46–51.

Henriques, J.G., S. Buziol, E. Stocker, A. Voogd, and J.C. Menezes. 2010. Monitoring mammalian cell cultivations for monoclonal antibody production using near-infrared spectroscopy. *Adv. Biochem. Eng. Biotechnol.* 116: 73–97.

Henry, C.S., L.J. Broadbelt, and V. Hatzimanikatis. 2007. Thermodynamics-based metabolic flux analysis. *Biophys. J.* 92: 1792–1805.

Hinz, D.C. 2006. Process analytical technologies in the pharmaceutical industry: The FDA's PAT initiative. *Anal. Bioanal. Chem.* 384: 1036–1042.

Hu, W.-S. and V.B. Himes. 1989. Stoichiometric considerations of mammalian cell metabolism in bioreactors. In *Bioproducts and Bioprocesses*, edited by A. Fiechter, H. Okada, and R.D. Tanner, pp. 33–45. Berlin: Springer-Verlag.

Hyde, R., P.M. Taylor, and H.S. Hundal. 2003. Amino acid transporters: Roles in amino acid sensing and signaling in animal cells. *Biochem. J.* 373: 1–18.

ICH. 2009. Pharmaceutical Development. Q8 (R2). Revision 2.

Ik-Hwan, K., M.H. Cho, and S.S. Wang. 1993. Measurement of hydrodynamic shear by using dissolved oxygen sensor. *Biotechnol. Bioeng.* 41: 296–302.

Illingworth, J.A. 1981. A common source of error in pH measurements. *Biochem. J.* 195: 259–262.

Irani, N., M. Wirth, J. van Den Heuvel, and R. Wagner. 1999. Improvement of the primary metabolism of cell cultures by introducing a new cytoplasmic pyruvate carboxylase reaction. *Biotechnol. Bioeng.* 66 (4): 238–246.

Jensen, E.B. and S. Carken. 1990. Production of recombinant human growth hormone in *Escherichia coli*: Expression of different precursors and physiological effects of glucose, acetate, and salts. *Biotechnol. Bioeng.* 36: 1–11.

Johnston, W., R. Cord-Ruwisch, and M.J. Cooney. 2002. Industrial control of recombinant *E. coli* fed-batch culture: New perspectives on traditional controlled variables. *Bioprocess Biosyst. Eng.* 25: 111–120.

Kao, Y.-H., D. Hewitt, M. Trexler-Schmidt, and M.W. Laird. 2010. Mechanism of antibody reduction in cell culture production processes. *Biotechnol. Bioeng.* 107 (4): 622–632.

Kao, Y.-H., M.T. Schmidt, M. Laird, R.L. Wong, and D.P. Hewitt. 2009. Prevention of disulfide bond reduction during recombinant production of polypeptides. Patent WO 2009/009523.

Kermis, H.R., Y. Kostov, P. Harms, and G. Rao. 2008. Dual excitation ratiometric fluorescent pH sensor for noninvasive bioprocess monitoring: Development and application. *Biotechnol. Prog.* 18: 1047–1053.

Kim, B.S., S.C. Lee, S.Y. Lee, Y.K. Chang, and H.N. Chang. 2004. High cell density fed-batch cultivation of *Escherichia coli* using exponential feeding combined with pH-stat. *Bioprocess Biosyst. Eng.* 26 (3): 147–150.

Kirdar, A.O., K.D. Green, and A.S. Rathore. 2008. Application of multivariate data analysis for identification and successful resolution of a root cause for a bioprocessing application. *Biotechnol. Prog.* 24: 720–726.

Kong, D., R. Gentz, and J. Zhang. 1998. Development of a versatile computer integrated control system for bioprocess controls. *Cytotechnology* 26: 227–236.

Kovács, R., F. Házi, Z. Csikor, and P. Miháltz. 2007. Connection between oxygen uptake rate and carbon dioxide evolution rate in aerobic thermophilic sludge digestion. *Period. Polytech., Chem. Eng.* 51 (1): 17–22.

Laird, M.W. 2010. Antibody disulfide reduction in CHO production processes. Biochemical Engineering XVI Conference, Vermont, USA.

Lam, H. and R. Kiss. 2007. Opportunities for applying PAT to biotech processes. Biochemical Engineering XV Conference, Québec City, Canada.

Langheinrich, C. and A. Nienow. 1999. Control of pH in large-scale, free suspension animal cell bioreactors: Alkali addition and pH excursions. *Biotechnol. Bioeng.* 10: 171–179.

Larson, T.M., M. Gawlitzek, H. Evans, U. Albers, and J. Cacia. 2002. Chemometric evaluation of on-line high-pressure liquid chromatography in mammalian cell cultures: Analysis of amino acids and glucose. *Biotechnol. Bioeng.* 77: 553–563.

Larsson, E.M., J. Alegret, M. Kall, and D.S. Sutherland. 2007. Sensing characteristics of NIR localized surface plasmon resonances in gold nanorings for application as ultrasensitive biosensors. *Nano Lett.* 7: 1256–1263.

Lee, S.Y. 1996. High cell-density culture of *Escherichia coli. Trends Biotechnol.* 14 (3): 98–105.

Lee, J., Y. Seo, T. Lim, P.L. Bishop, and I. Papautsky. 2007. MEMS needle-type sensor array for in-situ measurements of dissolved oxygen and redox potential. *Environ. Sci. Technol.* 41: 7857–7863.

Legmann, R., H.B. Schreyer, R.G. Combs, E.L. McCormick, A.P. Russo, and S.T. Rodgers. 2009. A predictive high-throughput scale-down model of monoclonal antibody production in CHO cells. *Biotechnol. Bioeng.* 104: 1107–1120.

Li, J., P.C. Toh, Y.Y. Lee, M. Yap, and A. Amanullah. 2010 Establishment of an on-line, dynamic feeding method and applications for high titer CHO processes. Cell Culture Engineering XII Conference, Banff, Canada.

Lin, A.A., R. Kimura, and W.M. Miller. 1993. Production of tPA in recombinant CHO cells under oxygen-limited conditions. *Biotechnol. Bioeng.* 42: 339–350.

Lin, H.Y., B. Mathiszik, B. Xu, S.O. Enfors, and P. Neubauer. 2001. Determination of the maximum specific uptake capacities for glucose and oxygen in glucose-limited fed-batch cultivations of *Escherichia coli. Biotechnol. Bioeng.* 73 (5): 347–357.

Ljunggren, J. and L. Häggström. 1994. Catabolic control of hybridoma cells by glucose and glutamine limited fed batch cultures. *Biotechnol. Bioeng.* 44: 808–818.

Lopes, J.A., P.F. Costa, T.P. Alves, and J.C. Menezes. 2004. Chemometrics in bioprocess engineering: Process analytical technologies (PAT) applications. *Chemom. Intell. Lab. Syst.* 74: 269–275.

Luan, Y.-T., T.C. Stanek, and D. Drapeau. 2005. Restricted glucose feed for animal cell culture. US Patent Application Publication US2005/0070013 A1.

Luypaert, J., D.L. Massart, and Y. Vander Heyden. 2007. Near-infrared spectroscopy applications in pharmaceutical analysis. *Talanta* 72: 865–883.

Mandenius, C.F. 2000. Electronic noses for bioreactor monitoring. *Adv. Biochem. Eng./Biotechnol.* 66: 65–82.

Martinelle, K. and L. Häggström. 1993. Mechanisms of ammonia and ammonium ion toxicity in animal cells: Transport across cell membranes. *J. Biotechnol.* 30: 339–350.

Mattes, R., D. Root, M.A. Sugui, F. Chen, X. Shi, J. Liu, and P.A. Gilbert. 2009. Real-time bioreactor monitoring of osmolality and pH using near-infrared spectroscopy. *BioProcess Int.* 7: 44–50.

McQueen, A. and J.E. Bailey. 1990a. Effect of ammonium ion and extracellular pH on hybridoma cell metabolism and antibody production. *Biotechnol. Bioeng.* 34: 1067–1077.

McQueen, A. and J.E. Bailey. 1990b. Mathematical modeling of the effects of ammonium ion on the intracellular pH on hybridoma cell metabolism and antibody production. *Biotechnol. Bioeng.* 35: 897–906.

McQueen, A. and J.E. Bailey. 1991. Growth inhibition of hybridoma cells by ammonium ion: Correlation with effects on intracellular pH. *Bioprocess Eng.* 6: 49–61.

McShane, M. and G.L. Cote. 1998. Near-infrared spectroscopy for determination of glucose, lactate and ammonia in cell culture media. *Appl. Spectrosc.* 52: 1073–1078.

Menezes, J.C., A.P. Ferreira, L.O. Rodrigues, L.P. Brás, and T.P. Alves. 2010. Chemometrics role within the PAT context: Examples from primary pharmaceutical manufacturing. Edited by S. Brown, R. Tauler, and B. Walczak. *Compr. Chemom.* 3: 313–357 (4 volumes).

Miller, W.M., H.W. Blanch, and C.R. Wilke. 1988. A kinetic analysis of hybridoma growth and metabolism in batch and continuous suspension culture: Effect of nutrient concentration, dilution rate, and pH. *Biotechnol. Bioeng.* 32 (8): 947–965.

Miller, W.M., C.R. Wilke, and H.W. Blanch. 1989. Transient responses of hybridoma cells to nutrient additions in continuous culture: 1. Glucose pulse and step changes. *Biotechnol. Bioeng.* 33: 477–486.

Minamoto, Y., K. Ogawa, H. Abe, Y. Iochi, and K. Mitsugi. 1991. Development of a serum-free and heat-sterilizable medium and continuous high-density cell culture. *Cytotechnology* 5: 35–51.

Mun, M., S. Khoo, M. Trexler-Schmidt, and L.W. Laird. 2010. Implementation of harvested cell culture fluid (HCCF) air sparging to prevent antibody disulfide bond reduction. SIM Conference, San Francisco, USA.

Naciri, M., D. Kuystermans, and M. Al-Rubeai. 2008. Monitoring pH and dissolved oxygen in mammalian cell culture using optical sensors. *Cytotechnology* 57: 245–250.

Navratil, M., A. Norberg, L. Lembren, and C.F. Madenius. 2005. On-line multi-analyzer monitoring of biomass, glucose and acetate for growth rate control of a *Vibrio cholerae*, fed-batch cultivation. *J. Biotechnol.* 115: 67–79.

Neermann, J. and R. Wagner. 1996. Comparative analysis of glucose and glutamine metabolism in transformed mammalian cell lines, insect and primary liver cells. *J. Cell. Physiol.* 166: 152–169.

Opel, C., J. Li, and A. Amanullah. 2010. Quantitative modeling of viable cell density, cell size, intracellular conductivity and membrane capacitance in batch and fed-batch CHO processes using dielectric spectroscopy. *Biotechnol. Prog.* 26: 1187–1199.

Ozturk, S.S. and B.O. Palsson. 1991. Growth, metabolic and antibody production kinetics of hybridoma cell culture. 2. Effects of serum concentration, dissolved oxygen concentration, and medium pH in a batch reactor. *Biotechnol. Prog.* 7: 481–494.

Paredes, C., E. Prats, J.J. Cairo, F. Azorin, L. Cornudella, and F. Godia. 1999. Modification of glucose and glutamine metabolism in hybridoma cells through metabolic engineering. *Cytotechnology* 30: 1–3.

Paul, E.L., V. Atiempo-Obeng, and S. Kresta. 2004. *Handbook of Industrial Mixing: Science and Practice*. Hoboken, NJ: John Wiley & Sons.

Phelps, M., J. Hobbs, D.G. Kilburn, and R.F.B. Turner. 1995. An autoclavable glucose biosensor for microbial fermentation monitoring and control. *Biotechnol. Bioeng.* 46: 514–524.

Potyryailo, R. and W.G. Morris. 2007. Multianalyte chemical identification and quantitation using a single radio frequency identification sensor. *Anal. Chem.* 79: 45–51.

Proell, T. and M.N. Karim. 1994. Nonlinear control of a bioreactor using exact and I/O linearization. *Int. J. Control* 60: 499–519.

Qin, S.J. and T.J. McAvoy. 1992. Nonlinear PLS modeling using neural networks. *Comput. Chem. Eng.* 16: 379–391.

Ramirez, O.T. and R. Mutharasan. 1990. Cell cycle- and growth phase-dependent variations in size distribution, antibody productivity, and oxygen demand in hybridoma cultures. *Biotechnol. Bioeng.* 36: 838–848.

Reuveny, S., D. Velez, J.D. Macmillan, and L. Miller. 1986. Factors affecting cell growth and monoclonal antibody production in stirred reactors. *J. Immunol. Methods* 86: 53–59.

Rhiel, M., M. Cohen, D. Murhammer, and M.A. Arnold. 2002a. Nondestructive near-infrared spectroscopic measurement of multiple analytes in undiluted samples of serum-based cell culture media. *Biotechnol. Bioeng.* 77: 73–81.

Rhiel, M., P. Ducommun, I. Bolzonella, I. Marison, and U. von Stockar. 2002b. Real-time in situ monitoring of freely suspended and immobilized cell cultures based on mid-infrared spectroscopic measurements. *Biotechnol. Bioeng.* 77: 174–185.

Riley, M.R. and H.M. Crider. 2000. The effect of analyte concentration range on measurement errors obtained by NIR spectroscopy. *Talanta* 52: 473–484.

Rodrigues, L.O., L.M. Vieira, J.P. Cardoso, and J.C. Menezes. 2008. The use of NIR as multiparametric in-situ monitoring technique in filamentous fermentation systems. *Talanta* 75: 1356–1361.

Roitt, I., J. Brostoff, and D. Male. 1993. *Immunology*, 3rd edition, edited by Linda Gamlin. Mosby: Elsevier Science.

Roth, E., G. Ollenschlager, G. Hamilton, A. Simmel, K. Langer, W. Fekl, and R. Jakesz. 1988. Influence of two glutamine-containing dipeptides on growth of mammalian cells. *In Vitro Cell. Dev. Biol.* 24 (7): 696–698.

Saucedo, V., C. Feng, B. Wolk, and A. Arroyo. 2011. Studying the drift of in-line pH measurements in cell culture. *Biotechnol. Prog.* 27: 885–890.

Saucedo, V.M. and M.N. Karim. 1997. Real-time optimization of fed-batch bioreactors using Markov decision processes. *Biotechnol. Bioeng.* 55: 317–327.

Saucedo, V., M. Milligan, N. Lewin-Koh, D. Coleman, B. Wolk, T. Larson, and A. Arroyo. 2009. Practical issues implementing an in-situ NIR for real time monitoring of cell culture bioreactors. ACS Annual Conference, Washington, DC, USA.

Scarff, M., S.A. Arnold, L.M. Harvey, and B. McNeil. 2006. Near infrared spectroscopy for bioprocess monitoring and control: Current status and future trends. *Crit. Rev. Biotechnol.* 26: 17–39.

Schiller, A., R.A. Wessling, and B. Singaram. 2007. A fluorescent sensor array for saccharides based on boronic acid appended bipyridinium salts. *Angew Chem.* 119: 6577–6579.

Schmid, G., G.H. Blanch, and C.R. Wilke. 1990. Hybridoma growth, metabolism, and product formation in HEPES-buffered medium: II. Effect of pH. *Biotechnol. Lett.* 12 (9): 633–638.

Schulman, S.G., S. Chen, F. Bai, M.J.P. Leiner, L. Weis, and O.S. Wolfbeis. 1995. Dependence of the fluorescence of immobilized 1-hydroxypyrene-3,6,8-trisulfonate on solution pH: Extension of the range of applicability of a pH fluorosensor. *Anal. Chim. Acta* 304: 165–170.

Scully, P.J., L. Betancor, J. Bolyo, S. Dzyadevych, J.M. Guisan, R. Fernandez-Lafuente, N. Jaffrezic-Renault, G. Kuncova, V. Matejec, B. O'Kennedy, O. Podrazky, K. Rose, L. Sasek, and J.S. Young. 2007. Optical fibre biosensors using enzymatic transducers to monitor glucose. *Meas. Sci. Technol.* 18: 3177–3186.

Shiloach, J. and U. Rinas. 2009. Glucose and acetate metabolism in *E. coli*—System level analysis and biotechnological applications in protein production processes. In *Systems Biology and Biotechnology of Escherichia coli*, edited by S.Y. Lee. Netherlands: Springer Science, Business Media.

Stuart, D.A., J.M. Yuen, N. Shah, O. Lyandres, C.R. Yonzon, M.R. Glucksberg, J.T. Walsh, and R.P. Van Duyne. 2006. In vivo glucose measurement by surface-enhanced Raman spectroscopy. *Anal. Chem.* 78: 7211–7215.

Sureshkumar, G.K. and R. Mutharasan. 1991. The influence of temperature on a mouse–mouse hybridoma growth and monoclonal antibody production. *Biotechnol. Bioeng.* 37 (3): 292–295.

Tamburini, E., G. Vaccari, S. Tosi, and A. Trilli. 2003. Near-infrared spectroscopy: A tool for monitoring submerged fermentation processes using an immersion optical-fiber sensor. *Appl. Spectrosc.* 57: 132–138.

Trexler-Schmidt, M., S. Sargis, J. Chiu, S. Sze-Khoo, M. Mun, Y.-H. Kao, and M.W. Laird. 2010. Identification and prevention of antibody disulfide bond reduction during cell culture manufacturing. *Biotechnol. Bioeng.* 106 (3): 452–461.

Triadaphillou, S., E. Martin, G. Montague, A. Norden, P. Jeffkins, and S. Stimpson. 2007. Fermentation process tracking through enhanced spectral calibration modeling. *Biotechnol. Bioeng.* 97: 554–567.

U.S. FDA. 2004. Guidance for Industry: PAT—A framework for innovative pharmaceutical development, manufacturing, and quality assurance. FDA (CDER, CVM, ORA).

Vallino, J.J. and G. Stephanopoulos. 1993. Metabolic flux distributions in *Corynebacterium glutamicum* during growth and lysine overproduction. *Biotechnol. Bioeng.* 41: 633–646.

Varma, A. and B.O. Palsson. 1994. Metabolic flux balancing: Basic concepts, scientific and practical use. *Nat. Biotechnol.* 12: 994–998.

Wang, W., S. Singh, D.L. Zeng, K. King and S. Nema. 2007. Antibody structure, instability and formulation. *J. Pharm. Sci.* 96 (1): 1–26.

Wen, Z., D. Lju, B. Ye, and X. Zhou. 1997. Development of disposable electrochemical sensor with replaceable glucose oxidase tip. *Anal. Commun.* 34: 27–30.

Willis, R.C. 2004. PAT pending: Government legislation is prompting drug companies to reevaluate their manufacturing processes. *Mod. Drug Discovery* 7: 29–32.

Wise, B. and N.B. Gallagher. 1996. The process chemometrics approach to process monitoring and fault detection. *J. Process Control* 6: 329–348.

Wold, S., N. Kettaneh-Wold, and B. Skagerberg. 1989. Nonlinear PLS modeling. *Chemom. Intell. Lab. Syst.* 7: 53–65.

Wolfe, A.J. 2005. The acetate switch. *Microbiol. Mol. Biol. Rev.* 69 (1): 12–50.

Wood, I.S. and P. Trayhurn. 2003. Glucose transporters (GLUT and SGLT): Expanded families of sugar transport proteins. *Br. J. Nutr.* 89: 3–9.

Xie, L., W. Pilbrough, C. Metallo, T. Zhong, L. Pikus, J. Leung, J.G. Auniņš, and W. Zhou. 2002. Serum-free suspension cultivation of PER.C6 cells and recombinant adenovirus production under different pH conditions. *Biotechnol. Bioeng.* 80: 569–579.

Xu, Y., A.S. Jeevarajan, J. Fay, T.D. Taylor, and M.M. Anderson. 2002. On-line measurement of glucose in a rotating wall perfused vessel bioreactor using an amperometric glucose sensor. *J. Electrochem. Soc.* 149: H103–H106.

Zawada, J. and J. Swartz. 2005. Maintaining rapid growth in moderate-density *Escherichia coli* fermentations. *Biotechnol. Bioeng.* 89 (4): 407–415.

Zeng, A.P. and W.D. Deckwer. 1995. Mathematical modeling and analysis of glucose and glutamine utilization and regulation in cultures of continuous mammalian cells. *Biotechnol. Bioeng.* 47 (3): 334–346.

Zhangi, J.A., A.E. Schmelzer, T.P. Mendoza, R.H. Knop, and W.M. Miller. 1999. Bicarbonate concentration and osmolality are key determinants in the inhibition of CHO cell polysialylation under elevated pCO_2 or pH. *Biotechnol. Bioeng.* 65: 182–191.

Zhou, W., C.-C. Chen, B. Buckland, and J. Aunins. 1997. Fed-batch culture of recombinant NS0 myeloma cells with high monoclonal antibody production. *Biotechnol. Bioeng.* 55: 783–792.

Zhou, W., J. Rehm, and W.S. Hu. 1995. High viable cell concentration fed-batch cultures of hybridoma cells through on-line nutrient feeding. *Biotechnol. Bioeng.* 46 (6): 579–587.

Zhou, Y.H. and J. Titchener-Hooker. 1999. Simulation and optimization of integrated bioprocesses: A case study. *J. Chem. Technol. Biotechnol.* 74: 289–292.

Zhu, J., Z. Zhu, Z. Lai, R. Wang, X. Guo, X. Wu, G. Zhang, Z. Zhang, Y. Wang, and Z. Chen. 2002. Planar amperometric glucose sensor based on glucose oxidase immobilized by chitosan film on Prussian blue layer. *Sensors* 2: 127–136.

7 Process Analytical Technology Applied to Raw Materials

Erik Skibsted

CONTENTS

7.1 INTRODUCTION

In the biotechnology industry, there is a tendency that process analytical technology (PAT) primarily concerns implementation of sophisticated analytical technologies and advanced feedback controls. It is, therefore, worthwhile to revisit the U.S. Food and Drug Administration (FDA) definition of PAT (US FDA 2004): "The Agency considers PAT to be a system for designing, analyzing, and controlling manufacturing through timely measurements (i.e., during processing) of critical quality and performance attributes of raw and in-process materials and processes, with the goal of ensuring final product quality."

The definition states that PAT is focused on monitoring not only a specific unit operation during manufacturing but also all critical performance attributes, and that these are not restricted to samples of intermediate product or the final drug product itself; they can also involve raw materials. Some of the largest potential benefits and savings can be achieved by understanding and handling of variability in raw material quality, given their impact in biopharmaceutical process yields and product quality. The purpose of this chapter is, therefore, to give examples of statistical methodologies and spectroscopic analytical techniques that in conjunction can be used to analyze raw materials and control variability originating from raw materials.

Focus on raw materials depends on the stage of the manufacturing process. In the research and development stage, focus is on development of the manufacturing process and analytical methods to control drug product quality. Time-to-market is a key driver in the research strategies. Secondly, the needed amount of drug substance for clinical trials may be supplied by one or few pilot batches. Therefore, very few different raw material batches are normally used during the development stage. Consequently, the effect of specific raw material lot-to-lot variability will not be fully captured/understood and accounted for during development. Raw material variability is considered an optimization problem that will be taken care of when the manufacturing process is transferred to full-scale commercial production. However, after transfer to full-scale production, it is often difficult to perform experiments due to scale, economics, and lack of time slots for experimental activities.

In the biopharmaceutical industry, there are many different types of raw materials. Some examples are listed in Table 7.1.

The main problems regarding raw material variability are

1. Little or no focus on raw material variability during development, time-to-market focus, and few batches
2. Lack of knowledge and training in statistical methods

TABLE 7.1
Raw Material Examples

Type of Raw Material	Examples	Analysis	Note
Media components for fermentation	Yeast extract, defined media, vitamin mixtures, hydrolysates, amino acids, sugar syrup	ID[a] analysis, supplier certificate	Complex and undefined mixtures, usual very broad specifications
Fine chemicals for purification and chemical modification processes	Ethanol, TEA[b], methanol, resins	ID, purity, supplier certificate	Typical off-the-shelf commodities
Excipients used in formulation of final drug product	Mannitol, sucrose, sodium chloride, WFI[c]	ID, supplier certificate	Typical off-the-shelf commodities
Specialized chemicals for API[d] modifications	Fatty acids	ID, purity	Manufactured for specific product either in-house or by CMO[e]

[a] Identification.
[b] Triethylamine.
[c] Water for injection.
[d] Active pharmaceutical ingredient.
[e] Contract manufacturing organization.

3. No resources to develop special analytical methods besides ID methods
4. The difficulty and expense of performing experiments at large scale
5. The difficulty of ensuring that all potential raw material variation has been covered during development

A key task during development is to identify and set specifications for raw materials used for manufacturing. Raw materials that are supplied off-the-shelf are typically delivered with a supplier certificate that declares that the material passes certain analytical assessments performed by the supplier and sometimes also in accordance with pharmacopeia requirements. In many cases, specifications are adopted from the supplier certificate with little or no experimental justification.

In the case that raw material variability is being investigated during process development, a major difficulty is to ensure that the investigated batches cover the expected variation in future manufacturing. A brute force strategy could be to purchase as many different batches as possible and perform experiment with all batches. This strategy is, however, very time and resource demanding and is no guarantee that the extremes are investigated. This strategy will provide a process robust to the investigated variability in the raw material batches used.

If different raw material qualities require different optimal settings, there is no single optimum. Instead, there are individual sets of optimal settings depending on the properties of the raw material being used. The second example in this text addresses this problem. A further consideration is that after the transfer of a process from pilot-scale to full-scale manufacturing, the factor settings are usually validated and locked within narrow operating intervals. Then when the process is running in full scale, it will be exposed to a far greater range of different qualities of raw materials, and there is a risk of receiving material from outside the tested range, which may result in producing off-spec product quality. Finally, if the process is being developed with, for example, only one or a few raw material batches from a specific supplier, there is a tendency to continue using that specific supplier for future full-scale manufacturing. Although it is a business risk to rely on one sole supplier of raw material, it is also an opportunity to collaborate closely with the supplier to minimize raw material variations. Therefore, the scientific work with characterizing and understanding raw material quality should be done in collaboration with strategic sourcing experts to find answers to questions such as "How many different suppliers can provide the raw material?" and "Is there a price advantage if we can use many different raw material qualities?" In any case, characterization of the raw materials and understanding the impact on process and product quality are a necessity to develop a well-understood manufacturing process that consistently produces high-quality products at the lowest possible price.

A simple and low-cost strategy to ensure that as much raw material variation is investigated during the development stage is to obtain analytical data with many different multivariate spectroscopic techniques such as near-infrared (NIR), nuclear magnetic resonance, or mass spectrometry of as many different raw material batches as can be purchased. The analytical data are then processed with multivariate statistical tools like principal component analysis (PCA), and the most different batches can be identified. Then experiments are performed with these vital few raw material batches. This strategy ensures with a high likelihood that a large part of the expected raw material variation has been investigated during the development stage.

To ensure process understanding and the impact of raw material variability on process performance and product quality, it is necessary to use statistical methods and analytical techniques (Lanan 2009). With spectroscopic methods, it is, for example, possible to correlate spectral fingerprints of complex raw materials such as yeast extract with the product yield in the fermentation (Kasprow et al. 1998). With the correct raw material measurements, it is possible to optimize the manufacturing process by using statistical methodologies (Mevik et al. 2001; Berget and Næs 2002a,b). Using advanced multivariate statistical data analysis, it is also possible to merge spectroscopic data obtained on raw materials with process setting data in a design of experiments framework and obtain a system that can handle uncontrolled raw material variation (Jørgensen and Næs 2004).

The following two examples will explore in depth how these methodologies can be applied to two very different types of raw materials, that is, yeast extract for a fermentation process and insulin crystals for a pulmonary insulin product.

7.2 EXAMPLES

7.2.1 Example 1. Yeast Extract Quality and Influence on Fermentation Process

One of the most important processes in the biotechnological industry is fermentation. In the fermentation process, genetically modified microorganisms are grown by feeding them with different carbon and nitrogen sources as well as vitamins and other essential nutrients. During the course of the fermentation, the microorganisms grow and produce specified peptides and proteins that are harvested and purified in later process stages.

Fermentation processes deal with living organisms, and the active pharmaceutical ingredient (API) is produced via complex metabolic pathways inside the organism. This means that the process and product quality is subject to large variability.

Many different types of raw materials are used for fermentation, ranging from simple chemicals to complex nutrients. For example, concentrated sugar solutions (e.g., glucose and maltose) are used as carbon sources, whereas yeast extract is a commonly used nutrient. Yeast extract is a common name for various forms of processed yeast products made by extracting the cell content; it is supplied to the culture media during the fermentation and acts as a nitrogen source as well as a source of vitamins and cofactors. The composition of yeast extract is not well defined, but it contains amino acids, peptides, water-soluble vitamins, and carbohydrate, and batch-to-batch variation is known to affect the fermentation process.

In this example, the batch-to-batch variation of yeast extract was investigated for an industrial bacterial fermentation process. Genetically modified *Escherichia coli* bacteria were used to produce a recombinant protein used in a pharmaceutical product, and the variability of the yeast extract was suspected to influence the tendency to foaming, cell growth, and product yield.

The experiments were done in a 5-L laboratory-scale fermentation reactor. The bacterial inoculums were added to starting media with salts, protein hydrolysates, and yeast extract. After an initial exponential growth phase of 5–7 h, addition of glucose syrup was started. The fermentations were terminated after a total run time of 34 h. The strategy for the experiments was simply to lock the settings of the process factors and run fermentations with different yeast extract batches and measure the process responses.

It was not known before the experiments what constituent in the yeast extract would influence foaming, growth, or yield; therefore, a wide array of analytical methods was used for characterization (Table 7.2). In total, six different analytical methods were used. The amino acid, trace element, and cation/anion analyses are all wet chemistry-based methods that require substantial sample pretreatment but result in specific quantitative data on sample constituents. In contrast, the three spectroscopic methods are rapid and measure the sample without pretreatment, but result in nonspecific spectral fingerprints of the sample constituents that have characteristic absorption profiles. The process response data are shown in Table 7.3.

The yeast extract came from two suppliers, identified as A and B. Different types of yeast extract were tested from each supplier, with some of the batches being tested in multiple fermentations. The different types were identified as 1, 2, etc., and different batches as a, b, c, etc. A batch label of "A1.a" means supplier A, type 1 yeast extract, and batch a, and "A2.c" means supplier A, type 2 yeast extract, and the third batch of this specific type. In total, 28 fermentations were performed with 19 different yeast extract batches.

TABLE 7.2
Analytical Methods Used to Characterize Yeast Extract

Analytical Method	Description	Data
Amino acid analysis	Total amount and amount of free amino acids (mM)	Quantitative results; amino acids
Trace element	Selected trace elements (ppm)	Quantitative results; Al, Co, Cr, Cu, Mn, Ni, V, and Zn
Cation and anion	Selected cations and anions (%)	Quantitative results; quinate, lactate, acetate, propionate, formate, pyruvate, chloride, bromide, succinate, phosphate, sulfate, citrate, potassium, ammonium, sodium
Mid-IR	Mid-infrared spectroscopic analysis (absorbance)	Spectral fingerprint; amino acids, carbohydrates, organic acids, vitamins
NIR	Near-infrared spectroscopic analysis (absorbance)	Spectral fingerprint; amino acids, carbohydrates, organic acids, vitamins
Fluorescence	2-D fluorescence spectroscopic analysis (intensity)	Spectral fingerprint; vitamins, tryptophan, cofactors

TABLE 7.3

Response Data

Response Name	Description
Foaming	Subjective measure of foaming tendency: no foaming (0), little foaming (1), significant foaming (2)
Acid	Total amount of added sulfuric acid during fermentation (relative units)
Base	Total amount of added sodium hydroxide during fermentation (relative units)
OD525 24h	Optical density at 525 nm after 24 h
OD525 34h	Optical density at 525 nm after 34 h
DCW 24h	Dry cell weight after 24 h (g/L)
DCW 34h	Dry cell weight after 34 h (g/L)
Viability 24h	Colony forming units per milliliter after 24 h (CFU/ml)
Peak CO_2	Maximum CO_2 value in exhaust gas during initial growth phase
Protein ELISA 34h	Total recombinant protein after 34 h by ELISA (relative units[a])
Protein HPLC 34h	Soluble recombinant protein after 34 h by ion-exchange HPLC (relative units)

[a] For proprietary reasons, the protein concentration values have been recalculated to relative units. The average protein concentration for the 28 fermentations is given the value 100 and the other values are relative to that.

7.2.1.1 Principal Component Analysis of Response Data

PCA was used to get an overview of the response data from the fermentation runs and to investigate reproducibility when using the same yeast extract batch. The response data from the 28 fermentations were autoscaled, and a PCA model was calculated with two principal components (PCs) explaining 81% of the variation. The two-dimensional (2-D) score plot for PC1 against PC2 showed a clear grouping according to foaming in the PC1 direction, whereas the PC2 direction explained the protein yield. Fermentations with a high yield had a high PC2 value. Batch A1.a was used in five fermentations, and their scores were all close, indicating good reproducibility (marked with the small tilted ellipse) (Figure 7.1).

7.2.1.2 Regression Analysis between Analytical Data and Response Data

The next step in the data analysis was to evaluate if the analytical data were correlated to the response data. Regression analysis was used to evaluate the correlation between the analytical data and three process responses, that is, *Foaming, DCW 34h*, and *Protein enzyme-linked immunosorbent assay (ELISA) 34h*. The coefficient of variation (RSD%) was calculated for each of the responses from the five fermentations made with the same yeast extract batch (A1.a). The *Foaming* response was a qualitative subjective evaluation from process operators (0 = no foaming, 1 = foaming, and 2 = significant foaming). All five fermentations with yeast extract batch A1.a were given 0 for *Foaming* response and three fermentations with A1.c 2, 2, and

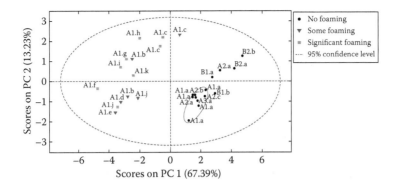

FIGURE 7.1 2-D score plot of response data. Markers are colored according to foaming tendency.

1. It is obviously easier to assign 0 to foaming than to judge if a foaming is 1 or 2. Therefore, in the regression analysis, RSD% for foaming response was not calculated (Table 7.4).

Partial least squares (PLS) regression modeling was used for all regression models. Leave-one-out cross validation was used to evaluate the regression models. For all models, the optimum number of latent variables (LVs) was found by comparing minima for the curves of root-mean-standard-error-of-calibration (RMSEC) and root-mean-standard-error-of-calibration-cross-validated (RMSECV) with different numbers of LVs. The different analytical methods can be compared by model statistics, that is, the correlation coefficients (R2 CV) and their prediction error (RMSECV); for example, if the fluorescence data best correlated with the foaming tendency (correlation coefficient = 0.841) (Table 7.5).

The regression models for foaming tendency showed that fluorescence, mid-IR, and NIR were capable of predicting the foaming tendency, whereas the wet chemistry-based methods were not usable (as the Y variance described by models based on wet chemistry was low to very low). The DCW was well correlated with the cation and anion and NIR measurements, and the best correlation was found with fluorescence and DCW. The final response yield after 34 h was measured by ELISA.

TABLE 7.4

Responses Used in Regression Analysis

Response	RSD%	Range
Foaming	na	0–2
DCW 34h	1%	19.9–30.5 (g/L)
Protein ELISA 34h	4.3%	67.2–134.9 (relative units)

Note: Coefficient of variation (RSD%) based on five repeated batch fermentations and range observed in all fermentations.

TABLE 7.5
PLS Model Results

Response		Foaming				
Analytical Method	# LV	Percentage Variance Captured X-Block	Percentage Variance Captured Y-Block	R2 CV[a]	RMSEC	RMSECV
Amino acid analysis	1	74	38	0.223	0.7	0.8
Trace elements	1	99	41	0.277	0.7	0.8
Cation and anion	2	83	81	0.687	0.4	0.5
Mid-IR	4	99.99	90	0.824	0.3	0.4
NIR	2	91.69	86.13	0.778	0.3	0.4
Fluorescence	3	98.67	90.11	0.841	0.3	0.4

Response		DCW 34h				
Analytical Method	# LV	Percentage Variance Captured X-Block	Percentage Variance Captured Y-Block	R2 CV	RMSEC	RMSECV
Amino acid analysis	1	75	62	0.515	2.1	2.4
Trace elements	2	99.97	65	0.497	2.0	2.5
Cation and anion	2	83	83	0.729	1.4	1.8
Mid-IR	5	99.75	93.20	0.653	0.9	2.2
NIR	4	99.67	90.10	0.753	1.1	1.8
Fluorescence	5	100	94.43	0.810	0.8	1.5

Response		Protein ELISA 34h				
Analytical Method	# LV	Percentage Variance Captured X-Block	Percentage Variance Captured Y-Block	R2 CV	RMSEC	RMSECV
Amino acid analysis	1	74	23	0.02	15.2	18.5
Trace elements	2	99.97	28	0.006	15.2	20.5
Cation and anion	4	99	41	0.054	13.8	18.6
Mid-IR	1	93.42	26.87	0.098	15.3	17.5
NIR	2	72.77	74.85	0.416	9.0	14.1
Fluorescence	5	99.99	61.68	0.187	10.7	17.7

[a] R2 CV. The correlation coefficient between measured and predicted values for the cross-validated result.

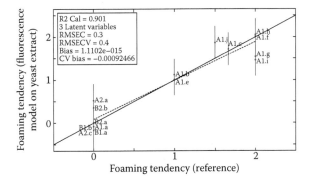

FIGURE 7.2 Fluorescence model for foaming tendency. R2 Cal is the correlation coefficient for the calibration model. Solid black line is perfect fit. Vertical bars through each calibration point are ±2 × RMSEC.

The only analytical data that showed a promising correlation were the NIR measurements, though the correlation was not very impressive. Nevertheless, all other methods were significantly worse compared with NIR. The best regression models are shown in Figures 7.2 through 7.4.

The example demonstrated how spectroscopic techniques can be used to measure a complex media such as yeast extract and how with multivariate regression models it is then possible to use the spectroscopic data to forecast the foaming tendency and DCW in the fermentation process. The NIR product model had a relative high prediction error, but it was useful for predicting if a given yeast extract batch would give a low, medium, or high yield. Compared to the wet chemistry-based methods, spectroscopic data are much harder to interpret, but in this example, they were superior as potential process control tools.

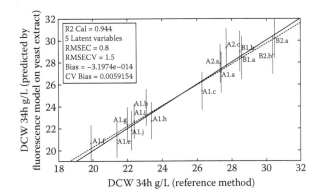

FIGURE 7.3 Fluorescence model for DCW 34h. R2 Cal is the correlation coefficient for the calibration model. Solid black line is perfect fit. Vertical bars through each calibration point are ±2 × RMSEC.

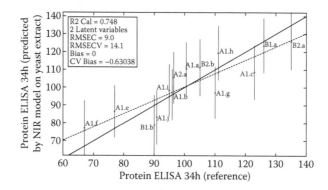

FIGURE 7.4 NIR model for ELISA protein 34h. R2 Cal is the correlation coefficient for the calibration model. Solid black line is perfect fit. Vertical bars through each calibration point are ±2 × RMSEC.

7.2.2 EXAMPLE 2. RAW MATERIAL INFLUENCE ON THE DESIGN SPACE OF MICRONIZATION PROCESS

This example explores how two different raw material qualities had a significant influence on the design space for the controllable process factors. The design space is defined by ICH (U.S. Department of Health and Human Services et al. 2009) as "The multidimensional combination and interaction of input variables (e.g., material attributes) and process parameters that have been demonstrated to provide assurance of quality."

The manufacturing process was a pharmaceutical micronization process. Insulin crystals were injected into the mill chamber of a spiral jet mill (Figure 7.5) through a feed channel. Compressed nitrogen was used as a grinding gas that was injected into the mill chamber via six grinding nozzles with adjustable inlet angles. The gas jets created a rotating gas vortex circulating the insulin crystals until they were micronized and left the mill chamber through an outlet in the center of the mill chamber.

The final micronized insulin was filled into special blisters and loaded into a special medical device. The medical device was then used for pulmonary administration of insulin for diabetic patients.

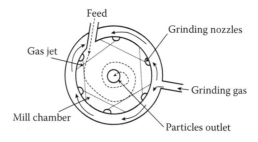

FIGURE 7.5 Spiral jet mill (top view of cross section).

TABLE 7.6
Design

Factor	Range	Abbreviation
Feed rate	The crystals were fed into the spiral jet mill using two different rates: 55 and 200 g/h	FR
Grinding air pressure	The nitrogen pressure was investigated at 6 and 13 bar	GAP
Grinding nozzle angle	Two different grinding nozzles were tested with 19° and 32° inlet angles, respectively	GNA
Insulin type	Two different insulin types named *molecule A* (an amorphous powder) and *molecule B* (a crystalline powder) were micronized	Type

Several critical quality attributes of the micronized insulin were important to control, but one of the most critical was the particle size. A thorough risk assessment exercise identified numerous factors that could influence the final particle size, and design of experiments methodology was used to investigate four factors that scored high in the risk assessment analysis. The four different factors were feed rate, grinding air pressure, grinding nozzle angle, and insulin type. A factorial design with two levels and three center point experiments was used, making a total of 19 experiments. After each experiment, the mass median aerodynamic diameter (MMAD*) was determined.

A multiple linear regression (MLR) model was developed between MMAD and the factors in Table 7.6. By inspection of the residuals, a single experiment was clearly identified as an outlier and was removed. The MLR model was recalculated with 18 experimental data points. The model fitted the MMAD very nicely (Figure 7.6) and explained 98% of the response variation. The cross-validated result was a fit of 90% and with a very high degree of reproducibility, assuring a valid model.

The model coefficient plot (Figure 7.7) showed that the grinding air pressure and insulin type were the most significant factors (coefficient with error bars that do not cross zero) influencing the MMAD. Furthermore, a significant interaction between the feed rate and insulin type was identified. It is this interaction effect specifically that causes the design spaces to change according to the insulin type.

The interaction between feed rate and insulin type is visualized in the interaction effects plot (Figure 7.8). The plot shows that the effect of the feed rate is dependent on which molecule is being micronized. When an amorphous insulin type like molecule A is micronized, the feed rate has a large effect on the final particle size, and increasing the feed rate will increase the size of the micronized particles. If, on the other hand, a crystalline insulin type like molecule B is micronized, the feed rate has a much lower effect, and increasing the feed rate will result in smaller particles.

---

* The *median* particle size (*mass median particle diameter*) is the particle diameter that divides the frequency distribution in half; 50% of the aerosol mass has particles with a larger diameter, and 50% of the aerosol mass has particles with a smaller diameter; unit is micrometers.

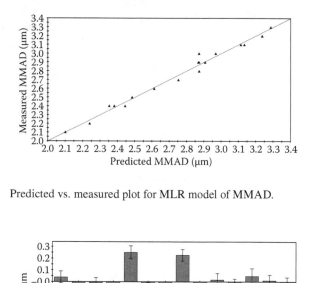

FIGURE 7.6 Predicted vs. measured plot for MLR model of MMAD.

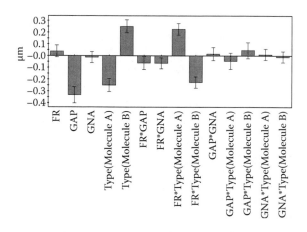

FIGURE 7.7 MLR model coefficient plot. Each coefficient has a whisker bar plotted, which is the error (95% confidence). The main effects are the grinding air pressure (GAP), the insulin type [Type(Molecule A) and Type(Molecule B)], and the interaction between insulin type and feed rate [FR*Type(Molecule A) and FR*Type(Molecule B)].

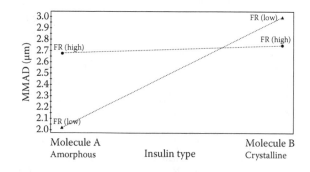

FIGURE 7.8 Interaction effects plot for feed rate and insulin type.

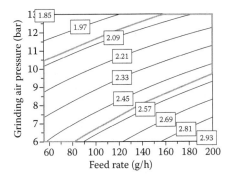

FIGURE 7.9 Contour plot for insulin type molecule A (amorphous powder). Area defined by light gray line shows the design space for a process that must manufacture micronized insulin with an MMAD between 2 and 2.5 μm.

The model can be used to visualize the design space. Contour plots show MMAD as a function of feed rate and grinding air pressure. Figure 7.9 shows the contour plot for amorphous insulin (molecule A), and Figure 7.10 shows the contour plot for crystalline insulin (molecule B), both shown for a grinding nozzle angle of 32°. The contour plots are very different due to the interaction effect between insulin type and feed rate. The design space is defined as the operating space within which the process can be run freely and still achieve a predefined quality. If the quality requirements for the micronized insulin, for example, are set to an MMAD between 2 and 2.5 μm, the design spaces (shown with the areas defined by light gray lines in the contour plots) differ a lot depending on the characteristics of the insulin type. If a raw material with the characteristics of molecule A (amorphous material) is to be micronized, the design space is large, whereas a more crystalline raw material with characteristics like those of molecule B (crystalline) has a far more restricted design space.

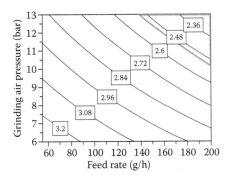

FIGURE 7.10 Contour plot for insulin type molecule B (crystalline powder). Area defined by light gray line shows the design space for a process that must manufacture micronized insulin with an MMAD between 2 and 2.5 μm.

7.3 CONCLUSIONS

The two examples demonstrate how raw material characteristics can influence the manufacturing process and drug product quality in two very different biopharmaceutical processes. In the first example, yeast extract was analyzed with six different analytical methods, both specific and also nonspecific spectroscopic methods. It was not possible to clearly identify chemical components in yeast extract that correlated to process performance parameters with the specific methods, but the nonspecific spectroscopic methods were superior, and statistical models could be developed, which were useful for predicting process performance by rapid measurement. With such techniques, it is possible to, for example, preselect yeast extract batches with predictable performance for manufacturing. In the process development phase, the technique can be used to select representative yeast extract batches, which are then used when a robust process is being developed that is capable of handling the variability of the yeast extract. In the second example, it was demonstrated how the design space for the manufacturing of pulmonary insulin powder changed significantly when the raw material characteristics changed. By applying a design of experiments methodology, the raw material impact on the design space could be found, and a process capable of handling raw material variability can be developed.

ACKNOWLEDGMENTS

Analytical work has been performed by Anette Behrens, John Lyngkilde, Anne Rodenberg, Jakob Gnistrup, and Erik Skibsted. Experimental work and scientific discussion have been done in collaboration with Mats Åkesson, Flemming Junker, Richard Smith, and Erik Skibsted, all employees at Novo Nordisk, Denmark. The 2-D fluorescence measurements were performed at the Faculty of Life Sciences, University of Copenhagen.

REFERENCES

Berget, I. and T. Næs. 2002a. Sorting of raw materials with focus on multiple end-product properties. *J. Chemom.* 16: 263–273.

Berget, I. and T. Næs. 2002b. Optimal sorting of raw materials, based on the predicted end-product quality. *Qual. Eng.* 14 (3): 459–478.

Jørgensen, K. and T. Næs. 2004. A design and analysis strategy for situations with uncontrolled raw material variation. *J. Chemom.* 18: 45–52.

Kasprow, R.P., A.J. Lange, and D.J. Kirwan. 1998. Correlation of fermentation yield with yeast extract composition as characterized by near-infrared spectroscopy. *Biotechnol. Prog.* 14: 318–325.

Lanan, M. 2009. QbD for raw materials. In *Quality by Design for Biopharmaceuticals*, edited by A.S. Rathore and R. Mhatre, pp. 193–209. Hoboken, NJ: John Wiley & Sons.

Mevik, B.-H., E.M. Færgestad, M.R. Ellekjær, and T. Næs. 2001. Using raw material measurements in robust process optimization. *Chemom. Intell. Lab. Syst.* 55: 133–145.

U.S. Department of Health and Human Services, Food and Drug Administration, Center for Drug Evaluation and Research (CDER), and Center for Biologics Evaluation and Research (CBER). 2009. Q8(R2) Pharmaceutical Development. Guidance for Industry.

Available online at http://www.ich.org/fileadmin/Public_Web_Site/ICH_Products/Guidelines/ Quality/Q8_R1/Step4/Q8_R2_Guideline.pdf.

US FDA. 2004. Guidance for Industry PAT—A Framework for Innovative Pharmaceutical Development, Manufacturing, and Quality Assurance. Available online at http://www .fda.gov/downloads/Drugs/GuidanceComplianceRegulatoryInformation/Guidances/ UCM070305.pdf.

8 Process Analysis Technology for Enhanced Verification of Bioprocess System Cleaning

Peter K. Watler, Keith Bader, John M. Hyde, and Shuichi Yamamoto

CONTENTS

8.1 INTRODUCTION

In the manufacture of therapeutic products, contamination control is of paramount importance to product quality, patient safety, and good manufacturing practice (GMP) compliance. Cleaning processes are the primary line of defense in preventing product adulteration from contaminants such as foreign matter, particulates, product residues, adventitious agents such as viral and microbial organisms, and processing reagents. Despite their critical nature, the control and monitoring of effective cleaning processes can be elusive. A recent survey found that 60% of Food and Drug Administration (FDA) warning letters cited Code of Federal Regulations 21 CFR 211.67 deficiencies relating to inadequate equipment cleaning procedures, monitoring, and documentation (European Compliance Academy 2006). These regulatory, patient safety, and product quality concerns make process analytical technologies (PATs) a powerful tool for enhancing the verification and validation of cleaning processes to ensure that they are well controlled and consistently remove contaminants every time they are used.

8.2 MECHANISMS FOR CLEANING BIOPROCESS EQUIPMENT SYSTEMS

Bioprocesses including recombinant biotherapeutics, plasma products, and vaccines all leave carbonaceous residues on the surface of process equipment. Typical process residues contain proteins, lipids, sugars, and salts that result from the host organism and media and buffer solutions used during processing. Accordingly, a majority of cleaning operations in the biotechnology industry are aqueous-based processes using formulated detergents and cleaning agents to degrade and dissolve the process residues. Bioprocess cleaning processes employ several mechanisms to break down and remove process residues. These mechanisms include (1) peptization—a cleavage reaction that degrades and disperses proteinaceous residues into smaller peptide chains that are soluble in aqueous solutions; (2) dissolution—a reaction that solubilizes residues such as non–heat-treated proteins, short-chain alcohols, excipients, and monovalent salts into aqueous cleaning solutions; (3) saponification—a hydration reaction where free hydroxide ions break the ester bonds between fatty acids and glycerol of triglycerides, resulting in degradants that are soluble in aqueous solutions; and (4) demineralization—a reaction where acidic conditions dissolve insoluble mineral salts. All of these mechanisms may be used to varying degrees in cleaning processes for bioprocess systems. Cleaning processes are controlled by specifying and monitoring four key operating parameters: cleaning solution temperature, cleaning agent concentration, cleaning solution turbulence (flow rate), and cleaning time.

8.3 METHODS FOR MONITORING CLEANING EFFECTIVENESS

8.3.1 TRADITIONAL MONITORING APPROACHES

After a cleaning process has been developed, the traditional approach to cleaning validation relies on characterizing the behavior of the cleaning process relative to a single set of operating parameters and critical quality attributes (CQAs). This characterization establishes that the process operates reproducibly within the specified range of the operating parameters. This is typically demonstrated through documented repetitions of the cleaning process applied to representative production residues.

In addition to controlling the cleaning parameters, it is equally important to monitor the CQAs that define the effectiveness of the cleaning process. Monitoring cleaning process CQAs typically involves multiple orthogonal off-line analytical measurements. Because of the complexity of the cleaning process and the analytical measurements, cleaning processes are typically validated to provide an added level of assurance of equipment cleanliness (Fordor and Hyde 2005). Cleaning validation provides documented evidence that a cleaning process consistently and effectively reduces potential product and cleaning agent residues to predetermined, acceptable limits. Once validated, cleaning processes are then typically monitored at a reduced level of analytical testing to verify the cleaning effectiveness. This "cleaning verification" provides documented evidence that the equipment has been successfully cleaned and can be released for processing.

8.3.2 PAT Monitoring Approaches

In 2004, the FDA provided industry with guidance to "encourage the voluntary development and implementation [of] process analytical technology" (FDA 2004). In the past decade, enabling advances in bioprocess system design, batch control, system automation, sensor technology, and data acquisition have been leveraged to create real-time, on-line PATs suitable for assessing the CQAs of bioprocess cleaning processes. However, the goal of PAT is not simply to measure parameters and attributes but "to understand processes such that quality is controlled & assured throughout production" (Watts and Clark 2006). The FDA has emphasized that PAT necessitates "an understanding of the scientific and engineering principles involved, and identification of the variables which affect product quality" (Brorson 2010). In practice, PAT requires the combination of three elements: (1) analytics to measure process characteristics, (2) a mechanistic understanding how these characteristics relate to performance and quality attributes, and (3) statistical process control techniques to evaluate and act on the PAT data.

Implementing PAT in this manner has enabled real-time, on-line techniques for assessing cleaning process effectiveness. For example, statistical analysis of the analytics and process mechanisms has enabled development of forward processing criteria (FPC) to aid in monitoring cleaning process effectiveness. FPC are specified criteria that must be met before feeding the process stream to the unit operation. During unit operation execution, PAT can be used to measure process CQAs that provide further evidence that the system was effectively cleaned. By establishing control limits for FPC and CQAs, PAT can control and assure cleaning efficacy throughout production. This more rigorous approach of real-time cleaning verification may lead to less reliance on, and ultimately a reduction in, the cleaning validation effort. These technologies and criteria will be reviewed for application in assessing the effectiveness of process vessel and chromatography cleaning processes.

8.4 APPLICATION OF PAT FOR MONITORING AND VERIFYING EQUIPMENT SYSTEM CLEANING

Critical cleaning process parameters that must be controlled are cleaning solution concentration, temperature, external energy, and cleaning process duration or contact time. To know whether a particular combination of these parameters is effective, system owners must understand the design space of the cleaning process for a particular residue set, thereby allowing active management of process variability. Some of the design space information may be determined empirically through bench-scale testing, whereas other data may be obtained *in situ* through the implementation of on-line monitoring instrumentation to determine such quantities as total organic carbon (TOC) and conductivity, both of which may be utilized as FPC. Additionally, flow rate may be monitored as a CQA because it relates to the application of external energy.

External energy is the mechanical force applied to process equipment surfaces to aid in the transport of cleaning solutions to soiled surfaces as well as the removal of cleaning solutions and residues from process equipment surfaces. Most often, external energy is imparted to process equipment surfaces in the form of turbulence, which is dependent on cleaning solution flow rate. Turbulence is described by the Reynolds number which is given as

$$Re = \left(\frac{\rho u D}{\mu} \right), \tag{8.1}$$

where

ρ = fluid density, u = fluid velocity, D = pipe diameter, and μ = viscosity.

At Reynolds number >2300, the flow becomes less ordered and more chaotic, tending toward becoming turbulent (Perry and Green 1997). This is the minimum flow regime that must be maintained to facilitate mass transfer to and from the residue to achieve an effective cleaning. Given that most process equipment will exhibit varying geometry within the equipment itself as well as in the clean-in-place (CIP) system piping, varying flow regimes will be evident throughout a cleaning circuit. With an understanding of the equipment design, the control system for the cleaning system can easily be programmed to not only monitor flow rate to the equipment but also calculate a Reynolds number for each portion of the equipment, thereby returning a parameter value more relevant to the cleaning process than a flow rate alone. Because the Reynolds number can be calculated in real time from the measured flow rate, this concept can be taken a step further, allowing active control of the cleaning cycle to maintain a turbulent flow regime throughout the cleaning circuit by adjusting the CIP supply flow rate. To ensure that the minimum Reynolds number is achieved, the value of the largest pipe diameter in the cleaning circuit should be used.

Conductivity and TOC may both be monitored on-line in the final rinse to determine if residual process material or cleaning agents are present. The degree to which active process control may be enacted using this information depends on the sensor technology used to monitor a specific forward processing criterion. For instance, conductivity may be monitored on-line without significant delay and, given appropriate circumstances, may be used as an endpoint determinant for the cleaning cycle.

8.4.1 ON-LINE TOC ANALYSIS FOR MONITORING SYSTEM CLEANING

Cellular debris is composed of lipids, polysaccharides, proteins, RNA, and DNA, and is up to 50% carbon on a dry basis (Vogel 1997). In addition, growth media and buffer solutions often contain carbon-containing molecules. Because of this prevalence of carbon in bioprocessing, TOC analysis is frequently used as a PAT to assess cleaning efficacy for the removal of host, product, and processing agent residues.

The cleaning process must be highly effective in removing process contaminants and must also include rigorous methods for assessing the efficacy of the cleaning processes. These measurements are generally derived from the collection of rinse water that is then subjected to off-line analysis. By demonstrating the absence of carbonaceous residues in rinse water, TOC analysis is an effective PAT for evaluating the efficacy of the cleaning processes.

Unfortunately, some on-line TOC technologies that perform destructive analyses on a captive sample can exhibit delays on the order of 1–5 min before the results are available, making them impractical for real-time process control. This lag behind real-time monitoring can also be further exacerbated by the configuration of the sampling equipment used to deliver captive samples to an analyzer. To wit, longer sampling lines lead to a greater residence time before the sample is delivered to the analyzer. Furthermore, many analyzers are not configured by the manufacturer specifically for use in CIP monitoring, and there is little opportunity to fully customize features to meet all the requirements for an on-line CIP application. Regardless, such instrumentation is still valuable for making critical to quality decisions at line or shortly after processing is concluded.

Although the basic operational principles for all TOC analyzers are much the same, the oxidation and sensor technologies vary between manufacturers. Matching the characteristics of CIP processes with an array of specific sensor and oxidation technologies compatible with those characteristics will yield a robust implementation of the on-line analyzer, enabling minimized operational and validation efforts with respect to cleaning processes. Accordingly, selection of an appropriate instrument requires knowledge of the analytical instrument operating principles, the CIP system operation, and cleaning process conditions to ensure that the CIP process itself does not interfere with the analytical results. For example, to obtain accurate results in CIP PAT applications, the on-line analyzer must not be confounded by interference from ionic species, variations in sample pressure, or changes in sample temperature. Because conductivity is used in some cases to quantify evolved CO_2, the ionic species in many cleaning agent formulations must be considered as a potential source of interference. These conductive species may be addressed through the use of a membrane conductometric sensor as in Figure 8.1, or through the use of photometric detection schemes that are insensitive to the presence of conductive ions. Membrane conductometric detectors allow selective permeability of CO_2 across a membrane without permitting other conductive ions into the measurement zone. The measured conductivity thus results entirely from inorganic carbon and total carbon, which has been oxidized to CO_2, effectively eliminating interference from conductive ionic species.

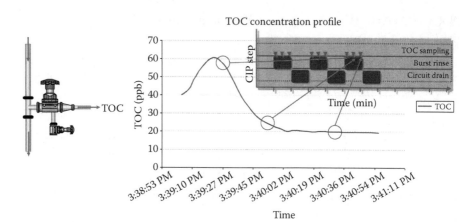

FIGURE 8.1 Continuous sampling arrangement.

For on-line TOC analyzers in which samples are directly introduced to the analyzer from the CIP return manifold, sample temperature and pressure are relevant parameters to consider. Sufficient pressure is required in the sample line to ensure that the analyzed sample concentration does not significantly lag that in the CIP return piping. Additionally, care should also be taken to protect the analyzer from pressures exceeding manufacturer's recommendations. In most cases, CIP pressures will not exceed the pressure specifications for an instrument; however, close attention must still be given to the configuration, size, and placement of automated sampling valves and associated sample lines drawing from CIP system return lines. Stabilization of analyzer inlet pressure and flow rate will allow for consistency in the residence time of fluid in the sample lines.

Temperature fluctuations are a relevant concern depending on the selected analyzer, especially if the analysis method is conductometric. Conductivity is a temperature-dependent measurement that each instrument manufacturer accommodates in a different manner. Temperature variations in the sample stream may be addressed through temperature-compensated conductivity sensors, or measurement of raw conductivity data with sampling apparatus that allows for temperature equilibration through ambient dissipation or active heat exchange. Alternatively, detection methods that are not temperature dependent (such as nondispersive infrared sensors) are also available; if critical specifications of the instrument (such as accuracy and limit of quantitation) meet the needs of a particular application, then concerns of interference from conductive species may be addressed.

Before installation of an on-line instrument for monitoring the cleaning process, some decision as to what information is to be collected and how it will be used must be made. The sampling equipment has a significant effect on the acquired data; for example, if a transient concentration profile is desired to evaluate the rinse down characteristics of a CIP circuit, a configuration such as that shown in Figure 8.1 would be appropriate.

Depending on the analysis time, a few points on a continuous profile may be generated. Conversely, if greater data integrity is desired for a single instance in time, a captive volume can be captured and isolated from the main CIP return path, and multiple samples can be drawn from the sample reservoir, which will presumably contain a relatively homogeneous sample (Figure 8.2).

The benefits and disadvantages of the arrangements shown in Figure 8.1 and 8.2 must be weighed against the goals for installing an on-line instrument. A continuous sampling arrangement provides more information about the behavior of the final rinse phase that can, for example, be used for the development of more efficient cleaning cycle rinse times. Conversely, a captive sample volume taken at the end of a rinse phase cannot convey any information about the rinse profile, but does offer adequate volume for replicate analyses and greater statistical certainty about the analytical result pertaining to cleaning efficacy.

Figure 8.3 shows the configuration of an on-line TOC analyzer installed on the return line of an educator-assisted CIP skid. Because the return line is under vacuum to aid in the return of cleaning solutions to the CIP skid, samples must be drawn from the return line to the TOC analyzer using a peristaltic pump. The sample is also drawn from the bottom of a horizontal return line through a zero static diaphragm valve to ensure that air is not drawn into the sample line.

The sampling strategy used for this implementation is consistent with the general approach shown in Figure 8.3 for continuous sampling. Collecting and analyzing the rinsate using this system produces results indicative of washout kinetics and can be used for at-scale optimization as well as evaluating rinse washout.

Examination of the shape of the final rinse concentration profile relays information about the cleaning circuit from which the information was collected. For example, the rinse profile shown in Figure 8.4 shows two distinct peaks. In this case, the second peak is a result of final rinse water being redirected through an alternate flow path on a bioreactor cleaning circuit. With each increase in TOC, there is also an increase in solution conductivity, indicating that the rise in TOC is likely a result

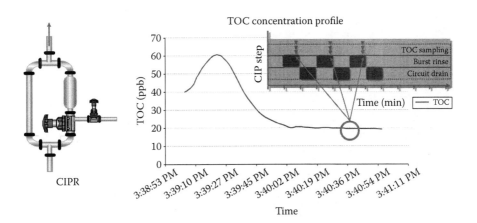

FIGURE 8.2 Captive volume sampling assembly placed in the CIP return (CIPR) line.

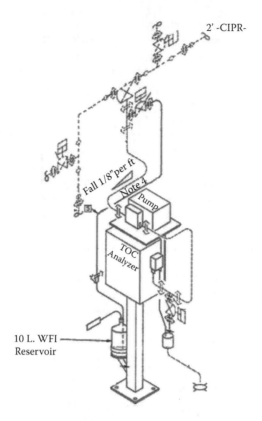

FIGURE 8.3 Configuration for on-line TOC analyzer to monitor CIP final rinse.

of the surfactants in the formulated cleaning agent used for the chemical wash, rather than from residual production residues, which typically exhibit lower conductivities and would not rinse from the system as readily.

Similarly, examination of the concentration profiles over time can also reveal interesting and useful information. Figure 8.5 illustrates how three separate cleaning operations can be graphically compared for consistency from run-to-run. Directly overlaying the three profiles provides a process "fingerprint" showing that the

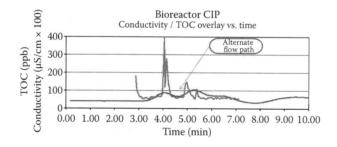

FIGURE 8.4 Conductivity and TOC profile for CIP final rinse.

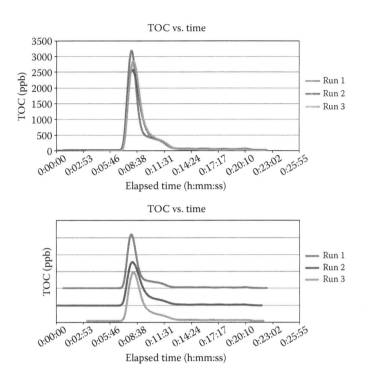

FIGURE 8.5   Concentration profile fingerprint.

general shape and magnitude of each peak is very similar. For comparison, these same peaks have also been stacked in the chart.

8.5 APPLICATION OF PAT FOR MONITORING AND VERIFYING CHROMATOGRAPHY COLUMN CLEANING

It is common industry practice to validate column reuse by assessing column suitability with a postcleaning integrity test, postproduction peak processing, and periodic post-use "mock" elutions (e.g., after every five production batches) lots (Rathore and Sofer 2005). This practice was initially adopted decades ago when column operations were less automated, only off-line analysis was available, and chromatographic mechanisms were less understood. Now, column operations are fully automated and on-line analysis is available, enabling column testing to be seamlessly integrated with real-time column operation including equilibration and elution phases. Several techniques such as height equivalent to a theoretical plate (HETP), asymmetry factor, back pressure, transition analysis, product peak retention time, and peak shape analysis exist for monitoring column operation (Molony and Undey 2009; Podgornik et al. 2005; Lendero et al. 2005). By using these techniques, FPC can be developed to provide physical and chemical measures to verify that the chromatography column is suitable for use. Evaluation of FPC before *every* production run provides a high level of assurance for successful purification before loading the product. FPC can be incorporated into batch records, SOPs, verification checks, or automated programs as column pre-use criteria. FPC must be satisfied before column reuse by requiring associated PAT measurements to meet predetermined acceptance criteria. PATs provide a means for measuring FPC, including column integrity, column back pressure, flow rate, capacity, and mock pool protein carryover for every manufacturing lot. In addition to FPC, critical performance attributes (CPAs) can be used to monitor column performance *during* use to further confirm acceptable separation operation. Although PAT enables real-time viewing of chromatography performance attributes such as pressure drop, flow rate, and ultraviolet (UV) absorbance (Figure 8.6), these FPC and CPA measures can also be trended and evaluated against Nelson (1984) or Western Electric Company (1958) control rules to assess and predict process control of the chromatography column.

In this case study of an ion exchange chromatography column, one complete column cycle consists of three phases: equilibration, production, and cleaning. FPC and CPA testing is fully integrated with the phases of column operation as shown in Figure 8.7. During the equilibration phase, the caustic storage buffer is neutralized and flushed with a high-concentration buffer, and then the column is equilibrated using a combination of the equilibration and elution buffers. Because these two buffers are of similar composition, the elution buffer can be used to partially equilibrate the column. During the final equilibration phase, a small volume of equilibration buffer is applied to displace the high-salt elution buffer and adjust the ligand counter-ion equilibria. With this strategy, a mock elution is incorporated into *every* equilibration phase. This allows for product/contaminant carryover to be measured with *every* column use.

During the production phase, column CPAs can be monitored to further verify that the column is operating within acceptable ranges.

During the cleaning phase, the UV sensor is used to monitor contaminant removal, providing a fingerprint of the cleaning operation (Figure 8.8).

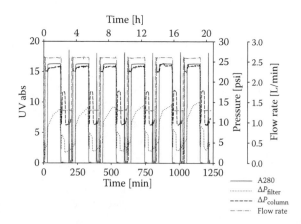

FIGURE 8.6 Chromatography PAT data profiles for pressure drop, flow rate, and UV absorbance comparing multiple production lots.

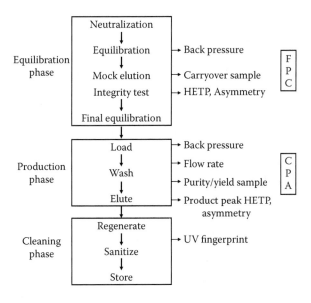

FIGURE 8.7   Operating phases and PAT testing of a chromatography cycle.

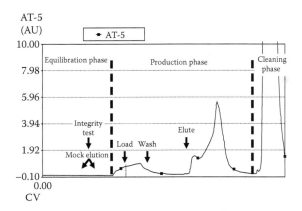

FIGURE 8.8   UV chromatogram showing equilibration, production, and cleaning phases.

8.5.1 COLUMN INTEGRITY TEST AS A PAT FORWARD PROCESSING CRITERION

The quality and integrity of the column packing are important factors for achieving reproducible chromatography column performance. The quality of the column packing can be tested before use to reduce process variability. The HETP is a generally accepted parameter for describing packing quality (Snyder 1972; Lettner et al. 1995). HETP values are commonly used to uncover gross problems with the setup of large-scale equipment and column packing (Jungbauer and Boschetti 1994).

A second measurement for assessing column packing is the asymmetry factor (A_f). A_f is a numerical expression of the peak skewness. The peak shape can help diagnose the cause of certain packing anomalies (Mitchell et al. 1997). Peak tailing can be caused by exponential washout kinetics to the flow profile. Peak fronting is normally caused by flow channeling and can indicate a low packing density. Magnetic resonance imaging has confirmed that undercompression of column packing can cause peak fronting, whereas overcompression can cause peak tailing (Dickson et al. 1998). HETP and A_f serve as quantitative FPC providing an assessment of column physical integrity before loading the product. These FPC aid in confirming that the hydrodynamic properties of the packed bed are maintained within qualifications by demonstrating that the packed bed has not shifted or compressed and is homogeneous (i.e., without channels or voids).

HETP and A_f can be calculated from tracer retention and dispersion data from either pulse response or frontal analysis (Watler et al. 2003; Podgornik et al. 2005). It is important that the test method measures zone spreading induced by the flow pattern through the packing while minimizing contributions from all other sources. To minimize unwanted contributions, the following practices are observed: (1) the mathematical calculation is standardized; (2) the test velocity is optimized to minimize contributions from diffusion and mass transfer resistances; and (3) the interactions of the tracer molecule with the stationary phase are minimized. When using NaCl as the tracer molecule, the exchange of sodium and chloride ions between the mobile phase and resin can be minimized by using an equilibration buffer and tracer consisting of the same co- and counter-ions (Helfferich 1962). HETP and A_f can be measured according to the following equations:

$$\text{HETP} = \frac{L}{5.545} \cdot \frac{w_{1/2}^2}{t_R^2} \tag{8.2}$$

$$A_f = \frac{B_{10}}{A_{10}}, \tag{8.3}$$

where
> L = packed column height, $w_{1/2}$ = peak width at half height, t_R = peak retention time, A_{10} = leading half peak width at 10% of peak height, and B_{10} = trailing half peak width at 10% of peak height.

Automated chromatography systems can calculate and report integrity test results and can activate alarms if values are beyond acceptable ranges (Figure 8.9).

Pre-use integrity test results can be plotted on statistical process control charts with control rules applied to identify deterioration of column integrity before use. The control charts shown in Figures 8.10 and 8.11 indicate that before each use, the column HETP and A_f were in statistical control and within the predetermined FPC acceptance criteria.

AT-3 post-column conductivity graph

Peak max	21.73
Half height	10.865
Baseline offset	10.43
10% height	2.173
Note: T1=0	
T2	735
T3	776
T4	816
A10	689
B10	863
HETP cm	0.0393
Asymmetry	1
Bed height	20
Column volume	62.34

Analysis type	Pulse

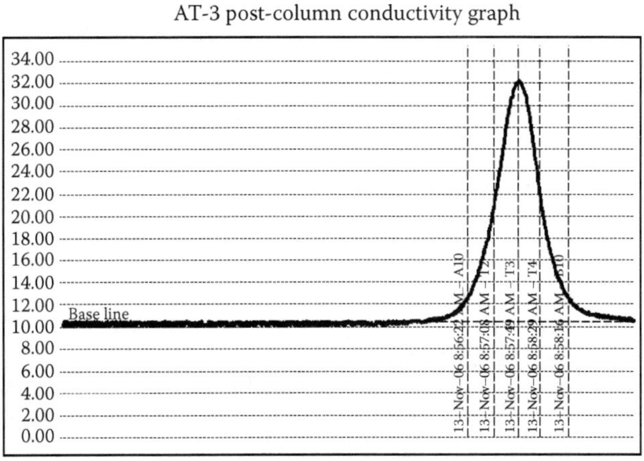

FIGURE 8.9   Automated chromatography column pre-use integrity test.

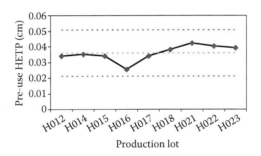

FIGURE 8.10   Pre-use HETP control chart indicating no control rule violations and within acceptance criteria of HETP <0.07 cm.

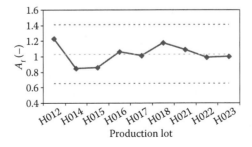

FIGURE 8.11   Pre-use A_f control chart indicating no control rule violations and within acceptance criteria of $0.8 < A_f < 1.6$.

In summary, the column integrity test is a PAT that can be used to monitor column cleaning and provides an indication that

1. Repeated use does not affect the packed bed integrity and hydrodynamic properties within pre-established acceptable limits.
2. The packed bed has not shifted or compressed.
3. The packed bed is homogeneous (i.e., without channels and voids).
4. Physical structure (size and strength) of the media has not been altered.

8.5.2 Column Pressure Drop as a PAT Forward Processing Criterion

Pressure drop is an important measure by which chromatography columns can be evaluated (Kaltenbrunner et al. 2000; Janson and Hedman 1982). Fluid flow through the packed bed causes a pressure drop due to friction forces (Giddings 1991; Bird et al. 1960; Dolejs et al. 1998). Bed compression occurs with many of the particles used in preparative chromatography because many are made of compressible materials such as cross-linked polymers of dextran, agarose, polyacrylamide, cellulose, and polystyrene. The Kozeny–Carman equation can be modified to account for the compressibility of the chromatography beads (Stickel and Fotopoulos 2001; Watler et al. 2003). The bed pressure drop equation was adjusted by modifying the void volume by a compression factor that is unique to each type of chromatography resin as described below:

$$\Delta P = u_o L \frac{150\mu}{d_p^2} \frac{(1-\varepsilon)^2}{\varepsilon^3} \tag{8.4}$$

$$\lambda = \lambda_c \frac{u}{u_c} \tag{8.5}$$

$$\varepsilon = \frac{\varepsilon_o - \lambda}{1 - \lambda}, \tag{8.6}$$

where
ΔP = packed bed pressure drop, u_o = superficial velocity, u_c = critical velocity, L = bed length, μ = viscosity, ε = void fraction, d_p = particle diameter, ε_o = initial void fraction, ε = compressed void fraction, λ = bed compression factor, and λ_c = critical bed compression factor.

These equations demonstrate that the column pressure drop follows a characteristic relationship as a function of flow rate, media type, and column diameter as shown in Figure 8.12.

It is known that cross-linking increases the rigidity of agarose gels (Janson and Ryden 1998), which affects bead compression and, hence, the pressure drop in chromatography columns. Because cross-linking density affects the compression modulus of the gels (Watler et al. 1988), changes in the physical structure of chromatography beads can be detected by monitoring column pressure drop for deviations from the characteristic pressure–flow curve for a given column system. Column pressure drop

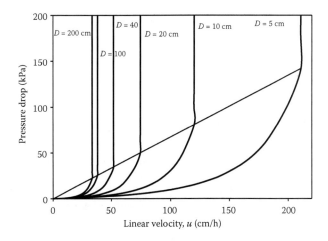

FIGURE 8.12   Predicted column pressure drop for Sepharose CL-6B as a function of column diameter and flow rate. Settled bed height = 20 cm, calculated from Kozeny–Carmen equation with void fraction adjusted for bed compression.

can be used as a PAT to monitor the physical stability of the chromatography resin as shown in Figure 8.13. This control chart demonstrates that the column pressure drop is in statistical control and is within the predetermined FPC acceptance criteria. An increase in back pressure beyond established limits can indicate bed fouling, at which point the column resin should be replaced.

In summary, the column pressure drop is a PAT that can be used to monitor column cleaning and provides an indication that

1. Chromatography beads maintain physical strength.
2. Media has not shifted or compressed.
3. Packed bed is homogeneous without channels or voids.
4. No product or contaminants has accumulated in the packed bed (fouling).
5. No product or contaminants has accumulated on the column frit and flow adapter

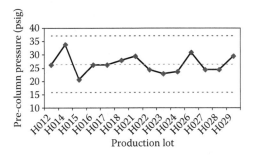

FIGURE 8.13   Pre-column pressure drop control chart for a 63-cm ion exchange chromatography column indicating no control rule violations and within <41 psig.

8.5.3 Product Peak HETP as a PAT Critical Quality Attribute and Forward Processing Criterion

The two factors governing a separation are peak retention volume and peak width (Schoenmakers 1986). Peak resolution is the ratio of the distance between two peaks to the width of the peaks and is indicative of the separation. Resolution is the result of two effects: (1) the separating power of the media (retention) and (2) the column efficiency (peak broadening). Peak broadening is influenced by system plumbing, media type, particle size, media size distribution, and conditions such as linear velocity, sample viscosity and diffusivity, and packing homogeneity (Sofer 1995; ASTM 1981).

When chromatography is used for high-resolution protein separation, retention and peak broadening can be very sensitive to operating conditions. This is especially true for isocratic elution where the separation is very sensitive to small changes in typical operating factors such as pH, ionic strength, and resin capacity (Kaltenbrunner et al. 1993; Ogez et al. 1991; Yamamoto et al. 1999). The number of theoretical plates, a measure of column performance, is defined by the degree of peak broadening and relates a peak's width to its elution volume. Product peak HETP is a useful measure for monitoring peak resolution and column separation performance. A loss of resin capacity would result in a reduction in peak retention time with a resultant increase in product peak HETP. Similarly, physical deterioration of the resin beads or packed bed homogeneity would increase band spreading (peak width) and result in an increase in product peak HETP. Because product peak HETP is influenced by resin capacity and resin bead physical characteristics, it provides an indication of column cleaning and column suitability for reuse. The product peak HETP is a PAT with application as a CPA for confirming chromatographic performance as well as an FPC to confirm suitability of the column for reuse.

Product peak HETP requires measurement of the peak retention time from the start of elution and measurement of the peak width. Such measurements can be automatically obtained from data acquired using chromatography control systems. Product peak HETP is calculated using the equation given in the section above. Results can be plotted on statistical process control charts with control rules applied to identify deterioration of separation performance before performance dropping below acceptable limits. The control chart shown in Figure 8.14 indicates that following each use, the product peak HETP was in statistical control and within the predetermined CPA/FPC acceptance criteria.

In summary, the product peak HETP is a PAT that can be used to monitor column cleaning, stability, and suitability for reuse by providing an indication that

1. The resin capacity has not deteriorated to unacceptable levels.
2. The physical structure (i.e., size and strength) of the media has not been altered.
3. The incoming resin quality is within acceptable limits (Wahome et al. 2008).

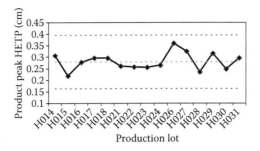

FIGURE 8.14 Product peak HETP control chart indicating no control rule violations and within acceptance criteria of HETP <0.50 cm.

8.5.4 PRODUCT PEAK ASYMMETRY FACTOR AS A PAT CRITICAL QUALITY ATTRIBUTE

In a similar fashion, the symmetry of the product peak shown in Figure 8.8 can be measured and monitored (Figure 8.15). In this application, A_f serves as a measure of the consistency and reproducibility of product elution. This PAT measurement serves as an indication that there is no nonspecific binding of the product due to solute buildup or media degradation. Consistent peak symmetry also demonstrates consistency of product loading, yield, and elution buffer conditions. The control chart shown in Figure 8.15 indicates that with each use, the product peak A_f was in statistical control and within the predetermined CPA acceptance criteria. Product peak A_f serves as a real-time CQA demonstrating consistent chromatography performance.

In summary, the product peak asymmetry factor is a PAT that can be used to monitor column cleaning and provides an indication that

1. There is no nonspecific binding due to solute buildup or media degradation.
2. Column equilibration, loading, and elution conditions are consistent.
3. The column continues to be fit for its intended purpose.

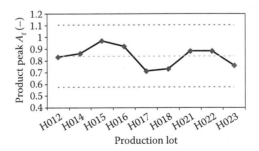

FIGURE 8.15 Product peak A_f control chart indicating no control rule violations and within acceptance criteria of $0.4 < A_f < 1.4$.

8.5.5 PRODUCT PEAK RETENTION AS A PAT FORWARD PROCESSING CRITERION

Chromatography mechanistic models are useful engineering tool to understand and control chromatography processes. Early models first appeared in the 1940s (see, for example, Martin and Synge 1941). More recently, ion exchange models have been developed based on the distribution coefficient by applying the law of mass action (Yamamoto and Ishihara 2000; Yamamoto 1995, 2005). The distribution coefficient (K) determines the retention time of a protein species. K is related to the ion exchange equilibria and provides quantitative information on the number of charges involved in the protein–ligand binding. Using this model, the distribution coefficient was related to the effective ionic capacity and the number of adsorption. The ion exchange reaction can be expressed from the law of mass action that relates K to the molarity of the resin ion exchange groups, the molarity of the ions in the mobile phase, the molarity of the bound protein, and the number of charges on the protein (Watler et al. 2003).

$$K = (K_e L^{Zp}) I^{-Zp} + K_{SEC} = A I^{-B} + K_{SEC} \tag{8.7}$$

where

L = effective total ion exchange capacity, I = ionic strength (counter ion concentration), K_e = equilibrium constant, Zp = protein valence (the number of charges involved in protein adsorption or effective charge), and K_{SEC} = distribution coefficient at nonbinding conditions (due to size exclusion).

The parameters $A = K_e L^{Zp}$ and $B = Zp$ are used for determining the values from linear gradient elution experiments (Yamamoto and Ishihara 2000; Yamamoto 1995, 2005).

The equation defining peak retention volume can be related to the distribution coefficient as

$$V_R = V_o(1 + HK), \tag{8.8}$$

where

V_R = peak retention time and H = the ratio of the stationary phase volume to the mobile phase volume = (1 − column bed void fraction)/(column bed void fraction).

Because K is directly related to the ion exchange capacity, it follows that peak retention volume is a direct measure of the total ionic capacity of the chromatography resin. This provides a simple method for monitoring the stability of the chromatography resin and assessing against predetermined criteria for acceptable ion exchange capacity. The control chart shown in Figure 8.16 indicates that with each use, the product retention volume was in statistical control and within the predetermined CPA acceptance criteria. Product peak retention volume is a PAT that serves as a real-time CQA demonstrating consistent chromatography performance and consistent resin capacity within acceptable ranges. Should the resin degrade with reuse, the resin capacity will be reduced and will result in a reduction in product peak retention volume.

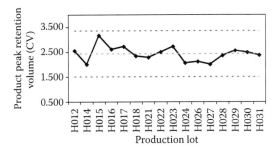

FIGURE 8.16   Product peak retention volume control chart indicating no control rule violations and within acceptance criteria of $1.5 < V_R < 3.5$.

In summary, the product peak asymmetry factor is a PAT that can be used to monitor column cleaning and provides an indication that

1. Media ligand capacity is maintained within acceptable ranges.
2. There is no buildup of contaminants blocking ligand-binding sites.
3. Media backbone is stable, with no nonspecific binding due to solute buildup or media degradation.

8.6   SUMMARY OF CHROMATOGRAPHY CLEANING PROCESS ANALYSIS TECHNOLOGY

To verify chromatography column suitability for reuse requires accessing the physical integrity of the packed bed, the stability of the resin chemical functionality, and the cleanliness of the packed column. The combination of PAT and process mechanistic knowledge can be combined to verify chromatography column acceptability for reuse. A summary of these PATs is given in Table 8.1.

TABLE 8.1
PATs for Verifying Chromatography Column Suitability for Reuse

	Analysis Method	Packed Bed Integrity	Media Stability	Carryover
Forward processing criteria (FPC)	HETP	✓	✓	
	A_f	✓	✓	
	Column back pressure	✓	✓	
	Mock pool protein analysis			✓
Critical performance attributes (CPAs)	Product pool purity	✓	✓	
	Product recovery		✓	
	Chromatogram reproducibility		✓	
	Product retention time	✓	✓	
	Product peak HETP	✓	✓	
	Product peak A_f	✓	✓	

8.7 CONCLUDING REMARKS

Advances in both sensor technology and automation have combined to provide a heightened level of monitoring and confirmation of equipment cleaning processes. As we have discussed, a wide array of orthogonal technologies can be applied to monitoring process criteria and quality attributes. This wide array of monitoring tools include on-line sensing of flowrate, temperature, TOC, conductivity, pressure drop, UV, HETP, and As. The benefits of employing these tools include demonstrating control and consistency of cleaning processes and providing greater assurance of adequate cleaning prior to equipment reuse. These online PATs can also enable cleaning fingerprints which are unique to each system and demonstrate cleaning consistency that would not be possible with a simple point sample. While providing heightened monitoring of process consistency, employing appropriate PATs can also reduce processing costs by verifying chromatography column cleaning prior to every reuse. Ultimately, these PAT techniques may obviate the burden of validating column cleaning over tens or even hundreds of cycles. Finally, employing PAT will improve cGMP compliance by serving as a useful tool to satisfy the Stage 3 'Continued Process Verification' expectations of the FDA's process validation guidance (FDA 2011).

REFERENCES

ASTM. 1981. *ASTM Standards on Chromatography*, First edition. Philadelphia, PA: ASTM.

Bird, R.B., W.E. Stewart, and E.N. Lightfoot. 1960. *Transport Phenomena*. New York: John Wiley & Sons.

Brorson, K. 2010. PAT and the future of biotechnology. *PDA J. Pharm. Sci. Tech.* 64 (2): 81.

Dickson, M.L., P. Leijon, L. Hagel, and E.J. Fernandez. 1998. Revealing packing heterogeneities and their causes using magnetic resonance imaging. Prep '98, Washington.

Dolejs, V., B. Siska, and P. Dolecek. 1998. Modification of the Kozeny–Carman concept for calculating pressure drop in flow of viscoplastic fluids through fixed beds. *Chem. Eng. Sci.* 53 (24): 4155–4158.

European Compliance Academy. 2006. Warning Letters Report 2005. Heidelberg, Germany. Available on-line at http://www.gmp-compliance.org/eca_news_702.html. Accessed on August 1, 2011.

FDA. 2004. Guidance for Industry, PAT—A Framework for Innovative Pharmaceutical Development, Manufacturing, and Quality Assurance. Rockville, MD. Available on-line at http://www.fda.gov/downloads/Drugs/GuidanceComplianceRegulatoryInformation/Guidances/ucm070305.pdf. Accessed on September 15, 2010.

FDA. 2011. Guidance for Industry - Process Validation: General Principles and Practices. U.S. Department of Health and Human Services. Rockville, MD. Available on-line at http://www.fda.gov/downloads/Drugs/GuidanceComplianceRegulatoryInformation/Guidances/UCM070336.pdf. Accessed on August 1, 2011.

Fordor, S. and J.M. Hyde. 2005. Increasing plant efficiency through CIP. BioPharm International. Available on-line at http://biopharminternational.findpharma.com/biopharm/GMPs%2FValidation/Increasing-Plant-Efficiency-Through-CIP/ArticleStandard/Article/detail/146346. Accessed on September 15, 2010.

Giddings, J.C. 1991. *Unified Separation Science*. New York: John Wiley & Sons.

Helfferich, F. 1962. *Ion Exchange*. New York: McGraw-Hill.

Janson, J.-C. and P. Hedman. 1982. Large scale chromatography of proteins. Advances in biochemical engineering. In *Chromatography*, edited by A. Fiechter. New York: Springer-Verlag.

Janson, J.-C. and L. Ryden. 1998. *Protein Purification*, Second edition. New York: Wiley-VCH.

Jungbauer, A. and E. Boschetti. 1994. Manufacture of recombinant proteins with safe and validated chromatographic sorbents. *J. Chromatogr. B.* 662: 143–179.

Kaltenbrunner, O., C. Tauer, J. Brunner, and A. Jungbauer. 1993. Isoprotein analysis by ion-exchange chromatography using a linear pH gradient combined with a salt gradient. *J. Chromatogr.* 639: 41–49.

Kaltenbrunner, O., P. Watler, and S. Yamamoto. 2000. Column qualification in process ion exchange chromatography. In *Bioseparation Engineering*, edited by I. Endo, T. Nagamune, S. Katoh, and T. Yonemoto, pp. 201–206. Maryland Heights, MO: Elsevier.

Lendero, N., J. Vidic, P. Brne, A. Podgornik, and A. Strancar. 2005. Simple method for determining the amount of ion-exchange groups on chromatographic supports. *J. Chromatogr. A* 1065: 29–38.

Lettner, H.P., O. Kaltenbrunner, and A. Jungbauer. 1995. HETP in process ion-exchange chromatography. *J. Chromatogr. Sci.* 33: 451–457.

Martin, A.J.P. and R.L.M. Synge. 1941. A new form of chromatogram employing two liquid phases. *Biochem. J.* 5: 1358–1368.

Mitchell, N.S., L. Hagel, and E.J. Fernandez. 1997. In situ analysis of protein chromatography and column efficiency using magnetic resonance imaging. *J. Chromatogr. A* 779: 73–89.

Molony, M. and C. Undey. 2009. PAT tools for biologics: Considerations and challenges. In *Quality by Design for Biopharmaceuticals*, edited by A.S. Rathore and R. Mhatre. Hoboken, NJ: John Wiley & Sons.

Nelson, L.S. 1984. The Shewhart control chart—Tests for special causes. *J. Qual. Technol.* 16 (4): 237–239.

Ogez, J., R. van Ries, N. Paoni, and S. Builder. 1991. Recombinant human tissue-plasminogen activator: Biochemistry, pharmacology and process development. In *Chromatographic and Membrane Processes in Biotechnology*, edited by C. Costa, J. Cabral, and J. Boston. Dordrecht, The Netherlands: Kluwer Academic Publishers.

Perry, R.H. and D.W. Green. 1997. *Perry's Chemical Engineers' Handbook*, Seventh edition. New York: McGraw-Hill.

Podgornik, A.A., J. Vidič, J. Jančar, N. Lendero, V. Frankovič, and A. Štrancar. 2005. Noninvasive methods for characterization of large-volume monolithic chromatography columns. *Chem. Eng. Technol.* 28 (11): 1435–1441.

Rathore, A. and G. Sofer. 2005. Life span studies for chromatography and filtration media. In *Process Validation in Manufacturing of Biopharmaceuticals: Guidelines, Current Practices, and Industrial Case Studies*, edited by A. Rathore and G. Sofer. Boca Raton, FL: Taylor & Francis.

Schoenmakers, P. 1986. *Optimization of Chromatographic Selectivity*. New York: Elsevier.

Snyder, L.R.J. 1972. *J. Chromatogr. Sci.* 10: 369–379.

Sofer, G. 1995. *Downstream Processing in Biotechnology Course # 9505103*. East Brunswick, NJ: The Center for Professional Advancement.

Stickel, J.J. and A. Fotopoulos. 2001. Pressure–flow relationships for packed beds of compressible chromatography media at laboratory and production scale. *Biotechnol. Prog.* 17 (4): 744–51.

Vogel, H.C. 1997. *Fermentation and Biochemical Engineering Handbook. Principles, Process Design, and Equipment*, Second edition. Park Ridge, NJ: Noyes Publications.

Wahome, J., W. Zhou, and A. Kundu. 2008. Impact of lot-to-lot variability of cation exchange chromatography resin on process performance. *Biopharm. Int.* 21 (5): 48–56.

Watler, P.K., C.H. Cholakis, and M.V. Sefton. 1988. Water content and compression modulus of some heparin–PVA hydrogels. *Biomaterials* 9 (2): 150–154.

Watler, P., O. Kaltenbrunner, and D. Feng. 2003. Engineering aspects of ion-exchange chromatography, in scale-up and optimization. In *Preparative Chromatography*, edited by A.S. Rathore and A. Velayudhan. New York: Marcel Dekker.

Watts, C. and J. Clark. 2006. PAT: Driving the future of pharmaceutical quality. *J. Process Anal. Technol.* 3: 6–9.

Western Electric Company. 1958. *Statistical Quality Control Handbook*, Second edition. New York: Western Electric Company.

Yamamoto, S. 1995. Plate height determination for gradient elution chromatography of proteins. *Biotechnol. Bioeng.* 48: 444–451.

Yamamoto, S. and T. Ishihara. 2000. Resolution and retention of proteins near isoelectric points in ion-exchange chromatography. Molecular recognition in electrostatic interaction chromatography. *Sep. Sci. Technol.* 35: 1707–1717.

Yamamoto, S., P.K. Watler, D. Feng, and O. Kaltenbrunner. 1999. Characterization of unstable ion-exchange chromatographic separation of proteins. *J. Chromatogr. A* 852 (1): 37–41.

Yamamoto, S. 2005. Electrostatic interaction chromatography process for protein separations: The impact of the engineering analysis of biorecognition mechanism on the process optimization. *Chem. Eng. Technol.* 28 (11): 1387–1393.

9 Cell Culture Process Analytical Technology Multiplexing Near-Infrared

Mariana L. Fazenda, Linda M. Harvey, and Brian McNeil

CONTENTS

9.1 NATURE OF CELL CULTURE SYSTEMS

Although there are several publications on cell culture processes and their optimization using spectroscopic tools in the context of process analytical technology (PAT), they do not address underlining details that affect the implementation of these systems at an industrial scale. This chapter discusses the intrinsic challenges of using near-infrared (NIR) as a PAT tool in fermentation cell culture processes.

Animal cell culture processes date back to the early 1950s when complex media using many animal-derived products (serum and blood-derived proteins) were used to grow highly specialized cell lines often anchorage dependent (existing on a support matrix) and nourished by the nutrient medium (Freshney 2000).

Nowadays, although some anchorage-dependent cell lines are still in use for manufacture of specialized bioproducts, for example, vaccines, most animal cell lines are cultured in chemically well-defined serum and animal product-free media (Paoli et al. 2010). The latter medium type has become the norm within the cell culture sector of the biotechnology industry, and there are several reasons why this is so. First, the use of medium components derived from animal sources implies the very real risk of introduction of adventitious viruses or even prions. This is a real concern because these viruses can potentially infect the cell line itself or be carried over into the product. Clearly, downstream processing of products manufactured from media containing animal-derived components could require costly downstream process cleanup and virus removal. The use of media with no animal-derived components avoids the risks and the costs. A further benefit is that these media are typically optically clear.

9.2 REACTOR TECHNOLOGY FOR ANIMAL CELL CULTURE

To differentiate from vessels used for cultivation of microbial cells, which are normally termed *fermenters*, the vessels used for animal cell culture are often referred to as *bioreactors* (Matthews 2008). Although over the years a vast range of reactor types have been used to culture anchorage-dependent and freely suspended cell lines, the workhorse of the cell culture industry is now the conventional stirred tank (STR) (Figure 9.1). As can be seen in Figure 9.1, the bioreactor STR is broadly similar to the fermenter STR, although there are some very important differences.

The design differences between animal cell reactors and microbial STRs are based on the fundamental differences between the structure and physiology of the cells involved. First, there is a considerable difference in size between a typical bacterial cell (2–3 μm in length by 1 μm in width) and a Chinese hamster ovary (CHO) cell, which can range from 10 to 15 μm in diameter (Pinto et al. 2008). Animal cells are far larger than microbial counterparts and have no rigid cell wall surrounding them; thus, they traditionally have been considered far more physically fragile and subject to damage by moderate hydrodynamic forces. This physical fragility is relative. However, there have been numerous studies into the shear sensitivity of these cells and their vulnerability to damage by the energy released by such physical phenomena as gas bubble disengagement (Wu 1995). The genetic complexity of animal cells causes the very slow growth rates normally seen when using these systems. A normal *Escherichia coli* fed-batch system can be completed in less than 24 h, whereas typical CHO batches may extend over 2 weeks (Arnold et al. 2003).

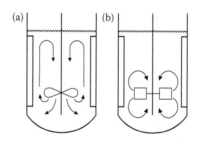

FIGURE 9.1 Bioreactor STR with (a) marine impeller and (b) turbine impeller. (Adapted from McNeil, B. and Harvey, L.A., *Fermentation: A Practical Approach*, IRL Press, Oxford, England, 1990.)

All of these structural and physical characteristics translate into design changes in the reactor. Slow growth means no great need for high oxygen transfer rates, and thus low gassing rates, and shear sensitivity means very low agitation rates of the stirrer and the use of marine impellers (low shear) instead of the microbial system's standard Rushton turbine (Figure 9.1). Animal cell systems also usually have a domed base (not flat as microbial) to reduce shear forces. pH control is usually affected in such systems by addition of CO_2 to the culture fluid to generate a bicarbonate buffer system mimicking what occurs in nature. Often, aeration is discontinuous with the process controller simply adding a burst of oxygen to the culture when the dissolved oxygen tension (DOT) approaches a predefined set point (often 20%–30% saturation). Cell density in such systems can be very low compared with microbial cell systems; typical batch densities in animal cell culture may be from 10^6 to 10^7 cells/ml, although some specialist cell lines such as PER-C6 can greatly exceed this. Overall dimensions of production-scale animal cell culture reactors are usually much less than equivalent microbial reactor systems. A typical microbial antibiotic production-scale fermenter would be around 150 m^3, whereas a CHO cell culture system for antibody-based drug manufacture could be around 10 m^3.

It is worth considering here what these differences (and similarities) between the two types of cells tell us about the application of techniques such as NIR, mid-infrared (MIR), and Raman within this context. Typical animal cell culture fluids are usually optically clear liquids showing simple Newtonian behavior; thus, the introduction of a probe-type sensor in these systems is inherently less challenging than in microbial systems, where there may be significant changes in process fluid rheology as time passes. As has been shown in numerous microbial studies, major changes in the physics of such systems can make spectroscopic analysis of the changes in levels of chemical analytes very challenging. This is not a problem we must face in the cell culture systems. The lack of a vigorous gas phase in cell culture also eases the challenges of using *in situ* probes because gas bubble intrusion into the measuring gap is less problematic.

Therefore, from first principles, the employment of potentially real-time monitoring techniques such as NIR or MIR should be much easier in cell culture systems because of the tractable nature of the process fluid and the modest mixing and gassing rates.

9.2.1 *Ex Situ* vs. *In Situ*

When a real-time sensor is applied to a reactor system, broadly speaking, there are two options on how it can be deployed: (1) *ex situ*, where the sensor is located outside the reactor or (2) *in situ*, where the sensor is located in the reactor (Vaidyanathan et al. 1999) (Figure 9.2). To date, nearly all applications of real-time monitoring of animal cells in culture have used the latter approach. The clear advantages are that the probe is within the sterile envelope, and for cell culture systems, this is especially important because maintenance of sterility (or more correctly monosepsis) is more challenging than for microbial systems. Physically robust fiber optic base sensors that can be deployed using standard fermenter/bioreactor ports (18 and 25 mm diameter) are available readily; thus, this sensor technology can easily be deployed

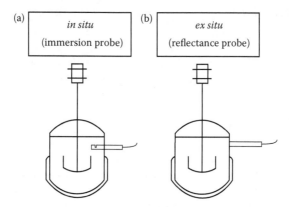

FIGURE 9.2 On-line sampling configuration. From left to right: (a) *in situ* measurements using an immersion probe; (b) *ex situ* measurements using a reflectance probe on the glass wall of the reactor. (Adapted from Cervera, A.E. et al., *Biotechnol. Prog.*, 25, 1561–1581, 2009.)

at lab to production scale. By contrast, the use of flow-through external loop systems generally finds even less favor with cell culture specialists because of the contents of the vessel having to flow through an *ex situ* loop before returning to the vessel. Furthermore, there is always the risk of introducing an artifact into the measurement or of physiological changes in the fluid in the loop (e.g., due to oxygen deprivation).

9.2.2 WHAT ARE WE INTERESTED IN MEASURING?

NIR, MIR, and Raman are all interesting spectroscopic methods, which have the potential to give real-time information about multiple analytes. NIR has been most widely applied to cell culture systems and is simpler to employ and to multiplex than MIR, but gives us less information than MIR due to the weak overlapping absorbances in the NIR region. This, of course, is a major advantage of using NIR in highly light-absorbing and light-scattering fermentation systems. So why has MIR been much less frequently examined in this context? The answer probably relates to the complexity and cost of the fibers (chalcogenide) and waveguides needed. It does have the potential to give real insights into cell culture medium composition as shown by Rhiel et al. (2002).

In many cell cultures, the analytes of major interest tend to be (1) the carbon and energy source, which is glucose in most cases; (2) the N source(s), which may include glutamine and a range of preformed amino acids; and (3) waste products such as ammonia and glutamate. These have been measured using *in situ* NIR but also by multiplexed NIR as will be discussed later. However, a recent report describes the use of NIR to assess medium osmolarity (Mattes et al. 2009). This is physiologically important in cell cultures because animal cells are really only comfortable in a relatively narrow range of osmolarity due to the absence of cell walls. Moreover, medium osmolarity increases due to pH control actions and feeding can lead to premature process cessation due to osmotic stress causing apoptosis (programmed cell death).

This is another measurement that is potentially translatable to multiplexed NIR, as are total cell number and viable cell number, which are two more vitally important process parameters (Paoli et al. 2010; Arnold et al. 2003). Therefore, although not widely reported, multiplexed NIR can give a great range of physiologically relevant information about what is occurring in cell culture systems.

9.2.3 A WORD OF CAUTION

The aforementioned spectroscopic methods (NIR, MIR, and Raman) and others such as ultraviolet and fluorescence have been widely used as PAT not only in pharmaceutical but also in oil and food industries for rapid and less expensive measurements of quality. Modeling methods such as the use of scores and residual statistics from principal component analysis (PCA) and partial least squares (PLS) are commonly used to summarize, monitor, optimize, and control a bioprocess. In order for these methods to be an integral part of the quality by design approach, the main challenge is the robustness of their calibration models to new acceptable sources of variability added to the process throughout their development and commercialization life cycle. The amount of effort while developing initial calibration models is enormous; therefore, it is necessary to consider all internal and external sources of variation. Aspects such as the evolution of the process design space (development, scale-up, and industrialization) and also, for example, the replacement or new added instruments can be critical sources of variation. Therefore, strategies for spectroscopic calibration model transfer, maintenance, and update should be in place before the implementation of real-time process monitoring. Different approaches have been applied, and most have been discussed in the literature (Fearn 2001; Feundale et al. 2002; Miller et al. 2008), such as dynamic orthogonal projection, direct cross prediction (based on slope/offset correction), and piecewise direct standardization methods.

In general, when using spectroscopic equipment in bioprocessing, several aspects need to be considered.

1. Sensors must be physically robust—capable of repeated sterilization.
2. Sensors must show matched performance.
3. Sensors should be able to monitor more than two process variables simultaneously and frequently.
4. Sensors must be scalable (operate at 2–10,000 L).
5. Ideally operating software should be designed with a process environment in mind.
6. Integrated data management.

Moreover, when using cell cultures, the common dangers of many biological systems for the spectroscopist interested in quantitative modeling are shared. Because the level of every analyte is linked to that of others via the process stoichiometry, there is a real danger that the use of a secondary analytical technique, such as NIR or MIR, might generate models without valid analytical basis. There are a number of approaches to dealing with the issue of collinearity of the data: (1) spiking with the analyte of interest completely breaks the link between analyte and matrix, thus eliminating collinearity, and (2) allowing clear assignment of analyte absorbance regions. Broadly speaking, adaptive calibration does something very similar, but also reduces the need for a large calibration dataset (a major drawback in calibration). However, both methods are difficult to employ *in situ* and especially in a multiplexed system, where the challenge is to deal with the matrix as it is, without any manipulation.

Defining the analytical basis of quantitative models (e.g., PCA) should of course help avoid collinearity because it allows us to examine the relative contribution of all sources of variance to the model, and it is good practice to do this as a routine.

9.3 MULTIPLEXING SYSTEMS

When calibration models are carried out using a single probe, multiple bioreactor runs are necessary (Figure 9.3). This is normally suitable for small vessels; however, this approach runs the real risk that the subsequent calibration models can also be probe and vessel specific. Moreover, it can be difficult to obtain sufficient quality spectral data from all process scales; for example, in early-phase process development, a full-scale bioreactor may only be run two to three times per product (which is not sufficient to develop a full-calibration model at this scale due to the small dataset). Using multiplexing systems comes at a time where sustainability awareness is at its peak, and it is necessary to consider each element of a process, in this case in a manufacturing process, as a potential factor of improvement from both an energy and cost-saving viewpoint. If it is possible to invest in a single piece of equipment, which will aid the understanding and control of an already established process, simultaneously reducing the overall cost and producing a more consistent and higher quality product, then it is extremely desirable. From this viewpoint, multiplexing can provide highly accurate measurement of constituents' concentrations, which is required for systematic modeling and control of complex processes, especially bioprocesses (Chen et al. 2011). It allows several fiber optic probes to be sequentially analyzed with one instrument, thereby decreasing the cost per point of analysis. It involves the use of multiple runs in parallel (Figure 9.4); thus, calibration data are

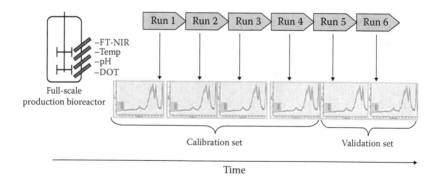

FIGURE 9.3 NIR calibration carried out in series.

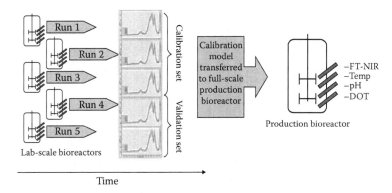

FIGURE 9.4 NIR calibration carried out in a multiplexing system.

collected from multiple fermentations incorporating a large degree of variability and expediting the calibration process and so reducing process timelines.

However, processing multiplex data can add additional challenges for the spectroscopist. There are not only the physicochemical variations picked by the spectra in each bioreactor used but also the optical differences in each of the probes. The optical differences in each probe can be seen as a physical factor, that is, breaking down the probe to its physical components and making sure that each of these is equivalent in terms of performance and quality, or, instead, deal with the contribution of these differences once the data have been collected and address them using advanced chemometrics analysis. Multivariate analysis including the use of PLS models or popular empirical preprocessing methods such as orthogonal signal correction (OSC), standard normal variate (SNV), and multiplicative signal correction (MSC) are commonly used while modeling and developing calibration methods, but may prove to be insufficient in dealing with multiplex systems. The following section will describe these two approaches.

9.3.1 Introducing "Variability"

When using multiplexing systems, it is necessary to evaluate the optical characteristics of the different probes, mirrors, and spectrometer channels to be used, because each individual optical component may add variability and thus affect spectra and calibration models. To tackle these critical important aspects (probe/mirror/spec channel), Roychoudhury et al. (2007) proposed a novel investigation where the impact of these factors was evaluated upon signal intensity. If the NIR multiplexing system (spectrophotometer channel, mirror optical properties, and probe design) performed consistently over the entire time course of the bioreactor runs, then the possibility of introducing error into the NIR signal would be excluded. Therefore, each of the above factors was varied in turn, and signal intensities were recorded in a range of "media," that is, air, methanol, and acetone (common reference standards used by manufacturers to check the optical performance of the probes). All the optical components contributed to some extent to variability (on the basis of the

F value); however, the greatest source of signal intensity variability was the differences between probes. They also observed that more variance occurred at lower wave numbers, which seems logical because these regions are less energetic. A single optical channel and a mirror were used, and a probe test (involving seven different probes) was conducted to evaluate the effect of probe variability on spectra. From the PCA analysis, it was possible to identify probe optical differences from the scores plot. The work described in this study highlights the potential impact of optical differences, despite the care taken during probe manufacture to ensure similar physical characteristics. Based on these results, it is possible to establish a maximum acceptance level for probe variance, and this could then be used by manufacturers as a guide to reduce issues regarding interprobe variability.

In terms of probe variability in multiplexed systems and its effect on modeling bioprocessing data, NIR calibration models from multiple small-scale bioreactors using CHO cell lines were then developed for prediction of concentrations of key analytes such as glucose, lactate, viable cell number, and total cell number. The idea was that these models could be used to monitor large-scale bioreactors facilitating the calibration process and reducing process timelines (Roychoudhury et al. 2007). Because mammalian cell cultures tend to be generally low-biomass processes with insignificant rheological changes grown in the completely soluble growth medium, the model-building exercise should be relatively straightforward (Arnold et al. 2003). Both single- and multiprobe models performed similarly on external validation (high R^2 and low size exclusion chromatography). The multiplex models were then used to predict analyte concentrations in cultures of different antibody-producing cell lines and in larger scale cultures (up to 150 L) of the original cell line. Surprisingly, given the high degree of variability (and, hence, robustness) built into the original calibration exercise, the overall model performance on different cell lines and at different scales was poor. The reasons for this were investigated, and it was clear that physical influences, specifically temperature fluctuations in the process plant, might have contributed significantly to spectral variability in the study. Even though the values and trajectory were in the right range, the models failed to predict accurately. The major issue was the sinusoidal variation in the room temperature, which was both seasonally and daily dependent (Figure 9.5). With the multiplexing system, the bioreactor and probe variabilities are incorporated into the models: data are collected simultaneously and during a certain period of time. This results in an unexpected uncontrolled variable, that is, temperature. Therefore, when the models are tested with spectral datasets captured at different climatic conditions, they fail to predict. One can argue that, with the conventional single-probe technique, this temperature variability could have been incorporated into the calibration models because the fermentation processes had to be operated in series during a longer period of time. Therefore, the choice of single probe or multiplex systems needs to be made according to the characteristics of the process to be analyzed.

A different approach to multiplex calibration development has been developed by Chen et al. (2011). In this study, a multiplex calibration algorithm is developed assuming that, in a multiplex system, variation only arises from the optical differences found in each probe and not from the matrix being analyzed. Therefore, Chen and coauthors anticipated the variance in the multiplexing system by developing a

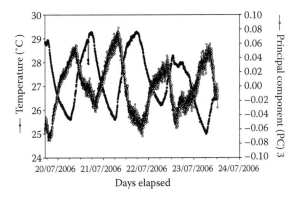

FIGURE 9.5 Temperature in the lab plotted alongside principal component PC3 of a single seed batch vessel of CHO cells vs. days elapsed.

suitable algorithm for that. This can be, however, debatable because each probe is in a different reactor; therefore, there is also an additional physicochemical variation picked by the spectra. The algorithm is based on Beer–Lambert's law, as the classic PLS algorithm, with the introduction of a multiplicative parameter for each optical probe to account for any spectral contribution. Once this parameter is determined (via an optical path-length estimation and correction method; Chen et al. 2006), the detrimental effects of each probe are effectively mitigated through a "dual calibration" strategy. Multiplex calibration models (MCMs) were then built for two completely different systems: a well-defined MIR data of ternary mixtures (acetone, ethanol, and ethyl acetate) and NIR data of the CHO cell process described previously. Their predictive performance was compared with that of PLS calibration models with and without data preprocessing methods such as SNV, MSC, and OSC using the root-mean-square error of prediction (RMSEP) as the performance criterion. The differences between the MIR spectra of the same ternary mixture sample recorded by two different probes were evident and could not be effectively modeled using PLS and preprocessing methods, even with combined datasets from the two different probes. The MCM calibration results, on the other hand, proved to give a much more accurate prediction in modeling the multiplex spectral data, achieving reductions of two to four times of the corresponding values achieved by PLS models. The MCM results obtained for the test samples from the multiplexed NIR probes in the CHO cell culture process also outperformed all the other methods investigated (approximately 54% reduction in the RMSEP values). However, it is extremely important to highlight the fact that it was not an overall improvement because the RMSEP for the calibration samples showed an increase. This is not at all surprising because the nature of such a system is extremely complex, and the MCM method only accounts for the optical probes differences, as mentioned initially. Although the present study seeks to provide an optimal calibration solution for such complex multiplex spectra and it is definitely a step forward in the development of suitable algorithms for such systems, the empiricality of this method is not sufficient, per se, to model the physicochemical variation picked by the spectra and intrinsic to the matrix.

9.3.2 Type of Spectrophotometers

The ability of NIR to use fiber optics to transmit light between the instrument and the sample over a considerable distance is one of the main advantages of process NIR over IR. This permits the instrument (a potential source of ignition) to be placed in safe areas where no flammable vapors exist. Only the fiber optic probe in contact with the sample and the optical fibers that connect to the instrument needs to be placed in the hazardous location. However, the most stable spectrometer will be useless if the fiber optic probe to which it is attached falls apart in the process. Process probes have to withstand the rigors of the manufacturing environment, including high temperatures, high pressures, aggressive chemicals, mechanical vibrations, and, often, combinations of the above. For example, another feature that may need to be considered in a multiplexed cell culture setup is the positioning and orientation of each probe within the appropriate bioreactor. It seems to surprise many new users of NIR *in situ* that the position of the probe within the vessel can affect spectral quality. Normally, the probes in a bioreactor would be close to the edge of the impeller zone, in a reasonably well-mixed region. However, probe orientation may also affect spectral collection quality. A typical NIR transflectance probe can be seen in Figure 9.6a. As can be seen from the figure, such probes normally have a gap (usually alterable before immersion). The orientation of the gap relative to flow patterns in the bioreactor itself can affect spectral quality. One way to minimize this, which is well understood within the industry, is to mark the probe casing and the bioreactor housing such that the probe orientation can be maintained from run to run (Figure 9.6b). This is a particular problem with dispersive instruments due to the mechanical scanning process involved. It is easy to see how additional variance can be contributed in a multiplexed setup if this is not taken into account. It is also possible that variations in the flexion angle of the fiber bundle itself might contribute to spectral variability. Although all of these aspects are very well known within the spectroscopy industry, they rarely, if ever, appear in scientific papers on the subject.

Ultimately, the success of an analyzer application is dependent on how well a spectrometer produces the spectra. Stability, reliability, wavelength precision, resistance to vibration, and temperature change are all factors that determine the performance of the analyzer and the application, thus leading to another variability factor, which

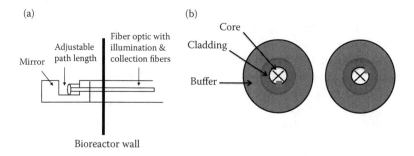

FIGURE 9.6 (a) Diagram of a transflectance NIR probe. (b) Top view of NIR probe highlighting how to mark the probe casing to maintain probe orientation from run to run.

is the type of spectrometer used. There are two distinct types of NIR spectrometers utilized: dispersive and Fourier transform near-infrared (FT-NIR). The theoretical advantages of FT-NIR over dispersive NIR systems are well known (Scarff et al. 2006; Peirs et al. 2002). Dispersive systems can only measure a small part of the spectrum at a time, so they usually scan the spectrum by moving the grating, the mirrors, or the detector. These early systems worked reasonably well in laboratories, where temperature variations and vibration could be controlled. In process environments, dispersive systems may be affected by variations in wavelength calibration, baseline stability, and mechanical reliability. FT-NIR spectroscopy was developed to overcome some limitations encountered with the dispersive instruments. In theory, FT instruments have some major advantages: (1) higher speed because all wave numbers are measured simultaneously; (2) simplified mechanics, with the moving mirror or quartz wedge being the only moving part in the instrument with a reduced risk of mechanical breakdown and less sensitivity to temperature variations and vibrations in field use; and (3) self-calibrating with a HeNe laser and wavelengths and, therefore, never needing calibration by the user (Peirs et al. 2002). However, nowadays, instrumentation is so developed that, for example, a dispersive process instrument with a microbundle optical fiber can be at least as robust as an FT-NIR system with a single optical fiber probe. One key question is how to objectively compare the two different approaches.

One of the best approaches to compare NIR probe function, and its effect on process monitoring, is to use PCA, rather than the more common, but simplistic approach of comparing heavily processed PLS model outputs. By contrast, PCA provides a direct measure of the NIR capability as a process fingerprinting technique, because it has the intrinsic ability to capture a sample's chemical and physical information (Henriques et al. 2009). PCA is a great tool for process monitoring, providing real-time information of process trajectories, such as growth patterns and growth deviations, that is, contaminations (Figure 9.7). This process "knowledge" becomes the "fingerprint" of the process at that specific phase, which can then be used for control decisions in the upcoming process phases. For example, an in-house

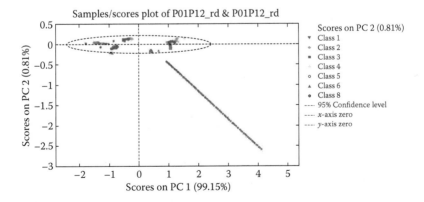

FIGURE 9.7 Scores plot of a PCA model from a production run of a CHO cell line. Identification of contamination by NIR spectroscopy, significantly before the process was actually terminated.

study conducted at Strathclyde University by the authors comparing an FT-NIR spectrophotometer with a dispersive spectrophotometer, with respect to its potential to monitor a series of consecutive CHO cell seed batch cultures in antibody production processes, was performed. PCA of the spectra of both instruments contained valuable information about both batch-to-batch variability and process time. The key finding was that despite the theoretical advantages of one system over the other, in practical usage, both systems had similar performances. However, to get the best from each system, the right chemometrics approach for that system had to be carefully selected.

9.3.3 DATA MANAGEMENT

To achieve the PAT goals of managing product quality and stepping into continuous process improvement involves, as discussed, both instrument and software development. However, with this come the expansion of on-line monitoring analyzers and, consequently, the rich data generated from them. Detailed chemical information on the ingredients and the process available in real time is needed; therefore, multivariate on-line instruments are more and more in demand. In a bioreactor, there can be an NIR probe, off gas analyzer, fermentation sensors (pH, DO, and temperature), etc., all collecting data. There can be a number of bioreactors used in a multiplex system where data is collected with these measurements. With this comes the extremely difficult task of managing the highly complex data acquired in an enormous flow of information including a mix of data formats (spectral, vector, and scalar data), before reaching the main and final goal of controlling (and limiting variability) in the process. The data acquired from NIR spectroscopy or any other instrumentation coupled to a bioprocess needs to be translated into useful information; and for that, it is necessary to interpret it. Of course, chemometrics deconvolutes data and transforms them into useful monitoring models, but the amount of data collected can overwhelm the system and the user if careful planning of a data management strategy has not been implemented in advance. The need to apply standard frameworks and accessible systems for transmission of the data generated by the analyzers involved in a specific process is extremely important. This can be achieved using interfaces that link external data sources (such as OPC, PI, and ORACLE), providing real-time connectivity to most distributed control systems. The data gathered from on-line monitoring cell culture in a single distributed database can then be interpreted and its relevance to controlling and limiting batch-to-batch variability can be understood, which leads to more consistent and predictable target product production with less loss or need for rework. It can also be used for audit purposes and be exchanged with other parties, which is extremely valuable at an industrial level.

9.4 CONCLUSIONS

In conclusion, in process monitoring, it is extremely important to have in place strategies for calibration model transfer, maintenance, and update before the implementation of real-time process monitoring. To handle this in terms of multiplexing systems, one can either (1) specify equipment performance standards to reduce interprobe variability and/or (2) accept probe-to-probe variance and develop algorithm-based model development to accommodate the variability. No doubt NIR spectroscopy can effectively reduce the development cycle and assist in scale and geographical transfer of a process, but it is unlikely that generic calibration models can be formulated unless intensive maintenance is taken into account from the start of the modeling exercise. In simple terms, model transferability should be built-in, perhaps using the approach outlined by Miller et al. (2008) or some similar calibration and instrument standardization methods. Ideally, operating software should be designed with a process environment in mind, because the immense data flow generated from all the sensors in place (in a PAT environment) can be a limiting step while implementing real-time bioprocessing. In saying this, the words of the FDA come to mind "quality cannot be tested into the product; it should be built-in or should be by design."

ACKNOWLEDGMENTS

The authors would like to acknowledge the support of GSK and the TSB Technology Programme of GSK.

REFERENCES

Arnold, S.A., J. Crowley, N. Woods, L.M. Harvey, and B. McNeil. 2003. In-situ near infrared spectroscopy to monitor key analytes in mammalian cell cultivation. *Biotechnol. Bioeng.* 84: 13–19.

Cervera, A.E., N. Petersen, A.E. Lantz, A. Larsen, and V.G. Krist. 2009. Application of near-infrared spectroscopy for monitoring and control of cell culture and fermentation. *Biotechnol. Prog.* 25: 1561–1581.

Chen, Z.P., J. Morris, and E. Martin. 2006. Extracting chemical information from spectral data with multiplicative light scattering effects by optical path-length estimation and correction. *Anal. Chem.* 78: 7674–7681.

Chen, Z.-P., L.-Z. Zhong, A. Nordon, D. Littlejohn, M. Holden, M.L. Fazenda, L.A. Harvey, B. McNeil, J. Faulkner, and J. Morris. 2011. Calibration of multiplexing fibre optic spectroscopy. *Anal. Chem.* 83 (7): 2655–2659.

Fearn, T. 2001. Standardisation and calibration transfer for near infrared instruments: A review. *J. Near Infrared Spectrosc.* 9: 229–244.

Feundale, R.N., N.A. Woody, H.W. Tan, A.J. Myles, S.D. Brown, and J. Ferre. 2002. Transfer of multivariate calibration models: A review. *Chemom. Intell. Lab. Syst.* 64: 181–192.

Freshney, R.I. 2000. *Introduction to Basic Principles in Animal Cell Culture: A Practical Approach*. Oxford, UK: Oxford University Press.

Henriques, J.G., S. Buziol, E. Stocker, A. Voogd, and J.C. Menezes. 2009. Monitoring mammalian cell cultivations for monoclonal antibody production using near-infrared spectroscopy. *Adv. Biochem. Eng./Biotechnol.* 116: 73–97.

Mattes, R., D. Root, M. Sugui, F. Chen, X. Shi, F. Jing, J. Liu, and P. Gilbert. 2009. Real-time bioreactor monitoring of osmolality and pH using near-infrared spectroscopy. *BioProcess Int.* 7: 44–50.

Matthews, G. 2008. Fermentation equipment selection: Lab scale bioreactor design considerations. In *Practical Fermentation Technology*, edited by B.A.H. McNeil, LM. Chichester: Wiley.

McNeil, B. and L.A. Harvey. 1990. *Fermentation: A Practical Approach*. Oxford, England: IRL Press.

Miller, C.E., R.T. Roglinski, N.B. Gallagher, and B.M. Wise. 2008. Combining calibration transfer and pre-processing: What steps, what order? Reno, NV: FACSS.

Paoli, T., J. Faulkner, R. O'Kennedy, and E. Keshavarz-Moore. 2010. A study of D-lactate and extracellular methylglyoxal production in lactate re-utilizing CHO cultures. *Biotechnol. Bioeng.* 107: 182–189.

Peirs, A., N. Scheerlinck, K. Touchant, and B.M. Nicolai. 2002. Comparison of Fourier transform and dispersive near-infrared reflectance spectroscopy for apple quality measurements. *Biosys. Eng.* 81: 305–311.

Pinto, R.C.V., R.A. Medronho, and L.R. Castilho. 2008. Separation of CHO cells using hydrocyclones. *Cytotechnology* 56: 57–67.

Rhiel, M., P. Ducommun, I. Bolzonella, I. Marison, and U. Von Stockar. 2002. Real-time in situ monitoring of freely suspended and immobilized cell cultures based on mid-infrared spectroscopic measurements. *Biotechnol. Bioeng.* 77: 174–185.

Roychoudhury, P., R. O'Kennedy, B. Mcneil, and L.M. Harvey. 2007. Multiplexing fibre optic near infrared (NIR) spectroscopy as an emerging technology to monitor industrial bioprocesses. *Anal. Chim. Acta* 590: 110–117.

Scarff, M., S.A. Arnold, L.M. Harvey, and B. McNeil. 2006. Near infrared spectroscopy for bioprocess monitoring and control: Current status and future trends. *Crit. Rev. Biotechnol.* 26: 17–39.

Vaidyanathan, S., G. Macaloney, J. Vaughn, B. McNeil, and L.M. Harvey. 1999. Monitoring of submerged bioprocesses. *Crit. Rev. Biotechnol.* 19: 277–316.

Wu, J.Y. 1995. Mechanisms of animal-cell damage associated with gas-bubbles and cell protection by medium additives. *J. Biotechnol.* 43: 81–94.

10 Process Analytical Technology for Bioseparation Unit Operations

Anurag S. Rathore, Vishal Ghare, and Rahul Bhambure

CONTENTS

10.1 INTRODUCTION

The pharmaceutical current good manufacturing practices (cGMP) for the 21st Century Initiative was implemented by the Food and Drug Administration (FDA) in 2002 to enhance and modernize the regulation of pharmaceutical manufacturing (FDA GMP Guidance 2002a). Some of the key concepts from this initiative emerged in the form of the guidance *PAT—A Framework for Innovative Pharmaceutical Manufacturing and Quality Assurance* (FDA PAT Guidance 2004). The scientific, risk-based framework outlined in this guidance, process analytical technology or PAT, is intended to support innovation and efficiency in pharmaceutical development, manufacturing, and quality assurance. The framework is founded on process understanding to facilitate innovation and risk-based regulatory decisions by industry and the agency. The approach is based on science and engineering principles for assessing and mitigating risks related to poor product and process quality. In addition to the PAT guidance, three important guidance documents were published as part of the International Conference on Harmonization (ICH) guidelines: ICH Q8 *Pharmaceutical Development*, ICH Q9 *Quality Risk Management*, and ICH Q10 *Pharmaceutical Quality System* (ICH Q8 Guidance 2009; ICH Q9 Guidance 2005; ICH Q10 Guidance 2008). PAT-related activities also occurred in other regulatory jurisdictions. A European Medicines Agency (EMEA) PAT team was created in November 2003 with the aim to review the implications of PAT and to ensure that the European regulatory framework and the authorities are prepared for and adequately equipped to conduct thorough and effective evaluations of PAT-based submissions. A reflection paper has been published that provides preliminary recommendations on how PAT-related information should be presented in regulatory applications (EMEA Reflection Paper 2006).

This chapter focuses on PAT applications for the various unit operations that are used in bioseparations. Special focus has been given on the research done in the last decade.

10.2 PAT APPLICATIONS IN THE PHARMACEUTICAL INDUSTRY

The successful introduction of the PAT guidance to the pharmaceutical industry in September 2004 by the FDA was a result of several observations by the agency. These included the need for improving the capability and efficiency of pharmaceutical manufacturing while maintaining product quality, decreasing the risk of releasing inferior quality products, and avoiding drug shortages and the risk of non-approval or delayed approval of the product due to low quality (FDA 2002b). From an economic point of view, it has been pointed out that the lack of scientific understanding of the process as well as product and early expenditure in the process before the phase III clinical studies are the key areas where improvement can be possible through the implementation of PAT (FDA 2002c). The overall expectation is that PAT can enable the pharmaceutical industry to overcome limitations in manufacturing in accordance with the regulations through the use of advances in the technology for fundamental understanding of the process and the impact of process variation onto the quality, productivity, and economics of the process.

As per the PAT guidance, some of the benefits to the pharmaceutical industry of implementing PAT are as follows (FDA PAT Guidance 2004; Molony and Undey 2009; Low and Phillips 2009):

1. Better scientific understanding of the process
2. Improvement in the yield due to prevention of the scrap, rejects, and reprocessing
3. Reduction in the production cycle time by using on-line, at-line, or in-line measurements and control
4. Decrease in energy consumption and improvement in efficiency by the conversion of the batch process into a continuous process

TABLE 10.1

Major Categories of PAT Tools Used in the Pharmaceutical Industry

Category	Examples of PAT Tools	Applications
Multivariate data analysis	• Design of the experiments • Principal component analysis • Multiple linear regression • Partial least square analysis • Pattern recognition • Soft independent modeling of class analogy	• Understanding multifactorial relationship between the process variables and product quality • Identification of interactions between the product and process variables
At-line/On-line/ In-line process analyzers	• Near-infrared (NIR) transmission spectroscopy • NIR reflectance spectroscopy • NIR microscopy • Raman spectroscopy • High-performance liquid chromatography	• Physical, chemical, and biological examination of the product • Qualitative and quantitative data collection • In-process testing
Process control	• Feed forward control • Feed backward control • Process modeling • Process simulation	• Real-time monitoring • Control of critical quality attributes of the product • Prediction of potential failure mode(s)
Knowledge management	• Electronic records • Laboratory information management system	• Data archival • Documentation of process understanding • Troubleshooting

5. Cost reduction due to the reduced waste and reduced energy consumption
6. Possibility of real-time release of the batches

Table 10.1 presents some of the key categories in which use of PAT has been proposed in the pharmaceutical industry. The major categories are statistical tools (chemometrics), modern analytical technologies (process analyzers), control schemes that can be applied (process control), and process documentation knowledge management. We will be discussing these applications pertaining to bioseparation unit operations later in this chapter.

10.3 DEFINITION OF PAT

PAT has been defined as "a system for designing, analyzing, and controlling manufacturing through timely measurements (i.e., during processing) of critical quality and performance attributes of raw and in-process materials and processes, with the goal of ensuring final product quality" (FDA PAT Guidance 2004). A desired goal of the PAT framework is to design and develop well-understood processes that will consistently ensure a predefined quality at the end of the manufacturing process. A

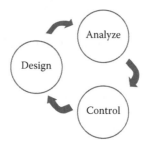

FIGURE 10.1 Illustration of the three major steps in creation and implementation of PAT.

process is generally considered well understood when (1) all critical sources of variability are identified and explained; (2) variability is managed by the process; and (3) product quality attributes can be accurately and reliably predicted over the design space established for materials used, process parameters, manufacturing, environmental, and other conditions.

From an implementation perspective, perhaps PAT can be visualized as a three-step process illustrated in Figure 10.1 (Read et al. 2010a,b). The *Design* phase is about identifying the critical quality attributes (CQAs) of the product that impact the product's safety and efficacy in the clinic and the critical process parameters (CPPs) of the process that significantly impact the CQAs (van Hoek et al. 2009; Seely 2005). The *Analyze* phase is about identifying the suitable analyzers and approaches to facilitate measurement of CQAs and CPPs via on-line, in-line, or at-line analysis (FDA PAT Guidance 2004; Molony and Undey 2009; Read et al. 2010a,b). Finally, the *Control* phase involves using the process understanding from the Design phase and the capabilities from the Analyze phase to create a control scheme that will ensure consistent process performance and product quality. This distinction is perhaps best clarified by the following example involving a relatively simple unit operation of sterile filtration (Sharma et al. 2008). The overall objective of this application was to prevent clogging of a sterile filter.

10.3.1 Design

The intermediate blocking law for fitting constant flow rate experimental curves is often used to model normal flow filtration (Hermia 1982; Hlavacek and Bouchet 1993). This is especially applicable to the application presented here involving sterile filtration with pore blocking due to aggregation (Hlavacek and Bouchet 1993). It assumes that particles can deposit on any part of the membrane surface and that any particle depositing on a pore plugs it completely. The decrease in free surface, dS, is proportional to the free surface, S, and the reduction of the free surface of pores is identical to the probability of a pore being blocked:

$$\frac{dS}{dV} = -\sigma \frac{S}{\varepsilon A}. \tag{10.1}$$

Integration of Equation 10.1 and application of Darcy's law would yield (Sharma et al. 2008)

$$\Delta P = \Delta P_0 \exp\left(\frac{\sigma V}{\varepsilon A}\right),\tag{10.2}$$

where ΔP is the pressure differential (psig), ΔP_0 is the initial pressure differential (psig), V/A is the throughput (L/m^2), ε is the porosity, and σ is the clogging coefficient (m^{-1}). Equation 10.2 can be rearranged to calculate maximum throughput as follows (Sharma et al. 2008):

$$\left(\frac{V}{A}\right)_{max} = \left(\frac{\varepsilon}{\sigma}\right)(\ln 30 - \ln P_0).\tag{10.3}$$

If data on filter performance are available, Equation 10.3 can also be used to predict the maximum throughput that can be achieved before the pressure drop reaches 30 psig.

Experiments were conducted to study the relationship between flow rate and clogging using an antibody solution and a tangential flow filtration system over varying temperatures (4°C–22°C), protein concentrations (40–100 g/L), and hold times (0–72 h). JMP software version 6 from SAS Institute was used to perform statistical analysis.

As shown in Figure 10.2, the predictions of Equation 10.2 were found to fit to experimental data with high accuracy ($R^2 > 0.9957$), confirming the validity of the model used. A key conclusion from the data analysis was that hold time exhibited a statistically significant effect on throughput. This is so because conditions that lead to filter fouling are similar to conditions that foster molecular aggregation. In

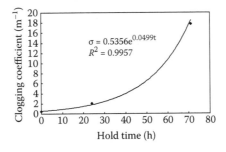

FIGURE 10.2 Clogging coefficient as a function of hold time for a 4°C, 70 g/L, pH 4.0 mAb sample filtered at 1000 LMH. (Adapted from Sharma, A., S. Anderson, A.S. Rathore, *BioPharm Int.* 21: 53–57, 2008. Copyright permission from Advanstar Communications.)

sterile filtration, aggregation is the most likely cause of clogging. The intermediate law states that each molecule can deposit on any part of the membrane surface and completely plug the pore (Hlavacek and Bouchet 1993). This is consistent with the protein monolayer theory proposed by Kelly and Zydney (1995) and Syedain et al. (2006).

Molecules prone to aggregation due to shear deformation or free thiol groups form small aggregates. These particles then randomly deposit anywhere on the membrane surface and serve as nucleation sites for growing aggregates. Passing molecules may then become attached to the aggregates at these nucleation sites. As these sites grow within the pore, the pore clogs up completely. When a protein solution is held for extended periods of time, aggregates can form and grow. When the filtering process commences, there are now more "seeds" to serve as nucleation sites for pore plugging.

Figure 10.3 illustrates the relationship between the theoretical maximum throughput and hold time under the chosen experimental conditions. It is seen that the maximum throughput follows an exponential decay with increasing hold time. For a given product, concentration, pH, and temperature of processing are fixed. To ensure successful execution of the sterile filtration step, the focus should be on ensuring that adequate filter area is available at the time of processing. The required area depends on the change in the clogging coefficient with hold time. This relationship can be stated in the form of Equation 10.4.

$$\left(\frac{V}{A}\right)_{max} = \left(\frac{V}{A}\right)_{max,0} \exp(-\delta t), \tag{10.4}$$

where the constant of proportionality, δ, is a decay constant (h^{-1}) and $(V/A)_{max,0}$ is the maximum throughput at a hold time of 0 h.

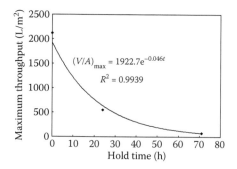

FIGURE 10.3 Maximum throughput as a function of hold time for a 4°C, 70 g/L, pH 4.0 mAb sample filtered at 1000 LMH. (Adapted from Sharma, A., S. Anderson, A.S. Rathore, *BioPharm Int.* 21: 53–57, 2008. Copyright permission from Advanstar Communications.)

10.3.2 Analysis

For this case, flux and the pressure across the filter need to be monitored to ensure that the control scheme works satisfactorily and the pressure stays below P_{max}. These data will also be useful in modifying the model if needed.

10.3.3 Control

The control scheme in this case will be based on the model illustrated in Equation 10.4. For every molecule coming into the facility, P_{max} tests will be performed at hold times of zero and a maximum allowable hold time that may be required during manufacturing. If the filterability of the product is not influenced by hold time, the data can be used for filter sizing. For cases where there is a significant impact of hold time on filterability of the product stream, it is recommended that a third P_{max} experiment be performed at an intermediate value of hold time. With three data points, the clogging coefficient can be determined using Equation 10.4 along with the relationship between the theoretical maximum throughput and hold time. The exponential fit can be used to generate the filter area that would be required to process a known amount of material for a given hold time. This allows us to control the filter area that is to be present before initiating processing of the product and ensuring successful sterile filtration.

10.4 PAT APPLICATION IN BIOSEPARATION UNIT OPERATIONS

Unit operations that are most commonly used for harvest and purification of biotech products include centrifugation, filtration, and chromatography (Molony and Undey 2009). However, several other unit operations, including flocculation, extraction, precipitation, and refolding, may also be used as required. Figure 10.4 illustrates the balance between the typical time that is available for making a decision for a step before proceeding to the next unit operation and the time required for analysis and decision making. It is evident that for steps (such as refolding) where the time

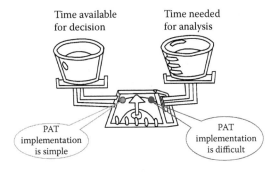

FIGURE 10.4 Illustration showing that the degree of difficulty in implementation of a PAT application depends on the time available for making the decision vs. the time needed for analysis.

window for decision making is wide enough to accommodate analysis of the sample and data evaluation to facilitate decision on the future course of action, implementation of PAT application is relatively straightforward (Rathore et al. 2006). However, in the case of unit operations such as process chromatography, where the time available for decision making may be smaller than the time for analysis, implementing a PAT application successfully requires changes to process, equipment, and analytical methods (Rathore et al. 2008a,b, 2009, 2010). In the following, we review PAT applications for each of these unit operations. Our focus will be on major developments that have occurred in the last 5 years (2004–2009).

10.4.1 CENTRIFUGATION

Centrifugation is commonly used for harvesting/separation of the cells from the fermentation or cell culture broth because of its robust performance and scalability (Rose 2008). The basic principle involves the use of the density difference between cells and surrounding liquid and the centrifugal force that yields the separation (Russell et al. 2007). Of the different varieties of centrifuges that are commercially available, disc stack centrifuges are commonly used for continuous operations (Rose 2008). Near-infrared (NIR) transmission technique has been used as a PAT tool for measurement of sedimentation velocity (Kuentz and Rothlisberger 2003). A particle separation analyzer was used for measurement of separation/sedimentation kinetics of bentonite/xanthan gum mixtures during centrifugation (Figure 10.5). This opto-electronic sensor system measured NIR transmission profiles of horizontally inserted samples tubes and the clarification data enabled optimization of concentration of excipients to be added in the final formulation.

10.4.2 FLOCCULATION

Flocculation is often used for clarification of cells, cell debris, host cell proteins, and nucleic acids. The flocculants reduce the turbidity of solutions by agglutination of cell debris and other fine particles (Riske et al. 2007). A variety of agents such as chitosan (Riske et al. 2007), polyethyleneimine (PEI) (Salt et al. 1995; Yeung

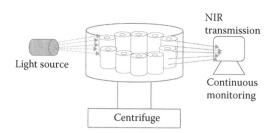

FIGURE 10.5 Rapid assessment of sedimentation stability in dispersions using near-infrared transmission measurements during centrifugation and oscillatory rheology. (Adapted from Kuentz, M. and D. Rothlisberger, *Eur. J. Pharm. Biopharm.* 56: 355–361, 2003.)

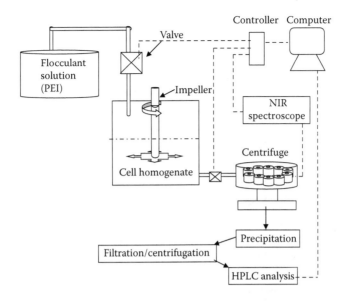

FIGURE 10.6   On-line monitoring of flocculant addition rate by using NIR spectroscopy in yeast homogenate. (Adapted from Yeung, K.S.Y., M. Hoare, F. Thornhill, T. Williams, J.D. Vaghjiani, *Biotechnol. Bioeng.* 63: 684–693, 1999.)

et al. 1999), acetonitrile copolymers (Shan et al. 1996), and other polyelectrolytes (Aspelund et al. 2008; Karim et al. 2003) have been used as flocculating agents. Measurement of flocculant strength and filter resistance have been used as PAT tools for controlling flocculation concentration in an application involving use of chitosan with different charge densities for flocculation of cell debris particles in *Escherichia coli* fermentation broth (Agerkvist 1992). Yeung et al. (1999) have developed a calibration model using multivariate data for concentration of contaminants removed after flocculation, using PEI as a flocculant in a yeast homogenate. The flocculant was added continuously and the flocculation solution was centrifuged. NIR spectroscopy was used as a PAT tool for monitoring contaminants after centrifugation (Figure 10.6). The data generated by NIR were sent to controller for adjustment of PEI addition rate, and this control strategy was used for optimization of the flocculant (PEI) addition. In another application illustrated in Figure 10.7, the authors used absorbance at 600 nm as a PAT tool for monitoring turbidity, which served as an indicator of chitosan concentration for flocculation (Riske et al. 2007). Control of the chitosan concentration in this manner yielded a sixfold to sevenfold improvement in clarification throughput without affecting recovery of the monoclonal antibody product.

10.4.3 Extraction

Extraction involves separation of the components on the basis of their relative solubility in the two different immiscible liquids (generally aqueous and organic phase).

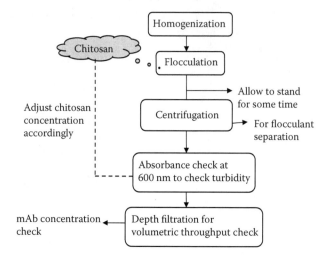

FIGURE 10.7 Flow chart for optimization of flocculant concentration by monitoring turbidity by UV spectrophotometry. (Adapted from Riske, F., J. Schroeder, J. Belliveau et al., *J. Biotechnol.* 128: 813–823, 2007.)

Traditionally, in a two-phase extraction process, parameters such as temperature or pH are monitored for process control. More recently, purification of proteins by organic solvent extraction has been shown to be possible by optimizing the proportion of organic solvent mixtures via monitoring with infrared (IR) spectroscopy and nuclear magnetic resonance (NMR) spectroscopy (Winstone et al. 2002). Moreover, NIR spectroscopy with partial least squares (PLS) has been suggested as a PAT tool for quantification of extract of *Ginkgo biloba* (Rosa et al. 2008). This approach enabled pattern recognition based on operating parameters such as type of solvent, time of extraction, and extractant concentration.

10.4.4 Precipitation

Precipitation is used when the product of interest or an intermediate with specific properties has to be removed from a process solution containing other contaminants. PAT approaches based on in-line/on-line spectroscopic methods have been applied for a selection of precipitants/reactants, solvents, additives, and reactant/precipitant concentrations (Holwill et al. 1997). The use of a microcentrifuge designed for auto-sampling, delivering sample for analysis, and washing out of solids from the centrifuge bowl has been suggested as a PAT tool for monitoring precipitation (Richardson et al. 1996). Their system, as shown in Figure 10.8, consisted of a bowl operated by an air-driven turbine. The feed pump was switched on to enter the sample into the centrifuge bowl. The vacuum solenoid valve was opened at the same time to clear the sample line from the preceding sample. When the sample line was cleared, the solenoid valve was closed and the new sample was fed into the bowl for a fixed time.

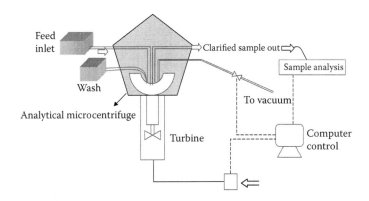

FIGURE 10.8   Automated microcentrifuge for autosampling and washing used for in-line monitoring of biological systems. (Adapted from Richardson, P., J. Molloy, R. Ravenhall et al., *J. Biotechnol.* 49: 111–118, 1996.)

A model for control of fractional precipitation based on data analysis by least squares and comparison by Kalman filter algorithm has been published (Holwill et al. 1997). IR and Raman spectroscopy have been suggested as tools for real-time measurement of solid–liquid suspensions, as found in precipitation (Nitari et al. 1989). Protein precipitation followed by filtration in 96-well plate format has been shown to enable rapid sample preparation technique for high-throughput analysis using liquid chromatography (Biddlecombe and Pleasance 1999). A similar approach has been used for implementing an on-line sample preparation technique using liquid–liquid extraction or precipitation before analysis by liquid chromatography–mass spectrometry to avoid signal modification (Marchi et al. 2007; Wen et al. 2009; Shin et al. 2008). Magnetoelastic sensors have been used to monitor kinetics of precipitation during *in situ* precipitation of salts (Bouropaulos et al. 2005). Figure 10.9 shows a setup where precipitate measurement was performed with attenuated total reflectance Fourier

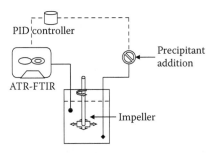

FIGURE 10.9   In-line control of precipitation process by measurement of precipitate formed during precipitation by ATR-FTIR spectroscopy technique; feedback process control done by PID controller to control precipitant addition. (Adapted from Qu, H., H. Alatalo, H. Hatakka, J. Kohonen, M. Louhi-Kulanen, *J. Cryst. Growth* 311: 3466–3475, 2009.)

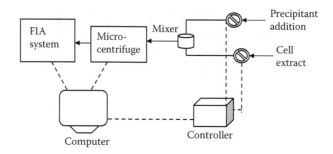

FIGURE 10.10 On-line monitoring of fractional precipitation by using microcentrifuge coupled with FIA analysis system; controller gives feedback control for addition of precipitant and thereby precipitation is controlled. (Adapted from Nitari, M., P. Richardson, R. Ravenhall et al., The modelling and control of fractionation processes for enzyme and protein purification, Proceedings of American Control Conference, Pittsburgh, PA, pp. 2436–2440, 1989.)

transform infrared (ATR-FTIR) and the information was used for feedback process control via proportional–integral–derivative (PID) controller (Qu et al. 2009; Pollanen et al. 2005a,b).

The feeding rate of the precipitant was adjusted on the basis of ATR-FTIR measurements and solubility data, enabling control of the precipitation process. The use of an automated centrifuge coupled with flow injection analyzer (FIA) for solubility profile measurement for purification of an enzyme/protein (alcohol dehydrogenase) has been suggested for on-line control of fractional precipitation (Nitari et al. 1989). Separation performance for a contaminant was described from the fractional diagram and used to decide on the fractionation cut and enable feedback control of precipitant addition (Figure 10.10).

10.4.5 FILTRATION

Filtration is the most commonly used unit operation in biotech processes, as normal flow filtration (depth filtration, nanofiltration, and sterile filtration) or tangential flow filtration (ultrafiltration, microfiltration, and diafiltration) applications fulfill a myriad of functions. Table 10.2 presents a summary of the various analytical tools that have been used to gather process information for filtration steps. It is evident that a variety of approaches have been proposed in the literature, each with associated pros and cons. Intermittent back flushing with compressed nitrogen gas avoids cake formation and blocking of the membrane and ensures constant flux thus yielding more than 99.5% retention and greater than 95% reduction in turbidity with respect to that of the feed material (Kuberkar and Davis 2001). As seen in Figure 10.11, process modeling of cross-flow microfiltration membranes for clarification of fermentation broths has been shown to be useful in predicting the timing of the above-mentioned back flushes and, thus, maintaining high flux and low turbidity (Carrere et al. 2001; Milcent and Carrere 2001). The possibility of using absorbance at 410 nm

TABLE 10.2

PAT Tools Used for Gathering Information during Filtration Unit Operation

Technique	Strength	Deficiencies	References
A. Monitoring of Concentration Polarization			
Light deflection techniques: shadowgraphy	Offers a method for measurement of concentration profile within medium with flexibility in choice of medium	Does not consider concentration polarization layer profile; limited to binary systems	Chen et al. 2004
Light deflection techniques: refractography	Can measure both concentration and concentration gradient	Technique limited to ultrafiltration (UF) of large molecules	Gowman and Ethier 1997
Magnetic resonance imaging	Measures concentration of colloidal suspension and polarization layer; Works with optically opaque samples	Cannot measure concentration gradient	Yao et al. 1995
Radioisotope labeling	Can monitor concentration polarization	Cannot measure polarization layer thickness	Mc-Donough et al. 1995
Electron diode array microscope	Measures concentration polarization in UF and microfiltration	Measurement is only above 20 μm of the membrane surface	Mc-Donough et al. 1992
Pressure sensors	Can monitor physicochemical parameters governing polarization	Pressure sensors may disturb concentration polarization layer	Zhang and Ethier 2001
B. Monitoring of Cake Formation and Membrane Fouling			
Optical microscope	Can estimate critical flux and lowest permeate flux to indicate onset of cake formation	Works with transparent membranes only; observation only until monolayer formation	Li et al. 1998
Image analyzers for direct visualization	Allows observation of deposition after monolayer formation	Cell deposition should be constant for correct analysis	Mores and Davis 2001
Laser triangulometry	Can track cake formation	Cannot measure cake porosity	Altmann and Ripperger 1997
Optic laser sensors	Measures cake porosity	Can only measure thickness up to 30 μm; turbidity obstructs photo detection	Hamachi and Mietton-Peuchot 2001
Ultrasonic time domain reflectometry	Measure membrane fouling during reverse osmosis	Observation not specific to porous structure or membrane skin layer	Mairal et al. 2000

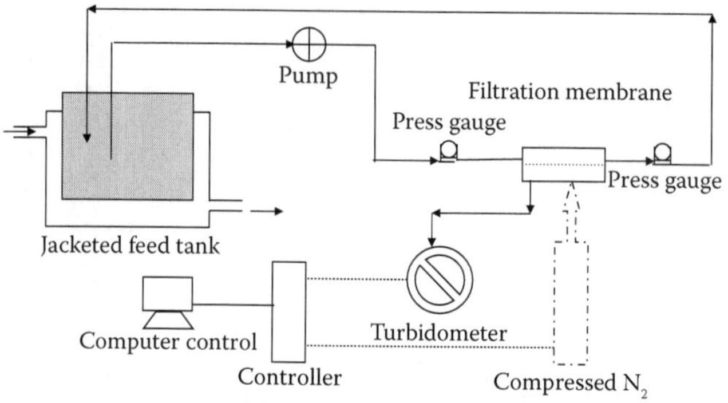

FIGURE 10.11 Plot of flux against time for on-line control of microfiltration; back flushing done intermittently to avoid the blockage of the membrane. (Adapted from Carrere, H., F. Blaszkow, H.R. Balman, *J. Membr. Sci.* 186: 219–230, 2001.)

for monitoring filtration of host cell proteins via depth filtration has been suggested (Yigzaw et al. 2006; Read et al. 2010a).

In a recent publication, the use of pH probe was suggested as a surrogate marker and a PAT tool for signaling completion of a diafiltration step (Rathore et al. 2006). Removal of the different species in the permeate after each diafiltration volume was monitored (see Figure 10.12). Species 1 was identified as the control species with respect to migration across the membrane, and a correlation between the concentration and pH of Species 1 was established. The correlation was then validated by testing at 0.5× and 2× of Species 1 concentration. It was shown that implementation of this PAT application is likely to result in process robustness (savings in buffer, process time, and plant capacity), and provides assurance that the product meets the predefined quality criteria. Using a setup illustrated in Figure 10.13, Maheshkumar et al. applied a PAT strategy involving continuous monitoring of turbidity to control fouling of the microfiltration membrane. The application was intended for processing of biomass-based power plant effluents via microfiltration (Maheshkumar et al. 2007).

10.4.6 Refolding

Refolding is generally used for proteins expressed via microbial cell systems and has been more recently also used for monoclonal antibody products to obtain improved efficacy. Use of high-performance liquid chromatography (HPLC) as a monitoring tool for a refolding step has been proposed (Rathore et al. 2006). The purity of the product as well as the levels of different product-related variants and impurities were monitored as a function of time. The data suggested that it is feasible to implement a PAT-based control scheme that allows ending the refold step on the basis of product quality data. This would ensure consistency in product quality of the refold end pool, as well as improve operational efficiency by keeping refolding time only to what is

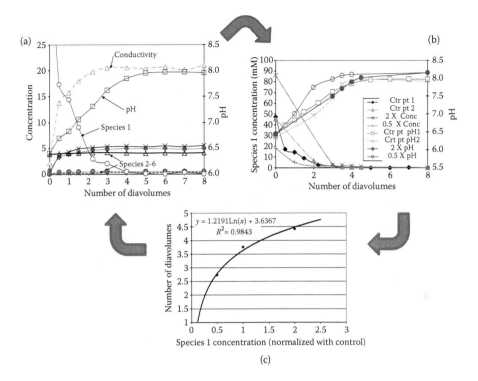

FIGURE 10.12 Illustration of PAT application diafiltration of a protein solution. (a) Concentration profiles of different excipients during diafiltration at 0.5×, 1×, and 2× the concentration of Species 1. (b) Relationship between pH and Species 1 concentration during diafiltration. (c) Relationship between number of diavolumes and Species 1 concentration. (Adapted from Rathore, A.S., A. Sharma, D. Chillin, *BioPharm Int.* 19: 48–57, 2006. Copyright permission from Advanstar Communications.)

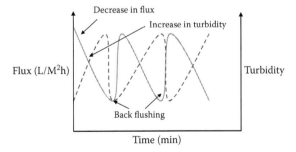

FIGURE 10.13 On-line control of microfiltration by turbidometry and back flushing done intermittently to avoid the blockage of the membrane. (Adapted from Maheshkumar, S., G.M. Madhu, S. Roy, *Sep. Purif. Technol.* 57: 25–36, 2007.)

really needed. In a more recent publication, Pizzaro et al. (2009) reported the use of % dissolved oxygen (DO) profile as a PAT tool to regulate process performance at the commercial manufacturing scale. An in-line DO sensor was used to monitor the oxygen levels in the reaction vessel, and a correlation was developed between the DO signal and the product quality attributes. The use of the PAT approach resulted in a more robust operation with the DO profiles enabling prediction of process performance.

10.4.7 CHROMATOGRAPHY

Chromatography remains the core of downstream processing due to the unique set of advantages it offers: robust operation, high resolution, and scalability. Pooling of chromatography columns is generally performed on the basis of absorbance at 280 nm, primarily for the operational simplicity that this approach offers. However, for protein products, absorbance at 280 nm is unable to differentiate between the product and other product-related impurities that chromatography steps are generally used to separate. Thus, use of a high-resolution method such as HPLC for making the pooling decision would be ideal and has been proposed in literature (Low 1986; Cooley and Stevenson 1992). Use of on-line reversed-phase HPLC for separating recombinant human insulin-like growth factor-I (IGF) and IGF aggregates to facilitate real-time pooling based on product quality has been successfully demonstrated (Fahrner et al. 1998). A similar application using chromatographic separation with protein A-immobilized media has been shown to reliably control antibody loading on a protein A column for a CHO-based recombinant antibody product (Fahrner and Blank 1999). Recently, an approach using a combination of on-line size-exclusion HPLC, differential refractometry, and multi-angle laser light scattering analysis has been shown to be useful for estimating real-time product quality of a mutant form of the human immunodeficiency virus. The accuracy of the approach was found to be >89% and the intra- and inter-day precision were found to be 0.9% and 3.6% relative standard deviation, respectively (Barackman et al. 2004). More recently, an application involving production of alcohol dehydrogenase in *Saccharomyces cerevisae* and purification by hydrophobic interaction chromatography (HIC) has been published (Chhatre et al. 2009). It was shown that stopped-flow analysis could be used as an at-line monitoring tool for HIC to provide a visual description of product levels and a real-time representation of yield. Rathore et al. recently published a series of case studies demonstrating the utility of a variety of analytical tools for performing analysis that would facilitate pooling of large-scale chromatography columns. The analytical tools that the authors used include HPLC (Rathore et al. 2008a,b, 2010) and fluorescence (Rathore et al. 2009). One such application is presented in Figure 10.14 (Rathore et al. 2010). The illustration shows the two critical steps for designing such an application. First, the feasibility of running the analytical method rapidly and the process slowly so as to minimize the lag between analysis and elution. Second, modeling of the data to allow for prediction of the pool purity with $n + 1$th fraction based on the pool purity with the nth fraction. It was shown that such a PAT approach could be used to generate a consistent product quality in the pools even with significant variations in the purity of the feed material.

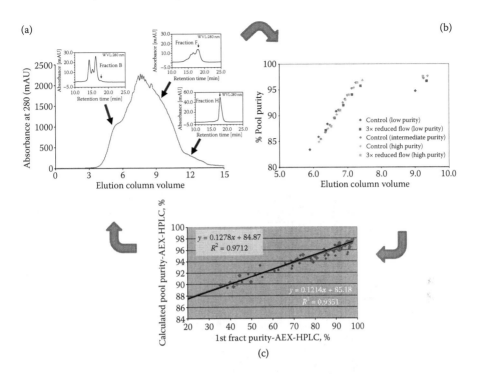

FIGURE 10.14 Illustration of PAT application involving pooling of chromatography column using on-line HPLC. (a) HPLC chromatograms corresponding to the three arbitrary fractions of the column elution profile. (b) Comparison of purity profiles for HIC process chromatography runs using high, intermediate, and low load purities. (c) Correlation between the anion exchange high-performance liquid chromatography (AEX-HPLC) purity of the first fractions to be included into the HIC pool vs. the AEX-HPLC purity of the HIC pool. Data has been obtained from pilot facility. Linear correlation is shown for the entire data set. (Adapted from Rathore, A.S., L. Parr, S. Dermawan, K. Lawson, Y. Lu, *Biotechnol. Prog.* 26 (2): 448–457, 2010.)

10.5 CONCLUSION

This chapter provides a review of PAT applications in the major unit operations involved in bioseparation. It can be concluded that while the ease of doing so may vary, in almost all cases, it is feasible to design control schemes that rely on measurement of product quality attributes and enable real-time decisions. Thus, implementation of these schemes is likely to result in a more consistent product quality and higher operational efficiency for most applications. However, these advantages are balanced by the requirement of a higher level of process understanding for designing these schemes and increased operational complexity for implementation. The slow adoption of PAT thus far in the biotech industry can be attributed not so much to the complexities of the biotech world but more so to the fact that implementing PAT requires changes in our approaches toward process and analytical development, manufacturing, quality assurance, and regulatory filings. It should also be emphasized that while significant advancements have been accomplished with respect to our ability to analyze/monitor key process and quality attributes in the biotech industry, more needs to be done with respect to using the collected data for subsequent control of the process so as to achieve optimal yield and product quality. Only the latter will allow us to achieve the most benefits from PAT implementation.

REFERENCES

Agerkvist, I. 1992. Mechanisms of flocculation with chitosan in *Escherichia coli* disintegrates: Effects of urea and chitosan characteristics. *Colloids Surf.* 69: 173–187.

Altmann, J. and S. Ripperger. 1997. Particle deposition and layer formation at the crossflow microfiltration. *J. Membr. Sci.* 124: 119–128.

Aspelund, M.T., G. Rozeboom, M. Heng, and C.E. Glatz. 2008. Improving permeate flux and product transmission in the microfiltration of a bacterial cell suspension by flocculation with cationic polyelectrolytes. *J. Membr. Sci.* 324: 198–208.

Barackman, J., I. Prado, C. Karunatilake, and K. Furuya. 2004. Evaluation of on-line high performance size-exclusion chromatography, differential refractometry, and multi-angle light scattering analysis for the monitoring of the oligomeric state of human immunodeficiency virus vaccine protein antigen. *J. Chromatogr. A* 1043: 57–64.

Biddlecombe, R.A. and S. Pleasance. 1999. Automated protein precipitation by filtration in the 96-well format. *J. Chromatogr. B* 734: 257–265.

Bouropaulos, N., D. Kouzoudis, and C. Grimes. 2005. The real-time, in situ monitoring of calcium oxalate and brushite precipitation using magnetoelastic sensors. *Sens. Actuators, B* 109: 227–223.

Carrere, H., F. Blaszko, and H.R. Balman. 2001. Modelling the clarification of lactic acid fermentation broths by cross-flow microfiltration. *J. Membr. Sci.* 186: 219–230.

Chen, J., Q. Li, and M. Elimelche. 2004. In situ monitoring techniques for concentration polarization and fouling phenomena in membrane filtration. *Adv. Colloid Interface Sci.* 107: 83–108.

Chhatre, S., G. Bou-Habib, M.P. Smith et al. 2009. Use of PAT principles for the open-loop control oflaboratory and pilot-scale chromatography columns. *J. Chem. Technol. Biotechnol.* 84: 1314–1322.

Cooley, R.E. and C.E. Stevenson. 1992. On-line HPLC as a process monitor in biotechnology. *Process Control Qual.* 2: 43–53.

European Medicines Agency (EMEA). 2006. Reflection paper: Chemical, pharmaceutical and biological information to be included in dossiers when process analytical technology (PAT) is employed. Doc. Ref. EMEA/INS/277260/2005.

Fahrner, R.A. and G.S. Blank. 1999. Real-time control of antibody loading during protein a affinity chromatography using an on-line assay. *J. Chromatogr. A* 849: 191–196.

Fahrner, R.A., P.M. Lester, G.S. Blank, and D.H. Reifsnyder. 1998. Real-time control of purified product collection during chromatography of recombinant human insulin-like growth factor-I using an on-line assay. *J. Chromatogr. A* 827: 37–43.

FDA, U.S. Department of Health and Human Services. 2002a. http://www.fda.gov/Drugs/DevelopmentApprovalProcess/Manufacturing/QuestionsandAnswersonCurrentGoodManufacturingPracticescGMPforDrugs/UCM071836.

FDA, U.S. Department of Health and Human Services. 2002b. Ajaz S. Hussain. The subcommittee on process analytical technologies (PAT): Overview and objectives. PhD Presentation. Available on-line http://www.fda.gov/ohrms/dockets/ac/02/slides/3841s1_01_hussain.ppt.

FDA, U.S. Department of Health and Human Services. 2002c. Productivity and the economics of regulatory compliance in pharmaceutical production. Doug Dean & Frances Bruttin Presentation. Available on-line at http://www.fda.gov/ohrms/dockets/ac/02/briefing/3841B1_07_PriceWaterhouseCoopers.ppt.

FDA, U.S. Department of Health and Human Services. 2004. Guidance for industry: PAT—A Framework for Innovative Pharmaceutical Development, Manufacturing, and Quality Assurance. Available on-line at http://www.fda.gov/downloads/Drugs/GuidanceComplianceRegulatoryInformation/Guidances/ucm070305.pdf.

Gowman, L. and C.R. Ethier. 1997. Concentration and concentration gradient measurements in an ultrafiltration concentration polarization layer. Part I: A laser-based refractometric experimental technique. *J. Membr. Sci.* 131: 95–105.

Hamachi, M. and M. Mietton-Peuchot. 2001. Cake thickness measurement with an optical laser sensor. *Chem. Eng. Res. Des.* 79: 151–155.

Hermia, J. 1982. Constant pressure blocking filtration laws—Application to power-law non-Newtonian fluids. *Trans. IChemE* 60: 183–187.

Hlavacek, M. and F. Bouchet. 1993. Constant flowrate blocking laws and an example of their application to dead-end microfiltration of protein solutions. *J. Membr. Sci.* 82: 285–295.

Holwill, I.J., S.J. Chard, M.T. Flanagen, and M. Hoare. 1997. A Kalman filter algorithm and monitoring apparatus for at-line control of fractional protein precipitation. *Biotechnol. Bioeng.* 53: 58–70.

ICH. Q8(R2): Pharmaceutical Development. 2009. International Conference on Harmonization of Technical Requirements for the Registration of Pharmaceuticals for Human Use, Geneva, Switzerland.

ICH. Q9: Quality Risk Management. 2005. International Conference on Harmonization of Technical Requirements for the Registration of Pharmaceuticals for Human Use, Geneva, Switzerland.

ICH. Q10: Pharmaceutical Quality System. 2008. International Conference on Harmonization of Technical Requirements for the Registration of Pharmaceuticals for Human Use, Geneva, Switzerland.

Karim, M.N., D. Hodge, and L. Simon. 2003. Data-based modeling and analysis of bioprocesses: Some real experiences. *Biotechnol. Prog.* 19: 1591–1605.

Kelly, S.T. and A.L. Zydney. 1995. Mechanism of BSA fouling during microfiltration. *J. Membr. Sci.* 107: 115–127.

Kuberkar, V.T. and R.H. Davis. 2001. Microfiltration of protein–cell mixtures with crossflushing or backflushing. *J. Membr. Sci.* 183: 1–14.

Kuentz, M. and D. Rothlisberger. 2003. Rapid assessment of sedimentation stability in dispersions using near infrared transmission measurements during centrifugation and oscillatory rheology. *Eur. J. Pharm. Biopharm.* 56: 355–361.

Li, H., A.G. Fen, H.G.L. Coster, and S. Vigneswaran. 1998. Direct observation of particle deposition on the membrane surface during crossflow microfiltration. *J. Membr. Sci.* 149: 83–97.

Low, D.K.R. 1986. The use of FPLC system in method development and process monitoring for industrial protein chromatography. *J. Chem. Tech. Biotechnol.* 36: 345–350.

Low, D. and J. Phillips. 2009. Evolution and integration of quality by design and process analytical technology. In *Quality by Design for Biopharmaceuticals*, edited by A.S. Rathore and R. Mhatre, pp. 255–286. Wiley Interscience, Hoboken, NJ.

Maheshkumar, S., G.M. Madhu, and S. Roy. 2007. Fouling behaviour, regeneration options and on-line control of biomass-based power plant effluents using microporous ceramic membranes. *Sep. Purif. Technol.* 57: 25–36.

Mairal, A.P., A.R. Greenberg, and W.B. Krantz. 2000. Investigation of membrane fouling and cleaning using ultrasonic time-domain reflectometry. *Desalination* 130: 45–60.

Marchi, I., S. Rudaz, M. Selman, and J. Veuthey. 2007. Evaluation of the influence of protein precipitation prior to on-line SPE–LC–API/MS procedures using multivariate data analysis. *J. Chromatogr. B* 845: 244–252.

Mc-Donough, R.M., H. Bauser, N. Stroh, and H. Chmeil. 1992. Separation efficiency of membranes in biotechnology: An experimental and mathematical study of flux control. *Chem. Eng. Sci.* 47: 271–279.

Mc-Donough, R.M., H. Bauser, N. Stroh, and U. Grauschopf. 1995. Experimental in situ measurement of concentration polarisation during ultra- and micro-filtration of bovine serum albumin and dextran blue solutions. *J. Membr. Sci.* 104: 51–63.

Milcent, S. and H. Carrere. 2001. Clarification of lactic acid fermentation broths. *Sep. Purif. Technol.* 22–23: 393–401.

Molony, M. and C. Undey. 2009. PAT tools for biologics: Considerations and challenges. In *Quality by Design for Biopharmaceuticals*, edited by A.S. Rathore and R. Mhatre, pp. 211–254. Wiley Interscience, Hoboken, NJ.

Mores, W.D. and R.H. Davis. 2001. Direct visual observation of yeast deposition and removal during microfiltration. *J. Membr. Sci.* 189: 217–230.

Niktari, M., P. Richardson, R. Ravenhall et al. 1989. The modelling and control of fractionation processes for enzyme and protein purification, pp. 2436–2440. Proceedings of American Control Conference, Institute of Electrical and Electronics Engineers, Pittsburgh, PA.

Pizzaro, S.A., R. Dinges, R. Adams, A. Sanchez, and C. Winter. 2009. Biomanufacturing process analytical technology (PAT) application for downstream processing: Using dissolved oxygen as an indicator of product quality for a protein refolding reaction. *Biotechnol. Bioeng.* 104: 340–351.

Pollanen, K., A. Hakkinen, S. Reinikainen et al. 2005a. ATR-FTIR in monitoring of crystallization processes: Comparison of indirect and direct OSC methods. *Chemom. Intell. Lab. Syst.* 76: 25–35.

Pollanen, K., A. Hakkinen, S. Reinikainen et al. 2005b. IR spectroscopy together with multivariate data analysis as a process analytical tool for in-line monitoring of crystallization process and solid-state analysis of crystalline product. *J. Pharm. Biomed. Anal.* 38: 275–284.

Qu, H., H. Alatalo, H. Hatakka, J. Kohonen, and M. Louhi-Kulanen. 2009. Raman and ATR FTIR spectroscopy in reactive crystallization: Simultaneous monitoring of solute concentration and polymorphic state of the crystals. *J. Cryst. Growth* 311: 3466–3475.

Rathore, A.S., X. Li, W. Bartkowski, A. Sharma, and Y. Lu. 2009. Case study and application of process analytical technology (PAT) towards bioprocessing. III: Use of tryptophan fluorescence as at-line tool for making pooling decisions for process chromatography. *Biotechnol. Prog.* 25: 1433–1439.

Rathore, A.S., L. Parr, S. Dermawan, K. Lawson, and Y. Lu. 2010. Large scale demonstration of process analytical technology (PAT) in bioprocessing: Use of high performance liquid chromatography (HPLC) for making real time pooling decisions for process chromatography. *Biotechnol. Prog.* 26 (2): 448–457.

Rathore, A.S., A. Sharma, and D. Chillin. 2006. Applying process analytical technology to biotech unit operations. *BioPharm Int.* 19: 48–57.

Rathore, A.S., M. Yu, S. Yeboah, and A. Sharma. 2008a. Case study and application of process analytical technology (PAT) towards bioprocessing: Use of on-line high performance liquid chromatography (HPLC) for making real time pooling decisions for process chromatography. *Biotechnol. Bioeng.* 100: 306–316.

Rathore, A.S., R. Wood, A. Sharma, and S. Dermawan. 2008b. Case study and application of process analytical technology (PAT) towards bioprocessing. II: Use of ultra performance liquid chromatography (UPLC) for making real time pooling decisions for process chromatography. *Biotechnol. Bioeng.* 101: 1366–1374.

Read, E.K., J.T. Park., R.B. Shah et al. 2010a. Process analytical technology (PAT) for bio-pharmaceutical products: Concepts and applications—Part I. *Biotechnol. Bioeng.* 105: 276–284.

Read, E.K., J.T. Park, R.B. Shah et al. 2010b. Process analytical technology (PAT) for bio-pharmaceutical products: Concepts and applications—Part II. *Biotechnol. Bioeng.* 105: 285–295.

Richardson, P., J. Molloy, R. Ravenhall et al. 1996. High speed centrifugal separator for rapid on-line sample clarification in biotechnology. *J. Biotechnol.* 49: 111–118.

Riske, F., J. Schroeder, J. Belliveau et al. 2007. The use of chitosan as a flocculant in mam-malian cell culture dramatically improves clarification throughput without adversely impacting monoclonal antibody recovery. *J. Biotechnol.* 128: 813–823.

Rosa, S.S., P.A. Barata, J.M. Martins, and J.C. Menezes. 2008. Near-infrared reflectance spec-troscopy as a process analytical technology tool in *Ginkgo biloba* extract quantification. *J. Pharm. Biomed. Anal.* 47: 320–327.

Rose, P. 2008. Biopharmaceutical technology: Cell harvesting getting cultural. *Filtr. Sep.* 45: 29–31.

Russell, E., A. Wang, and A.S. Rathore. 2007. Harvest of a therapeutic protein product from high cell density fermentation broths: Principles and case study. In *Process Scale Bioseparations for the Biopharmaceutical Industry*, edited by A.A. Shukla, M. Etzel, and S. Gadam, pp. 1–58. CRC Press, Boca Raton, FL.

Salt, D.E., S. Hay, O.R.T. Thomas, M. Hoare, and P. Dunnill. 1995. Selective flocculation of cellular contaminants from soluble proteins using polyethyleneimine: A study of several organisms and polymer molecular weights. *Enzyme Microb. Technol.* 17: 107–113.

Seely, J. 2005. Process characterization. In *Process Validation in Manufacturing of Biopharmaceuticals*, edited by A.S. Rathore and G. Sofer, pp. 31–68. Taylor & Francis, Boca Raton, FL.

Shan, J., J. Xia, Y. Guo, and X. Zhange. 1996. Flocculation of cell, cell debris and soluble pro-tein with methacryloyloxyethyl trimethylammonium chloride–acrylonitrile copolymer. *J. Biotechnol.* 49: 173–178.

Sharma, A., S. Anderson, and A.S. Rathore. 2008. Filter clogging issues in sterile filtration. *BioPharm Int.* 21: 53–57.

Shin, J., D.F. Pauly, J.A. Johnson, and R.F. Frye. 2008. Simplified method for determination of clarithromycin in human plasma using protein precipitation in a 96-well format and liquid chromatography–tandem mass spectrometry. *J. Chromatogr. B* 871: 130–134.

Syedain, Z.H., D.M. Bohonak, and A.L. Zydney. 2006. Protein fouling of virus filtration mem-branes: Effects of membrane orientation and operating conditions. *Biotechnol. Prog.* 22: 1163–1169.

van Hoek, P., J. Harms, X. Wang, and A.S. Rathore. 2009. Case study on definition of process design space for a microbial fermentation step. In *Quality by Design for Biopharmaceuticals: Perspectives and Case Studies*, edited by A.S. Rathore and P. Mhatre, pp. 85–109. Wiley Interscience, Hoboken, NJ.

Wen, X., P. Wu, K. Xu, J. Wang, and X. Hou. 2009. On-line precipitation–dissolution in knot-ted reactor for thermospray flame furnace AAS for determination of ultratrace cadmium. *Microchem. J.* 91: 193–196.

Winstone, T.L., K.A. Duncalf, and R.J. Turner. 2002. Optimization of expression and the purification by organic extraction of the integral membrane protein EmrE. *Protein Expression Purif.* 26: 111–121.

Yao, S., M. Costello, A.G. Fane, and J.M. Pope. 1995. Non-invasive observation of flow pro-files and polarisation layers in hollow fibre membrane filtration modules using NMR micro-imaging. *J. Membr. Sci.* 99: 207–216.

Yeung, K.S.Y., M. Hoare, F. Thornhill, T. Williams, and J.D. Vaghjiani. 1999. Near-infrared spectroscopy for bioprocess monitoring and control. *Biotechnol. Bioeng.* 63: 684–693.

Yigzaw, Y., R. Piper, M. Tran, and A.A. Shukla. 2006. Exploitation of the adsorptive properties of depth filters for host cell protein removal during monoclonal antibody purification. *Biotechnol. Prog.* 22: 288–296.

Zhang, W. and C.R. Ethier. 2001. Direct pressure measurements in a hyaluronan ultrafiltration concentration polarization layer. *Colloid Surf.* 180: 63–73.

11 Process Analytical Technology Use in Biofuels Manufacturing

Pedro Felizardo, José C. Menezes,
and M. Joana Neiva-Correia

CONTENTS

11.1 BIOFUELS TECHNOLOGY AND USES

Biofuels are defined as renewable fuels derived from biological feedstocks (biomass), and they include liquids such as bioethanol or biodiesel and gaseous fuels such as biogas (e.g., methane) or hydrogen (Koh and Ghazoul 2008). In this chapter, the focus will be on liquid biofuels for transport because that sector is a major consumer of petroleum-based fuels (Demirbas 2007).

Liquid biofuels can be classified as either first or second generation (Zabaniotou et al. 2008; Naik et al. 2010). First-generation biofuels are produced from conventional raw materials by using well-established and economical production technologies. That is the case of bioethanol produced from sugar or starch-containing materials and of biodiesel produced from soybean or rapeseed oils. Those fuels can be blended with petroleum-based fuels (ethanol in gasoline and biodiesel in diesel), burned by existing internal combustion engines, and distributed through the existing logistics systems. They are available today, with almost 50 billion liters produced annually (Naik et al. 2010). On the contrary, second-generation biofuels, such as cellulosic ethanol and Fischer–Tropsch fuels, are derived from lower-grade and non-food bio-materials such as lignocellulosic materials, as opposed to oil seeds or grains, and their production processes still have several technological barriers to be overcome (Naik et al. 2010).

The large-scale production of first-generation biofuels can offer some CO_2 reduction benefits but raises several problems. In fact, besides the social impacts and the doubts related with the benefits of their production and use in terms of the net carbon balance, the production of fuels from food crops may contribute to increasing the price and availability of food, which is obviously an important drawback (Reijnders and Huijbregts 2008). Even so, the use of first-generation fuels is of great interest nowadays and will remain until second-generation fuels are commercially available and economic.

11.2 BIOFUELS FOR THE TRANSPORT SECTOR

It is known that the transport sector is one of the major consumers of petroleum-based fuels such as diesel, gasoline, liquefied petroleum gas, and jet fuel. Several environmental aspects mainly related with the necessity of limiting CO_2 emissions, together with the issues of limited reserves and rising prices of petroleum, led to the search for feasible alternatives in this sector. The use of bioethanol in gasoline engines, and vegetable oils or biodiesel in diesel engines, is at the moment feasible and a proven alternative.

Several factors must be taken into account before using biofuels on a large scale. A biofuel must (1) be easy to use in today's combustion engines; (2) have a favorable economics in plant building, operational costs, and all associated feedstock and biofuel distribution logistics; (3) have an environmental benefit; and (4) represent no additional cost to the end user (Agarwal 2007).

11.2.1 BIOETHANOL

Bioethanol is a fuel produced from renewable sources such as sugar-containing crops (sugar cane, beet root, fruits, palm juice), starch-containing grains (wheat, barley, rice, sweet sorghum, corn), and cellulosic biomass (wood and wood waste, cedar, pine, agriculture residues, fibers) through the fermentation of sugars. This fuel can be used with no engine modification in 5% (v/v) blends with gasoline (EN 228 2004), whereas higher blends (up to 85% of ethanol) require engine modifications (Demirbas 2007).

The use of ethanol as fuel was first suggested in the 1930s. However, because of its relatively low cost, gasoline was used as the main transport fuel until the 1970s (Agarwal 2007). The use of bioethanol has several benefits due to its higher octane number and higher oxygen content than gasoline, which lead to a more complete combustion and to a decrease in the emissions of several pollutants (Agarwal 2007). However, ethanol has a lower heating value and is toxic and corrosive (Balat et al. 2008).

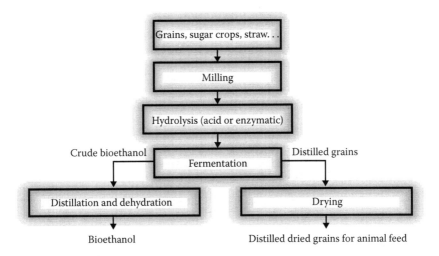

FIGURE 11.1   Illustration chart for the production of bioethanol. (Adapted from Agarwal, A.K., *Prog. Energy Combust. Sci.* 33: 233–271, 2007.)

First-generation bioethanol may be produced from sugar crops such as sugar beets or sugar cane, or starch materials such as corn and wheat. When sugars are used, the material can be directly fermented by different microorganisms, such as the yeast *Saccharomyces cerevisiae*, which hydrolyzes disaccharides to simple sugars that can be fermented to bioethanol. However, starch is a long-chain glucose polymer that cannot be directly converted into ethanol by conventional fermentation technology (Naik et al. 2010). Therefore, starch has to be broken into sugar monomers by acid or enzymatic hydrolysis (Reaction 11.1), which is then fermented into alcohol (Reaction 11.2).

$$(CH_2O)_n \rightarrow C_6H_{12}O_6 + C_6H_{12}O_6 \qquad \text{(Reaction 11.1)}$$
$$\text{Starch} \qquad \text{Glucose} \quad \text{Fructose}$$

$$C_6H_{12}O_6 \rightarrow 2\ C_2H_5OH + 2\ CO_2 \qquad \text{(Reaction 11.2)}$$

After fermentation, the crude alcohol is submitted to a distillation and dehydration process to yield anhydrous bioethanol. Figure 11.1 shows the flowchart for the production of bioethanol.

11.2.2 Vegetable Oils

The direct use of vegetable oils in diesel engines, while feasible, can be problematic. In fact, the use of high-viscosity and low-volatility fuels in compression ignition engines can be a problem especially in cold climates (Vyas et al. 2010). Several studies, such as the one conducted by Knothe et al. (1997), showed that the direct combustion of various vegetable oils can lead, among other problems, to cocking (carbonization of the fuel on the injector head), to problems in the injection system and in the starting of the cold engine, to the dilution of crankcase oil, and to the contamination of the lubricant oil. To solve those problems, it is possible to introduce changes in the engine, which involve modifying the fuel injection systems, or to preheat the oil before injection. Alternatively, it is also possible to change the characteristics of the oil that may include its dilution with fossil diesel and their modification by microemulsion, pyrolysis (Agarwal 2007; Vyas et al. 2010), and by transesterification. The latter alternative will be described in Section 11.2.3.

The main component of vegetable oils and animal fats are triacylglycerols that have a glycerol backbone with three short to long fatty acid chains as shown in Figure 11.2.

The fatty acids composition can vary significantly from oil feedstock. Nevertheless, the five main fatty acids in crops are saturated palmitic acid (C16:0), saturated stearic acid (C18:0), oleic acid (C18:1), linoleic acid (C18:2), and linolenic acid (C18:3) and, as described below, their relative abundance in each type of oil determines its properties and the possibility of its use as fuel or as raw material for biodiesel production (Table 11.1).

The standardization of the vegetable oils to be used as fuels in diesel engines follows several national or multinational standards, such as the DIN V 51605 standard for rapeseed oil in Germany, which prescribes quality specifications that must be met and their acceptable ranges: density, flash point, kinematic viscosity (40°C), heat of combustion, low-temperature behavior, cetane number, iodine number, acid value, oxidation stability, water content, carbon residue, sulfur content, contamination, and ash content.

As mentioned above, the oil's viscosity is extremely important and increases with the length of the fatty acid chains, whereas the degree of unsaturation of those chains determines the oil's iodine number. In fact, the iodine number is a measure of the degree of unsaturation of the methyl ester molecules and an indicator of how easily these molecules polymerize. Thus, for example, oils with higher values of the iodine number have lower oxidative stabilities, low values of the cetane number, and worse low-temperature behavior (Knothe et al. 2005; Saraf and Thomas 2007). The flash point of such oils is much higher than that of diesel fuel, which makes their transport and handling safer. On the other hand, the use of oils with high content of free fatty acids and water favors corrosion processes and the formation of engine deposits.

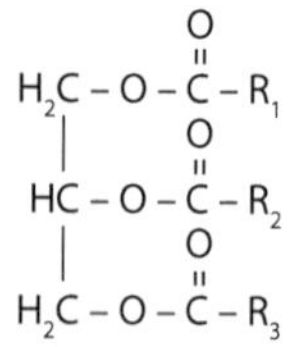

$$
\begin{array}{l}
\quad\quad\quad\quad\; O \\
\quad\quad\quad\quad\; \| \\
H_2C - O - C - R_1 \\
\quad| \quad\quad\quad\; O \\
\quad\quad\quad\quad\; \| \\
HC - O - C - R_2 \\
\quad| \quad\quad\quad\; O \\
\quad\quad\quad\quad\; \| \\
H_2C - O - C - R_3
\end{array}
$$

FIGURE 11.2 Schematic representation of a triglyceride molecule in which R_1, R_2, and R_3 represent the short to long fatty acid chains.

TABLE 11.1
Fatty Acids Composition of Vegetable Oils and Fats

Oils	C12:0	C14:0	C16:0	C18:0	C18:1	C18:2	C18:3	C20:0	C22:0	C22:1	C24:0
Corn	0	0	12	2	25	6	0	0	0	0	0
Cotton	0	0	28	1	13	58	0	0	0	0	0
Linseed	0	0	5	2	20	18	55	0	0	0	0
Peanut	0	0	11	2	48	32	1	1	1	0	1
Rapeseed	0	0	3	1	64	22	8	0	0	0	0
Safflower	0	0	9	2	12	78	0	0	0	0	0
Safflower with high oleic content	0	0	5	2	79	13	0	0	0	0	0
Sesame seed	0	0	13	4	53	30	0	0	0	0	0
Soybean	0	0	3	3	23	55	6	0	0	0	0
Sunflower	0	0	3	3	17	74	0	0	0	0	0
Palm	0	1–2	41–46	4–6.5	37–42	8–12	<0.5	<0.5	0	0	0
Lard	<0.5	<1.5	24–30	12–18	36–52	10–12	<1	<0.5	0	0	0
Tallow	0	2–4	23–29	20–35	26–45	2–6	<1	<0.5	0	0	0
Jatropha	0	1.4	15.6	9.7	40.8	32.1	0	0.4	0	0	0
Coconut	41–46	18–21	9–12	2–4	5–9	0.5–3	0	0	0	0	0

Source: Srivastava, A., R. Prasad, *Renewable and Sustainable Energy Rev.* 4, 111–133, 2000. With permission.
Elvers, B., F. Ullmann: *Ullmann's Encyclopedia of Industrial Chemistry.* 1992. Copyright Wiley-VCH, Verlag GmbH & Co. KGaA. Reproduced with permission.
Singh, S.P., D. Singh, *Renewable Sustainable Energy Rev. 14:* 200–216, 2010. With permission.

Note: C*n:m*—*n*, number of carbon atoms of the chain; *m*, number of unsaturated bonds.

As described in Section 11.2.3, if the vegetable oil is used for biodiesel production, the quality of the final product will be greatly influenced by the type of feedstock used in the production process, namely its composition, that is, water content, free fatty acids, and its iodine value (Saraf and Thomas 2007; Baptista et al. 2008b).

11.2.3 BIODIESEL

Biodiesel is one of the main alternatives to fossil fuels used in the transport sector because it can be used in conventional diesel engines without significant modifications. Biodiesel has a higher oxygen content than fossil diesel, and it is generally accepted that its use leads to a significant reduction in the emission of particulate matter as well as carbon monoxide, sulfur, polyaromatics, hydrocarbons, engine smoke, and noise, but, on the other hand, to an increase in NOx emissions (Bhatti et al. 2008; Reijnders and Huijbregts 2008; Lapuerta et al. 2008). However, as to emissions of greenhouse gases, the life cycle analysis of the emissions of biodiesel produced from virgin vegetable oils is also an issue and the expansion of biodiesel production may also lead to biodiversity losses (Reijnders and Huijbregts 2008). Additionally, the production of first-generation biodiesel from expensive high-quality virgin oils is the main reason for the higher price of biodiesel compared with fossil diesel. Therefore, the use of low-cost feedstocks and non-edible oils and fats will help overcome some of these problems, making biodiesel competitive in price and decreasing the negative impacts of biofuels on food markets (Singh and Singh 2010). Non-edible vegetable oils, such as jatropha oil, soapstocks and waste frying oils, and oils of non-vegetable origin, such as animal fats and oils from microalgae, are therefore alternatives that are being studied (Felizardo et al. 2006, 2008; Singh and Singh 2010).

Biodiesel is a mixture of mono-alkyl esters of long-chain fatty acids derived from renewable lipid feedstock. This fuel is mostly produced through a transesterification reaction (Figure 11.3) between a vegetable oil or fat and a short-chain alcohol (mainly methanol), to produce an ester and glycerol as a by-product (Felizardo et al. 2006). This reaction occurs stepwise, with mono- and diglycerides as intermediate products (Knothe et al. 2005; Marchetti et al. 2007). Recently, Vyas et al. (2010) published a comprehensive review on biodiesel production processes by transesterification.

The transesterification (alcoholysis) reaction of triglycerides occurs in the presence of acid or basic catalysts. Basic homogeneous catalysts have been the most used materials industrially and have surpassed acid catalysts because they provide

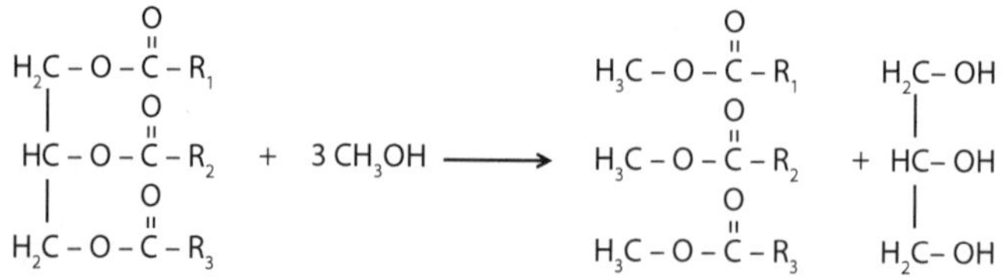

FIGURE 11.3 Schematic transesterification reaction with methanol. R_0, R_2, and R_3 represent the fatty acid chains.

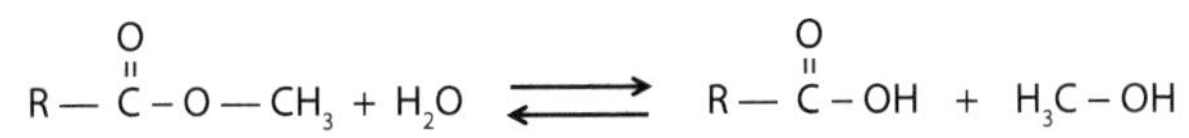

$$R - \overset{\overset{\displaystyle O}{\|}}{C} - O - CH_3 + H_2O \; \rightleftharpoons \; R - \overset{\overset{\displaystyle O}{\|}}{C} - OH + H_3C - OH$$

FIGURE 11.4 Hydrolysis reaction of the methyl esters with production of free fatty acids.

faster reaction rates and lead to the same yield of transesterification (Schudardt et al. 1998). Homogeneous catalysts, although effective, require the implementation of good downstream separation and product purification protocols, thus increasing production costs. To overcome these negative aspects, many types of solid catalysts have been tested in esterification and transesterification reactions, which may have several advantages such as the reduction of production costs and catalyst consumption, greater simplicity of the purification steps of biodiesel and glycerol phases, and the elimination of contaminated waste salt streams that require disposal (Vyas et al. 2010). However, for most of these catalysts, there are still several problems to solve, especially those related to catalyst stability and reuse.

As to homogeneous basic catalysts, sodium and potassium hydroxides and sodium and potassium methylates are widely used (Felizardo et al. 2006; Vyas et al. 2010). Methylated salts are more reactive than the corresponding hydroxides, but are expensive and require water-free reagents (Schuchardt et al. 1998; Meher et al. 2006). However, sodium and potassium hydroxides react with the alcohol to give water. This reaction is undesirable because water can react with the triglycerides and esters in hydrolysis reactions (Figure 11.4) and subsequent saponification of the free fatty acids occurs (Figure 11.5) (Schuchardt et al. 1998).

Moreover, soaps lead not only to a decrease in esters yield but also to the formation of emulsions, which do not allow an efficient separation of the esters and glycerol phases after the transesterification reaction. Because of that problem, for oils with a high concentration of free fatty acids (FFA > 1%), a pretreatment step is recommended. In fact, for oils and fats with an FFA content between 1% and 4%, an extra amount of basic catalyst in the transesterification reaction is necessary to balance the acidity of the oils. When the FFA concentration is greater than 4%, it is necessary to carry out a previous esterification reaction with methanol (Figure 11.6a) or with glycerol (Figure 11.6b) to convert the FFAs into esters (Canakci 2007; Felizardo et al. 2008). Afterward, the glycerides contained in the oil/ester phase can be transesterified by basic catalysis.

At the end of the transesterification reaction, the glycerol-rich phase is separated from the ester layer by decantation or centrifugation. The resulting ester phase contains several contaminants, such as methanol, glycerides, soaps, catalyst, and glycerol, and has to be purified, which may cause severe engine problems. Therefore, the ASTM and European Standards (ASTM D-6751 and EN 14214) impose maximum limits for those products, only attainable after purification. The purification

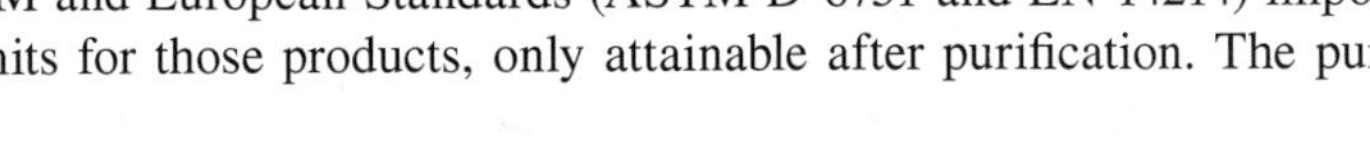

$$R - \overset{\overset{\displaystyle O}{\|}}{C} - OH + NaOH \; or \; NaOCH_3 \; \rightleftharpoons \; R - \overset{\overset{\displaystyle O}{\|}}{C} - ONa + H_2O \; or \; CH_3OH$$

FIGURE 11.5 Saponification reaction of the free fatty acids.

(a)
$$R-\overset{\overset{\displaystyle O}{\|}}{C}-OH + CH_3OH \xrightarrow{\quad H^+ \quad} R-\overset{\overset{\displaystyle O}{\|}}{C}-OCH_3 + H_2O$$

(b)
$$\begin{array}{c}
H_2C-OH \\
| \\
HC-OH \\
| \\
H_2C-OH
\end{array}
+ \; 3 \; R-\overset{\overset{\displaystyle O}{\|}}{C}-OH \longrightarrow
\begin{array}{c}
H_2C-O-\overset{\overset{\displaystyle O}{\|}}{C}-R \\
| \\
HC-O-\overset{\overset{\displaystyle O}{\|}}{C}-R \\
| \\
H_2C-O-\overset{\overset{\displaystyle O}{\|}}{C}-R
\end{array}
+ \; 3 \; H_2O$$

FIGURE 11.6 Methanolysis and glycerolysis reactions: (a) with methanol, (b) with glycerol.

processes usually include the removal of the excess methanol by distillation, washing steps, and a final drying to remove the water (Felizardo et al. 2006). Figure 11.7 presents a global overview of biodiesel production process.

In conclusion, biodiesel can be produced using several different feedstocks and technologies that may lead to different properties of the final product. Furthermore, the final quality of the biodiesel (such as the methyl esters, water, methanol, and glycerol contents; acid, iodine, and cetane numbers; cold filter plugging point; and oxidation stability) is related to raw material properties (water content, and acid and iodine numbers) and to process efficiency.

The quality control of raw materials and biodiesel is extremely important and, for example, EN 14214 requires the determination of 25 quality parameters for biodiesel. However, the full characterization of raw materials and biodiesel by the conventional analytical methods specified in the quality standards (volumetric titrations and gas chromatography) is very expensive and time consuming. Therefore, techniques such as mid- or near-infrared (MIR/NIR) spectroscopy have recently started to be applied as cheaper and faster alternative methods in quality control of both feedstock oils and biodiesel (Knothe 1999, 2000, 2001; Baptista et al. 2008a,b,c; Felizardo et al. 2007a,b; Felizardo 2010). Additionally, those analytical techniques allow the implementation of process analytical technology (PAT) strategies in biodiesel production to better understand and control in advance—before or during production—the effect of different feedstocks or seasonal variation in particular feedstocks. The development of PAT approaches in biodiesel production using NIR spectroscopy as analytical tool is described in the remainder of this chapter.

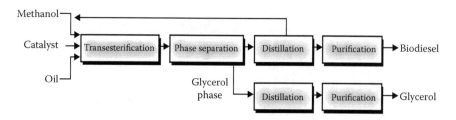

FIGURE 11.7 Biodiesel production.

11.3 PAT IN BIOFUELS

11.3.1 INTRODUCTION

A process is considered well understood when all critical sources of variability are identified and explained, the variability is managed by the process, and product quality attributes can be accurately and reliably predicted over the design space established for the materials, process parameters, and manufacturing conditions used (FDA 2004). In fact, the ability to predict the quality of the final product based on raw materials and process information reflects a high degree of process understanding, and it is therefore the main goal of any process engineering development effort. That goal requires the use of analytical techniques that can collect the necessary information in real time to evaluate the process and product, analyzing simultaneously relevant variables as they are collected. In addition, it is also necessary to use mathematical and statistical multivariate methods to treat all information collected (chemometrics).

NIR spectroscopy, together with chemometrics, is an appropriate analytical tool to attain that goal because it enables analyzing the process based on the overall spectrum of a sample's matrix, and not on a single chemical component, and relating the information collected to the state of raw materials, process, or product. Furthermore, as NIR spectroscopy allows the use of fiber-optic probes, it is easily multiplexed and implemented at an industrial scale.

11.3.2 NIR SPECTROSCOPY

NIR spectroscopy is a well-established analytical technique based on the absorption of electromagnetic energy in the region from 780 to 2500 nm (12,820–4000 cm^{-1}), typical of the vibrational transitions in polyatomic molecules, which allows multicomponent analysis in a fast and nondestructive way without requiring complex pretreatments. In the NIR region, a component typically absorbs at more than one wavelength (Næs et al. 2002; Felizardo et al. 2007b). On the other hand, for complex samples as those obtained in biofuel production processes, absorbance at a given wavelength may have contributions from more than one analyte in chemically complex matrices and thus the analytical information contained in NIR spectra is usually not selective and is influenced by physical, chemical, or structural sample attributes (Felizardo et al. 2007b). Thus, to relate spectral variation with the physical and chemical characteristics of a sample, it is necessary to use statistical and mathematical methods (i.e., via chemometrics) to extract from the spectrum the relevant information for the analysis of the parameter under study. The analysis of the NIR spectra requires the use of chemometric tools, which include principal components analysis (PCA) to perform the qualitative analysis of the spectra, or the projection into latent structures (PLS) regression to develop calibration models between spectral and analytical data for quantitative analysis (Otto 1999). As described elsewhere, the performance of the calibration models has to be analyzed by calculating the root mean square errors of calibration (RMSEC), cross-validation (RMSECV), and prediction (RMSEP) (Næs et al. 2002). The correlation between the NIR spectra and the analytical reference data can be improved through the use of specific spectra preprocessing and variable selection methods. Preprocessing of spectra reduces variations not directly related to the analyte concentration, such as random noise, baseline drift, and light scattering. Furthermore, variable selection methods allow determining the spectral region/s where variations are specifically related to changes in the analyte concentration, so that uninformative variables can be excluded, thus enabling the construction of simpler and more robust models for routine analysis with NIR (Felizardo et al. 2007b). NIR spectroscopy is an undisputed PAT monitoring technology given its intrinsic capability of combining different quality attributes (i.e., chemical, physical, and biological) into a multivariate sample assessment (see Section 11.3.3.1). By being able to capture information of different nature present in a sample's matrix, NIR spectroscopy is capable of producing a precise and reproducible fingerprint of a process state over time. These characteristic features explain why NIR spectroscopy is a de facto PAT monitoring tool with all the desired requirements (Menezes et al. 2010).

11.3.3 CASE STUDY—PAT IN BIODIESEL PRODUCTION

Biofuels are produced in a sequence of large batch operations involving multiple phases, a bio/chemical reaction, and several separation/purification steps. Just as in other bioprocesses, the composition variability of the raw materials can have a significant impact on process performance. Furthermore, despite the variability in the raw materials and the complexity in the unit operations used in their processing, the final product must comply with multiple quality specifications. Thus, the use of PAT throughout (1) raw material qualification and characterization, (2) production process monitoring and supervision, and (3) end-product multiparametric release is strongly recommended.

The application of PAT approaches to biodiesel production is illustrated in Figure 11.8. In that case, the logistics associated with the production of large quantities of a commodity-type product by batch operations, with strict quality specifications, increases the need of using PAT to reduce end-product variability as well to use fast multiparametric quality control to achieve safe and fast release and reduce inventory.

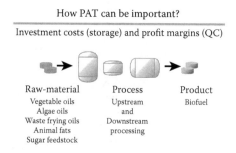

FIGURE 11.8 Biofuel production from a systems engineering PAT perspective. (From Menezes, J.C., P. Felizardo, M.J. Neiva Correia, *Spectroscopy*, Sept.: 30–35, 2008. With permission.)

The methodology illustrated in Figure 11.8 has been applied to biodiesel production with the aim of (1) accessing raw material variability by studying complex raw materials, developing raw material supervisory systems, and formulating mixed lots of raw materials for target specifications; (2) handling raw material variability during process by monitoring quality attributes in-process, at-line, or on-line process supervision and controlling critical issues to quality parameters; and (3) end-product quality control by carrying out the biodiesel quality control and blending operations for target specifications.

11.3.3.1 Raw Material Characterization

The final quality of biodiesel is greatly influenced by the type of feedstock used in the production process, namely its iodine value and contents of water and free fatty acids. In fact, as mentioned above, high moisture content in the oil can lead to the hydrolysis of the methyl esters with formation of free fatty acids, which lead to the decrease in esters yield and to an inefficient separation of the esters and glycerol phases after the transesterification reaction (Baptista et al. 2008a). The iodine value of the feedstock oil used for biodiesel production, which is a measure of the degree of unsaturation of the methyl esters molecules, must also be closely monitored because the iodine value of biodiesel is similar to the iodine value of the oil used in its production. Therefore, usually to comply with the limits imposed by the quality standards (EN 14214 and ASTM D-6751), mixtures of different oils (e.g., soybean, rapeseed, and palm oils) have to be used. Moreover, the iodine value of biodiesel, which is mostly dependent on the iodine value of the oil/mixture of oils used for its production, is a quick indicator of some important quality parameters such as its content of linolenic acid methyl ester, its cold properties (the cold filter plugging point, CFPP, and the cloud point, CP), and its stability (oxidation stability).

11.3.3.1.1 Principal Component Analysis

PCA was used to perform qualitative analysis of the spectra for sample classification. Figure 11.9 shows the score plots of the first and second principal components of the PCA model computed after analysis of different oils using an ABB BOMEM MB160

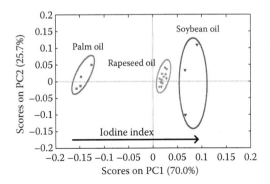

FIGURE 11.9 Scores on PC1 versus scores on PC2 using PCA for the calibration of the iodine value in vegetable oils.

(Zurich, Switzerland) spectrometer equipped with an InGaAs detector and a transflectance probe from SOLVIAS (Basel, Switzerland) with a 5-mm optical path length. From this figure, it is possible to conclude that this PCA model distinguishes the samples according to the type of oil and confirms that it is possible to use NIR as a process fingerprinting technique. In fact, without the need for calibration development, NIR spectroscopy is able to capture a sample's chemical and physical information.

The analysis of the scores plot presented in Figure 11.9 also shows that PC1, which captures 70.0% of the data variance, contains relevant information concerning the oil's composition, namely its iodine value. In fact, palm oils, with the lowest iodine values (around 60 g I_2/100 g), are positioned at the left, whereas soybean oils, with the highest iodine values (around 130 g I_2/100 g), are at the right side of the figure (Baptista et al. 2008c; Menezes et al. 2008).

11.3.3.1.2 Development of Calibration Models: Regression Analysis

PLS regression was used to develop calibration models between spectral data and the weight percentage of soybean, palm, and rapeseed oil in oil mixtures, and the oil's iodine values, fatty acids composition, density, kinematic viscosity, acid number, and water content. This calibration allows the controlled monitoring of all raw materials used for biodiesel production and can reduce the cost of biodiesel processing because raw materials represent a large part of biodiesel production costs.

Table 11.2 summarizes the performance of models derived to predict the percentage (w/w) of soybean, palm, and rapeseed oils in oil mixtures in the range of 0%–100%. NIR spectroscopy is able to accurately predict different oil contents in blends with good results in the entire analytical range (0%–100%) of ternary mixtures.

Table 11.3 presents the prediction errors of the fatty acids composition of oils using NIR spectroscopy. Prediction errors are similar to reference method errors (gas chromatography). Determining the fatty acids composition of oils and fats is very time consuming because they are measured as methyl esters by gas chromatography after a transesterification reaction. Therefore, the possibility of performing the analysis by NIR spectroscopy in minutes is of great significance.

TABLE 11.2

Calibration Results for the Prediction of the Composition of Oil Mixtures

Parameter	Soybean Oil (% w/w)	Palm Oil (% w/w)	Rapeseed Oil (% w/w)
R2 calibration	0.998	0.999	0.995
R2 validation	0.988	0.998	0.994
RMSEC (%)	3.3	1.6	3.0
RMSECV (%)	3.7	1.8	3.3
R2 prediction	0.997	0.998	0.993
RMSEP (%)	2.4	1.2	3.2

Source: Baptista, P., P. Felizardo, J.C. Menezes, M.J. Neiva Correia, *J. Near Infrared Spectrosc.* 16: 445–454, 2008. With permission.

TABLE 11.3
Calibration Results for the Fatty Acids Profile in Vegetable Oils

Parameter	Calibration Range	RMSEP	Analytical Error
C14:0 %(m/m)	0.06–1.21	0.02	0.01
C16:0 %(m/m)	4.9–39.0	0.8	0.9
C18:0 %(m/m)	1.6–4.1	0.2	0.1
C18:1 %(m/m)	26.4–66.1	2.1	2.0
C18:2 %(m/m)	10.8–49.7	1.8	1.7
C18:3 %(m/m)	2.2–8.3	0.2	0.6

Source: Felizardo, P., Near infrared spectroscopy and chemometrics in biodiesel production from vegetable oils and animal fats, PhD Thesis, IST—Technical University of Lisbon, Portugal, 2010. With permission.

Table 11.4 summarizes NIR spectroscopy calibration models developed to predict the oil's iodine value, viscosity, density, acid number, and water content. The prediction error for acid number is higher than the maximum reference method error. However, that model still allows the prediction of the sample's acid number because the prediction error (around 0.2 mg KOH g^{-1}) corresponds only to a 5% error considering the complete calibration range covered. For all other quality parameters, NIR spectroscopy models lead to prediction errors similar to the reference methods' errors (Baptista et al. 2008c).

The results in Tables 11.2 through 11.4 confirm NIR spectroscopy as a simple, fast, and reliable method for qualification and characterization of raw materials in a multiparametric manner (i.e., multiple quality specifications obtained from a single analytical measurement, that is, a spectrum).

11.3.3.2 Process Monitoring and Process Supervision
Process supervision takes into consideration several types of variables at once (i.e., it is multivariate) as opposed to monitoring. Process supervision is not only monitoring in perspective (i.e., the incoming monitored data point for a variable is plotted and compared to that variable's history along the batch); it takes up a higher level by comparing the running batch against previous batches (e.g., mapping the running

TABLE 11.4
Calibration Results of Several Quality Parameters for Oils

Parameter	Calibration Range	RMSEP	Reference Method Error
Iodine value (g I$_2$/100 g)	60–140	1.1	3
Dynamic viscosity (cP)	48.7–72.3	0.5	0.5
Density (kg m^{-3})	905–925	1.1	3
Acid number (mg KOH/g)	0.13–6.56	0.22	≈0.04
Water content (mg/kg)	478–2496	135	≈100

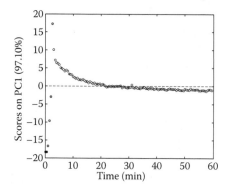

FIGURE 11.10 Scores on PC1 versus reaction time (black points—initial oil; gray points—methanol addition; open points—transesterification reaction).

batch trajectory over the nominal batch trajectory) (MacGregor and Bruwer 2008). The capabilities of in-process NIR for both process monitoring and supervision at industrial-pilot processes of biodiesel production are demonstrated below.

Figure 11.10 summarizes a process trajectory as seen via *in situ* NIR spectroscopy during one transesterification reaction (Felizardo 2010). The PCA map built from preprocessed NIR spectra is sensitive to the reaction stage. In fact, methanol addition leads to the increase of the PC1 score up to its maximum value, whereas during the transesterification reaction, it decreases up to a constant value that indicates the end point of the reaction.

NIR spectroscopy is sensitive to the chemical differences between feedstock oils and end-product biodiesel (Figure 11.11). The main differences are in the regions of $4425–4430$ cm^{-1} and around 6000 cm^{-1}, whereas the methyl ester displays a sharp peak in biodiesel while the feedstock oil exhibits a shoulder. Therefore, NIR spectroscopy can be used to monitor oil to biodiesel transesterification reactions.

Figure 11.12 presents the scores plot of the first two principal components of the PCA model developed after analyzing by NIR spectroscopy samples of soybean oil,

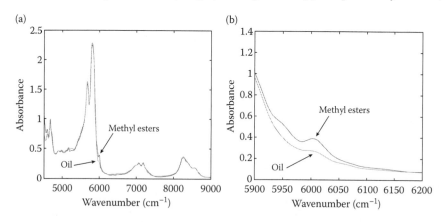

FIGURE 11.11 NIR spectra of soybean oil and soybean biodiesel.

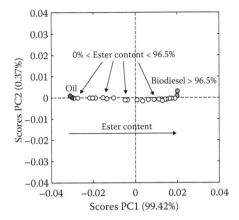

FIGURE 11.12   PCA for process fingerprinting based on the methyl esters content (spectral region 6200–4900 cm⁻¹).

biodiesel, and of the final products obtained in several batches carried out using heterogeneous catalysts, which have methyl esters content between oil (0%) and biodiesel (>96.5%). PC1 scores increase directly with the methyl esters content of the samples. This feature enables monitoring transesterification reaction completion of feedstock oil into biodiesel. Therefore, this result anticipates the potential of NIR spectroscopy to identify process trajectories during biodiesel production and to detect reaction end points.

Another interesting aspect is that NIR spectroscopy can also detect the influence of the operating variables on the reaction trajectories and final stages of reactions (Figure 11.13). In fact, process trajectories can be plotted directly using PCA performed without the need for any reference method and using only mean centering as

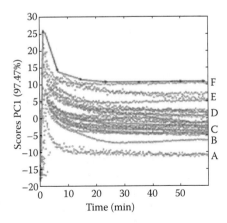

FIGURE 11.13   Scores on PC1 versus the reaction time for several transesterification batches.

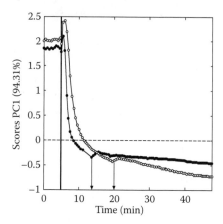

FIGURE 11.14 Scores on PC1 versus time during phase separation. The NIR probe was immersed in biodiesel phase. Stage 1—end of the transesterification reaction; Stage 2—stirring off-phases separation; Stage 3—crude biodiesel.

spectra preprocessing. As seen in the figure, the end points of the different batches are differentiated according to PC1, due to the different methanol content of the reaction mixtures. Reactions with end points near point A presented lower methanol contents, whereas the increase of methanol content leads to the increase of the PC1 scores from point A through F.

Besides the ability to monitor the transesterification reaction, NIR spectroscopy can also be used to follow phase separation on-line. In fact, Figure 11.14 shows that NIR spectroscopy distinguishes the three different stages of phase separation: Stage 1—end of the transesterification reaction (0–5 min); Stage 2—stirring off (5 min)— initial stage of phase separation (until 15–20 min); and Stage 3—slow part of phase separation inside the crude biodiesel phase (after 20 min). Furthermore, PCA also indicates that different separation conditions, in this example, different methanol contents in the reaction mixtures, lead to different end results.

Thus, these results indicate that NIR spectroscopy can be used for process monitoring and supervision of critical biodiesel production steps.

11.3.3.3 Biodiesel Quality Control

As mentioned above, the ASTM and European Standards (ASTM D-6751 and EN 14214) impose maximum limits for several quality parameters of biodiesel (25 parameters in EN 14214 and 20 parameters in ASTM) and their analyses are expensive and time consuming. However, after calibration, NIR spectroscopy is also able to perform the qualification of biodiesel and allows the establishment of expeditious methods for the release of biodiesel lots according to the required specifications. The following examples aim to demonstrate these aspects.

11.3.3.3.1 Principal Component Analysis

As for feedstock oils (Figure 11.9), the PCA model of Figure 11.15, capturing almost all relevant variance (98.75%) in the data, allowed grouping of the biodiesel samples

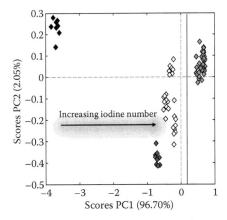

FIGURE 11.15   PC scores plot of several biodiesel samples produced from soybean oil (lighter gray), rapeseed oil colza (darker gray), palm oil (black), and oil mixtures (white). Black vertical line indicates iodine value = 120.

according to the type of oil used in their production. In fact, the scores on PC1 explain the iodine number of biodiesel (i.e., the unsaturation of its fatty acid chains), whereas PC2 appears to be related with the C18:1 and C18:2 content (Felizardo 2010).

Figure 11.15 confirms the usefulness of NIR spectroscopy in the establishment of the desired process design space for targeting biodiesel specifications. EN 14214 imposes an iodine value lower than 120 g I_2/100 g, and because the iodine value of biodiesel is similar to the iodine value of the oil used in its production, to obey the limits it is usually necessary to use mixtures of oils as raw materials or to mix biodiesel lots produced using different raw materials.

Thus, the collection of the NIR spectra and the use of PCA models allow verifying if the sample complies with the quality standards and/or if the type of mixture that has to be prepared is within the desired design space for its multiple quality specifications. The same conclusion was obtained earlier for raw materials, which means that NIR spectroscopy can link quality specifications for the feedstock oils with quality specifications of the final product, thus opening the road to PAT approaches to feed-forward control strategies and in the end to quality by design processes.

11.3.3.3.2   Development of PLS Calibration Models

PLS regression models were developed for all biodiesel quality parameters measurable by NIR spectroscopy. Table 11.5 shows that NIR spectroscopy in combination with multivariate calibration techniques allow for the quantitative analysis of biodiesel. After being properly developed and validated, NIR-based prediction models can, in less than 1–2 min of sample presentation and acquiring a spectrum, give most of the quality specifications required for any modern biodiesel plant to operate efficiently.

As an example, the performance of one of the developed NIR spectroscopy models is shown in Figure 11.16 for the determination of CFPP.

TABLE 11.5

NIR Spectroscopy Calibration Models for Biodiesel Quality Control

Property	Calibration Range	Calibration Error (RMSEP)	Reference Method Error	Reference
Esters content (%)	78.4–99.3	0.9	1.5	Baptista et al. 2008a
Monoglycerides content (ppm)	2641–11,738	654	1000	Felizardo 2010
Diglycerides content (ppm)	82–7846	389	400	Felizardo 2010
Total glycerol (ppm)	1371–4230	178	250	Felizardo 2010
Methanol content (ppm)	2–2864	73	50	Felizardo et al. 2007a,b
Linolenic acid ME content (C18:3) (%)	0.3–8.9	0.18	–	Baptista et al. 2008a
C14:0	0.05–1.05	0.02	–	
C16:0	4.0–36.0	0.79	–	
C18:0	1.8–4.5	0.22	–	
C18:1	25.0–62.0	1.79	–	
C18:2	13.0–54.0	2.5	–	
Iodine value (g I_2/100 g)	62–132	0.8	3	Baptista et al. 2008a
Density at 15°C (kg/m³)	877–922	0.9	0.5	
Kinematic viscosity at 40°C (mm²/s)	3.67–4.91	0.08	0.04	
Water content (ppm)	218–1859	80	50	Felizardo et al. 2007a,b
Acid value (mg KOH/g)	0.09–0.68	0.05	0.04	Baptista et al. 2008a
Oxidative stability	0.7–17.8	0.6	0.5	Felizardo 2010
CFPP (°C)	–13 to 5	1.0	1	Baptista et al. 2008a
CP (°C)	–3 to 9	0.4	2	Felizardo 2010

All calibrations developed in this work can be used industrially, with the purpose of process monitoring, as a simpler, faster, and more affordable alternative to classical quality control of the final product, thus forming the basis of more competitive manufacturing platforms.

11.4 CONCLUSIONS

The potential of PAT in manufacturing areas other than the pharmaceutical industry is far from properly exploited mainly due to the insufficient use of intrinsically multi-parametric monitoring tools (e.g., NIR), little use of available process information (e.g., data on historical batches or on raw material lots), and normally not taking a process/plant-wide perspective for the proposed PAT strategy. The application of

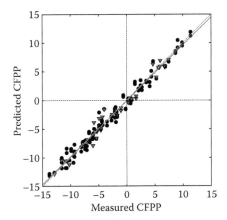

FIGURE 11.16 NIR spectroscopy models for determination of CFPP in biodiesel. Circles, calibration samples; inverted triangles, external prediction samples.

PAT concepts offers different opportunities depending on the regulatory oversight of specific industries. Here we tried to show the advantages of using PAT tools in biofuels production, namely biodiesel. Future work can be directed to ensuring, by means of real-world examples such as those shown here, that tools and approaches already in use in the production of medicines and widespread in the pharmaceutical industry become accepted by the authorities regulating biofuels production. PAT tools, especially those based on spectroscopy, are versatile and their advantages are diverse in terms of business benefits and technical advantages of all sorts. At current affordable prices of several at-line and portable NIR or MIR spectroscopy solutions, it makes little sense to invest in fully instrumented chemical laboratories to analyze feedstocks and the end product of biodiesel manufacturing plants. With current ASTM standards addressing MIR and NIR spectroscopy use for quantitative (ASTM 1655) and qualitative (ASTM 1790) analysis, the road is open to the generalized use of these monitoring techniques as PAT tools in biofuels manufacturing, as described here.

ACKNOWLEDGMENTS

We thank all our former students involved in the PAT and Biodiesel Program at IST-UTL in the past 6 years, especially Margarida Sousa-Uva, Pedro Fiolhais, and Patricia Baptista. We also thank many vendors, namely ABB (represented in Portugal by Paralab) and Yokogawa (represented in Portugal by Mr. P. Calado) for all the support with laboratory and process NIR instrumentation throughout the years. P.F. thanks the Portuguese National Science Foundation (Fundação para a Ciência e a Tecnologia) for financial support (PhD Grant SFRH/BDE/15566/2005).

REFERENCES

Agarwal, A.K. 2007. Biofuels (alcohols and biodiesel) applications as fuels for internal combustion engines. *Prog. Energy Combust. Sci.* 33: 233–271.

Balat, M., H. Balat, and C. Oz. 2008. Progress in bioethanol processing. *Prog. Energy Combust. Sci.* 34: 551–573.

Baptista, P., P. Felizardo, J.C. Menezes, and M.J. Neiva Correia. 2008a. Multivariate near infrared spectroscopy models for predicting the methyl esters content in biodiesel. *Anal. Chim. Acta* 607: 153–159.

Baptista, P., P. Felizardo, J.C. Menezes, and M.J. Neiva Correia. 2008b. Multivariate near infrared spectroscopy models for predicting the iodine value, CFPP, kinematic viscosity at 40°C and density at 15°C of biodiesel. *Talanta* 77: 144–151.

Baptista, P., P. Felizardo, J.C. Menezes, and M.J. Neiva Correia. 2008c. Monitoring the quality of oils for biodiesel production using multivariate near infrared spectroscopy models. *J. Near Infrared Spectrosc.* 16: 445–454.

Bhatti, H.N., M.A. Hanif, M. Qasim, and A. Rehman. 2008. Biodiesel production from waste tallow. *Fuel* 87 (13–14): 2961–2966.

Canakci, M. 2007. The potential of restaurant waste lipids as biodiesel feedstocks. *Bioresour. Technol.* 98: 183–190.

Demirbas, A. 2007. Progress and recent trends in biofuels. *Prog. Energy Combust. Sci.* 33: 1–18.

Elvers, B. and F. Ullmann. 1992. *Ullmann's Encyclopedia of Industrial Chemistry*, Weinheim, Germany: Wiley-VCH.

EN 228. 2010. European Standard EN 228: Automotive Fuels - Unleaded Petrol–Requirements and Test Methods. CEN - European Committee for Standardization, Brussels, Belgium.

FDA. 2004. Guidance for Industry. PAT—A framework for innovative pharmaceutical development, manufacturing, and quality assurance, September. Available online at http://www.fda.gov/. Accessed in May 2009.

Felizardo, P. 2010. Near infrared spectroscopy and chemometrics in biodiesel production from vegetable oils and animal fats. PhD Thesis, IST—Technical University of Lisbon, Portugal.

Felizardo, P., P. Baptista, J.C. Menezes, and M.J. Neiva Correia. 2007a. Multivariate near infrared spectroscopy models for predicting methanol and water content in biodiesel. *Anal. Chim. Acta* 595: 107–113.

Felizardo, P., P. Baptista, M. Sousa Uva, J.C. Menezes, and M.J. Neiva Correia. 2007b. Monitoring biodiesel fuel quality by near infrared spectroscopy. *J. Near Infrared Spectrosc.* 15: 97–105.

Felizardo, P., M.J.N. Correia, I. Raposo, J.F. Mendes, R. Berkmeier, and J.M. Bordado. 2006. Production of biodiesel from waste frying oils. *Waste Manage.* 26: 487–494.

Felizardo, P., J. Machado, D. Vergueiro, M.J.N. Correia, and J.M. Bordado. 2008. High free fatty acids oils for biodiesel production. Proceedings of the 10th International Chemical and Biological Engineering Conference—CHEMPOR 2008, Braga, Portugal, 4–6 September, pp. 1220–1225.

Knothe, G. 1999. Rapid monitoring of transesterification and assessing biodiesel fuel quality by NIR spectroscopy using a fiber-optic probe. *J. Am. Oil Chem. Soc.* 76 (7): 795–800.

Knothe, G. 2000. Monitoring the turnover of a progressing transesterification reaction by fiber-optic NIR spectroscopy with correlation to 1H-NMR spectroscopy. *J. Am. Oil Chem. Soc.* 77 (5): 489–493.

Knothe, G. 2001. Determining the blend level of mixtures of biodiesel with conventional diesel fuel by fiber optic NIR spectroscopy and 1H-NMR spectroscopy. *J. Am. Oil Chem. Soc.* 78: 1025–1028.

Knothe, G., R.O. Dunn, and M.O. Bagby. 1997. Biodiesel: The use of vegetable oils and their derivatives as alternative diesel fuels. *ACS Symp. Ser.* 666: 172–208.

Knothe, G., J. Van Gerpen, and J. Krahl. 2005. *The Biodiesel Handbook*. Champaign, IL: American Oil and Chemists Society Press.

Koh, L. and J. Ghazoul. 2008. Biofuels, biodiversity, and people: Understanding the conflicts and finding opportunities. *Biol. Conserv.* 141: 2450–2460.

Lapuerta, M., J.M. Herreros, L.L. Lyons, R. García-Contreras, and Y. Briceño. 2008. Effect of the alcohol type used in the production of waste cooking oil biodiesel on diesel performance and emissions. *Fuel* 87: 3161–3169.

MacGregor, J.F. and M.J. Bruwer. 2008. A framework for the development of design and control spaces. *J. Pharm. Innov.* 3: 15–22.

Marchetti, J.M., V.U. Miguel, and A.F. Errazu. 2007. Possible methods for biodiesel production. *Renewable Sustainable Energy Rev.* 11: 1300–1311.

Meher, L.C., D.V. Sagar, and S.N. Naik. 2006. Technical aspects of biodiesel production by transesterification—A review. *Renewable Sustainable Energy Rev.* 10: 248–268.

Menezes, J.C., P. Felizardo, and M.J. Neiva Correia. 2008. The use of process analytical technology (PAT) tools in biofuels production. *Spectroscopy*, Sept.: 30–35.

Menezes, J.C., A.P. Ferreira, L.O. Rodrigues, L.P. Brás, and T.P. Alves. 2010. Chemometrics role within the PAT context: Examples from primary pharmaceutical manufacturing. In *Comprehensive Chemometrics* (4 volumes), edited by S. Brown, R. Tauler, and B. Walczak, volume 3, pp. 313–357.

Naik, S.N., V.V. Goud, P.K. Rout, and A.K. Dalai. 2010. Production of first and second generation biofuels: A comprehensive review. *Renewable Sustainable Energy Rev.* 14: 578–597.

Næs, T., T. Isaksson, T. Fearn, and T. Davies. 2002. *A User Friendly Guide to Multivariate Calibration and Classification.* Chichester, UK: Nir Publications.

Otto, M. 1999. *Chemometrics: Statistics and Computer Application in Analytical Chemistry.* New York: Wiley-VCH.

Reijnders, L., and M.A.J. Huijbregts. 2008. Biogenic greenhouse gas emissions linked to the life cycles of biodiesel derived from European rapeseed and Brazilian soybeans. *J. Cleaner Prod.* 16: 1943–1948.

Saraf, S. and B. Thomas. 2007. Influence of feedstock and process chemistry on biodiesel quality. *Process Saf. Environ. Prot.* 85: 360–364.

Schuchardt, U., R. Sercheli, and R.M. Vargas. 1998. Transesterification of vegetable oils: A review. *J. Braz. Chem. Soc.* 9: 199–210.

Singh, S.P. and D. Singh. 2010. Biodiesel production through the use of different sources and characterization of oils and their esters as the substitute of diesel: A review. *Renewable Sustainable Energy Rev.* 14: 200–216.

Srivastava, A. and R. Prasad. 2000. Triglycerides-based diesel fuels. *Renewable Sustainable Energy Rev.* 4: 111–133.

Vyas, A.P., J.L. Verma, and N. Subrahmanyam. 2010. A review on FAME production processes. *Fuel* 89: 1–9.

Zabaniotou, A., O. Ioannidou, and V. Skoulou. 2008. Rapeseed residues utilization for energy and 2nd generation biofuels. *Fuel* 87: 1492–1502.

12 Application of Microreactors for Innovative Bioprocess Development and Manufacturing

Seth T. Rodgers and A. Peter Russo

CONTENTS

12.1 INTRODUCTION

Bioprocess design engineers face a set of challenges unlike those faced by process designers in most other industries. Perhaps the most important of these is that process designers cannot interact with the "machinery" of bioproduction directly, and so the quality attributes of the product are manipulated primarily by selection of the production cell line and by the optimization of the process and conditions under which those cells are cultured (Anderson and Reilly 2004; Butler 2005; Chun et al. 2003; Dowd et al. 1999; Farid 2007; Jain and Kumar 2008; Kunkel et al. 2000). A second challenge is posed by the fact that the "machinery" is complex in that a large number of process parameters, for example, pH, dissolved oxygen (DO), temperature (T), and nutrient levels can influence cellular processes, thereby impacting the quality attributes of the protein produced (Byrne et al. 2007). This multivariate optimization of a multivariate parameter set is clearly a more demanding task than that of simple optimization toward a single objective. Furthermore, there are, as yet, no rigorous models to explain the impact of process design choices on the quality attributes of the resulting bioproducts, so most of this process understanding must be developed empirically.

These challenges become more acute when considered in light of the quality by design (QbD) initiative of the Food and Drug Administration (FDA) (Cogdill and Drennen 2008). Under this paradigm, each process development decision represents an opportunity to reduce process variability and to "design quality" into the process. This is, of course, best done by identification of process sensitivities early and defining a multidimensional "design space" within which the process is as robust to changes in input parameters as possible (Harms et al. 2008).

The most straightforward strategy to develop this understanding would be the use of multifactorial, statistically designed experiments (DOEs) (Chun et al. 2003; Montgomery 2009; Swalley et al. 2006). Unfortunately, the set of potentially important parameters is large, and these culture parameters typically interact with each other. A simplified experimental design that does not involve variation of all important process parameters will miss some of these important interactions and leave open the possibility of unexpected process variations (Montgomery 2009). Because the number of trials (runs) in an experimental design grows roughly geometrically with the number of parameters varied, the requirements on experimental "bandwidth" for complete understanding can be very large (Rao et al. 2009). Although large DOE experimental campaigns can be broken into a series of smaller ones, process development experimentation can only be serialized to a limited extent before impacting the commercialization timeline.

It is natural to ask why potential bioprocess conditions and inputs could not be screened in high throughput just as candidate molecules are screened in drug discovery campaigns. At first glance, it would seem that the adoption of high-throughput experimentation for process development is a certain next step for the field of bioprocess design. Certainly, several authors have commented on the need for high-throughput experimentation in this field (Betts and Baganz 2006; Isett et al. 2007; Moran et al. 2000; Puskeiler et al. 2005; Rao et al. 2009; Reis et al. 2006; Willoughby 2006; Chen et al. 2009), and the application of automation to related areas such as cell culture and drug discovery has increased productivity in these areas dramatically. It is surprising, then, that a similar revolution has not taken place in bioprocess development.

There is certainly no "industry standard" high-throughput cell culture technology today, where various high-throughput systems are used sporadically alongside the "traditional" shaken flask and small bioreactor platforms that have been in use for more than 30 years. With this paradox in mind, it is worthwhile to consider some of the challenges that must be faced before high-throughput experimentation can play the same transformational, enabling role that it has in other fields.

Drawing on the literature as well as experience, it can be said there are at least three components needed to execute cell culture experiments in very high throughput: (1) a scale-down bioreactor model; (2) an automated system for culture maintenance, data acquisition, and control; and (3) a data management system capable of aggregating multivariate cell culture data sets and indexing them according to the multifactorial DOE, so that the effects of parameter manipulations may be readily discerned. In practice, a fourth component is usually needed in the form of a high-throughput analytical system to examine the quality attributes of protein products, as it is these data sets upon which decisions are made.

12.2 CHALLENGE ONE: CONTROLLED CELL CULTURE IN A MICROBIOREACTOR

The first challenge encountered is that none of the scale-down models in wide use today is ideally suited for simply "numbering up" to create a solution for high-throughput experimentation. The traditional platforms may be presented as offering a set of trade-offs between ease of achieving high throughput and the quality of the experimental data obtained.

Figure 12.1 shows that no traditional scale-down model can be readily used to provide high-quality data in high throughput at reasonable expense. The need for a new scale-down model is further attested to by the fact that there are currently several microreactor platforms available on the market, in addition to many academic reports that have been reviewed in the literature (Betts and Baganz 2006; Marques et al. 2009; Zhang et al. 2006). Platforms such as the DASGIP Parallel Bioreactor System and the Fluorometrix Cellstation, as well as others, use miniaturized stirred tank bioreactors operated in parallel. These designs are typically limited to 16 reactors or less. Higher throughputs are obtained with shaken plate platforms such as the Applikon M24 and the m2p-Labs BioLector. The Applikon M24 uses a shaken 24-well plate with individual gas and temperature controls for each well, whereas the BioLector uses 48- or 96-well plates with a common gas and temperature profile. Both of these systems lack liquid handling, so that fed-batch processes or base additions for pH control must be performed manually or by third-party liquid handlers. More recently, The Automation Partnership released the Advanced Microscale Bioreactor (ambr®), which uses 24 miniaturized stirred vessels with independent gas controls and two temperature zones. This platform contains an integrated liquid handler for adding or removing fluids from each bioreactor.

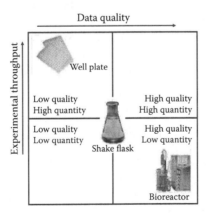

FIGURE 12.1 Fully instrumented and controlled bioreactors provide the best quality data but are too expensive to use in all but the smallest factorial designs. Flasks and well plates allow more experimentation but permit very limited control of culture conditions.

Certainly, all of these "new" model systems aim at providing the right combination of cost, throughput, and data quality for a given purpose. Although cost and throughput are clearly defined, the definition for "quality" of data obtained from small-scale bioreactor models is less clear. Fortunately, however, the creation and validation of scale-down models of production-scale bioreactors is a well-understood discipline in the biopharmaceutical industry (Willoughby 2006; Yang et al. 2007). Model designers typically divide the important culture variables into two groups: scale-independent variables, for example, temperature, pH, and DO; and scale-dependent variables, for example, liquid level, sparge rates, and impeller speed. Engineers typically strive to make the scale-independent parameters of the small system match those of the larger system exactly, and then adjust the scale-dependent variables based on dimensional analysis or heuristics.

The first design challenge for a new scale-down model is to create a culture vessel or microbioreactor that offers control of the important scale-independent variables and a sensible treatment of those which are scale dependent. Although no standard approach has emerged, some noteworthy efforts to create new scale-down model systems have already been made. These models can be separated into two groups. The first group follows the most "bioreactor-like" design paradigm. These solutions use dedicated gas and or fluid lines to supply each growing culture with nutrients and gasses, as well as taking data via dedicated electrical connections or optical channels such as those from DASGIP and Fluorometrix. These systems offer a high degree of control over the cell culture environment and thus have the potential for very high data quality, but due to the need to manage continuous fluid and gas connections, electronic probes, and mechanical impellers, these systems typically are not scaled below a few hundred milliliters and are thus difficult to automate. This limited degree of automation in turn means that these systems generally achieve only a modest increase in throughput over conventional systems. The second group follows a more "flask like" design paradigm, such as the Tubespin (DeJesus et al. 2004) and the SimCell (Legmann et al. 2009), where no continuous connections for fluids, gasses, agitation, or data acquisition are used. In this paradigm, the culture vessel interacts periodically with external (frequently automated), liquid handling, measurement, agitation, and control apparatus to execute the various actions required for cell culture. This flask-like paradigm has the obvious advantage of avoiding the tangle of dedicated lines for each reactor and liquid combination allowing different cell culture conditions to be examined in very high throughput. The trade-off is that only intermittent control actions are possible in such a system, potentially compromising the ability of the microreactor model to illustrate the impact of variation in process inputs such as pH, temperature, and media composition on process outcomes of interest such as cell viability, titer, and product quality.

The SimCell microbioreactor used to generate all data in the case studies presented here is illustrated in Figure 12.2. The reactors are arranged in arrays of six and presented in a well-plate footprint in the X and Y dimensions. Each reactor has a total volume of 800 μl and working volume of about 700 μl, with 100 μl reserved for gas headspace to provide for agitation as shown in Figure 12.2.

The scheme presented in Figure 12.2 allows several hundred cultures to be agitated simultaneously with only one axis of rotation. Computational fluid dynamics

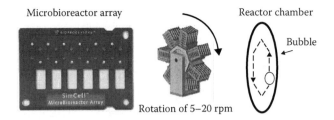

Microbioreactor array   Reactor chamber

Rotation of 5–20 rpm

Bubble

FIGURE 12.2 Agitation is provided by preserving a gas headspace and mounting the microbioreactor on a rotating wheel, forcing the bubble to traverse the chamber perimeter.

(CFD) studies were undertaken, and an average shear stress of 0.18 dyn/cm^2 was calculated under typical operating conditions. In addition, the system can be used to simulate a variety of shear rates by adjusting bubble size and rotation rate. Other solutions for different bioreactor functions are shown in Table 12.1.

The microbioreactors are designed to exchange carbon dioxide and oxygen with the external environment via diffusion through the reactor walls. Diffusion through the polymer is the rate-determining step, as predicted by the mass transfer model presented in Equation 12.1 through 12.6 and verified in the step test experiment shown in Figure 12.3.

Equation 12.1 is a transient mass balance on oxygen, with total oxygen in the reactor represented by the sum of the oxygen dissolved $C(t)^{in}$ in the liquid volume, V_l, and that in the gas headspace in equilibrium with the liquid. This is expressed by using the ideal gas law and Henry's law as $C(t)^{in}*HV_b/RT$, where $C^{in}(t)$ is the liquid phase oxygen concentration, H is Henry's law constant, V_b is the volume of the gas phase or "bubble," R is the gas constant, and T is temperature. The rate of diffusive exchange through the oxygen permeable reactor walls is described by the first term on the right-hand side, where A is the area available for gas exchange, Π is the wall permeability, l is the wall thickness, and $C(t)^{out}$ is the oxygen concentration in the environment outside the reactor. The rate of oxygen consumption with the reactor is represented as the product of cell density $\rho(t)$, specific oxygen demand per cell $R(t)$, and volume of the liquid phase V_l.

TABLE 12.1

Comparing Design Choices in Microbioreactors and Conventional Bioreactors

Bioreactor Function	Bioreactor Method	Microbioreactor Method
Agitation	Impeller	Bubble and rotation
Gas transfer	Active sparge and stir	Diffusion through walls
pH and DO measurement	Electrodes	Fluorescent sensors
Temperature control	Heated jacket	Equilibration with environment
pH control and feeding	Hard plumbed liquid lines	Intermittent addition through septum

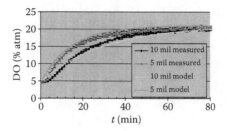

FIGURE 12.3 Doubling the polymer film thickness doubles the equilibration time constant in a static (i.e., unagitated) microbioreactor, in agreement with predictions of diffusion-controlled mass transfer.

$$\left(V_1 + \frac{HV_b}{RT} \right) \frac{dC(t)^{in}}{dt} = \frac{A\Pi H}{l} \left[C(t)^{out} - C(t)^{in} \right] - \rho(t)R(t)V_1 \qquad (12.1)$$

Because the amount of biomass changes slowly relative to the system equilibration time, a pseudo steady-state assumption is made and deviation variables are introduced, where $C'(t)$ represents the difference between the actual oxygen concentration inside the reactor and the steady-state value represented by $\overline{C}$.

$$C'(t) = C(t)^{in} - \overline{C}. \qquad (12.2)$$

Equation 12.3 represents the Laplace transform of Equation 12.1 after expressing the oxygen concentration in deviation variables to facilitate casting the transient mass balance in transfer function form as shown in Equation 12.4.

$$\frac{\left(V_1 + \dfrac{HV_b}{RT} \right) l}{A\Pi H} SC'(s)^{in} = \left[C'(s)^{out} - C'(s)^{in} \right] - \rho'(s)R'V_1 \frac{l}{A\Pi H} \qquad (12.3)$$

$$C'(s) = \frac{C'(s)^{out}}{\tau S + 1} - \frac{K\rho'(s)}{\tau S + 1}, \qquad (12.4)$$

where model parameters have been grouped to form an equilibration time constant as shown in Equation 12.5 and a "gain" shown in Equation 12.6. Physically, Equation 12.5 predicts the time constant for response to a step change in external oxygen concentration, whereas Equation 12.6 gives the "gain" of the system, in this case, the amount of change in the internal oxygen level for a given change in cellular respiration.

$$\tau = \frac{\left(V_1 + \dfrac{HV_b}{RT} \right) l}{A\Pi H} \qquad (12.5)$$

$$K = \frac{RV_1 l}{A \Pi H}.$$ (12.6)

These equations predict that a doubling of film thickness would double the time constant of the response to a step change.

With the mass transfer analysis in mind, it seems reasonable to conclude that the inside of the microbioreactor is homogeneous, and to expect that if the scale-independent variables are replicated correctly, the microbioreactor can be a good model of a "well-mixed" cell culture system at any scale. It is, however, unreasonable to expect the microreactor to model mixing effects or inhomogeneities at the production scale any better than a homogeneous bench top reactor. However, when combined with a CFD model of heterogeneities in gas or chemical species in a large-scale system, a microreactor could provide an attractive platform to study the likely impact of these variations on process outcomes.

Now that reactor materials and dimensions have been chosen to provide shear rates, gas transfer rates, and mixing representative of a bench-scale bioreactor, measurements and controls must be put in place. A minimal set of measurements would be those that simply support feedback control of the most important culture parameters, and because the cells require round-the-clock maintenance, it is desirable that these measurements and control actions be on-line and automated. In this work, and most other high-throughput systems, pH and DO were taken to be the minimal set of on-line variables and are measured optically by using fluorescent sensors printed in each microbioreactor. Cell growth was also monitored by forward light scattering. Of course, measurements beyond this minimal set could be added, and ideally, provision could be made for additional measurements and early deployment of appropriate process analytical technologies (PATs) such as NIR for glucose and lactate levels or high-throughput mass spectrometry or capillary electrophoresis assays to examine protein quality.

12.3 CHALLENGE TWO: ADDING AUTOMATION FOR THROUGHPUT

Once a suitable microbioreactor has been developed, the next step is to add automation to achieve high throughput. At first glance, there seems to be a direct parallel to high-throughput screening, amounting to integration of liquid handling, optical measurements, and incubation/agitation. This is done by creating automated "modules," dedicated to each of these unit operations, and integrating them around a robot arm. The automated culture system used in this work is presented in Figure 12.4.

Although the unit operations are very similar to those used in high-throughput screening, there are some differences between the requirements of high-throughput screening and high-throughput process development. The most important is that while high-throughput screening is frequently a "one pass" activity, animal cell culture experiments in bioreactors typically run for 10 days or more. This means that the microbioreactors must be repeatedly cycled through the incubation, measurement, and liquid handling modules to support the growing cells and maintain control over culture conditions until the cultures are ready to harvest.

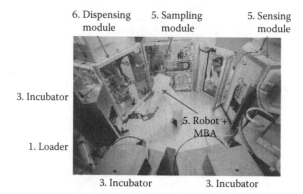

FIGURE 12.4 Automated cell culture system contains four unit operations: measurement, sampling, liquid dispensing, and incubation.

In a typical mammalian cell culture experiment, the robotic system will measure each reactor at 6- to 12-h intervals and add feeds and/or base for pH control as needed. The system will also adjust incubator oxygen levels in response to DO measurements and execute shifts in parameter set points as programmed. This cycling frequency dictates the frequency of measurement data points and control actions. The cycling frequency is the central trade-off of a system based on intermittent fluid connections: very good control of culture conditions can be achieved by frequent measurement and control actions, but only a few reactors can be serviced; on the other hand, a very large number of cultures can be run with a low cycling frequency (long cycle time), but with less control over culture conditions. A conceptual diagram of a cycle is presented in Figure 12.5.

Although a "round trip" for a specific microbioreactor array typically takes about 20 min, the sensing, sampling, and liquid dispensing modules each processes their reactor arrays in parallel leading to an overall throughput of about 6 reactor arrays (36 cultures) per hour of cycle time; a typical 12-h cycle time would then allow 72 arrays or 432 individual cell cultures.

Feedback control actions for pH and DO are performed based on measurements of pH and DO in individual culture chambers. pH control is straightforward, and a block diagram of the pH control loop is presented in Figure 12.6. Here, the base addition is driven by comparing the pH result to the desired set point, and adjusted by taking into account the CO_2 level in the incubators as well as the acidic/basic properties of the feed expected. Dissolved oxygen levels are controlled by manipulating incubator oxygen levels to compensate for cellular respiration as the cultures grow.

There is clearly some complexity in programming a robotic system to manage several hundred individual cell cultures over the course of 2 weeks even if all were processed in exactly the same way and thus subjected to the same culture condition. Given that a single multifactorial experiment might involve several hundred reactors, representing a hundred or more culture conditions, an efficient method for programming the robotic system to execute a given experimental campaign was required.

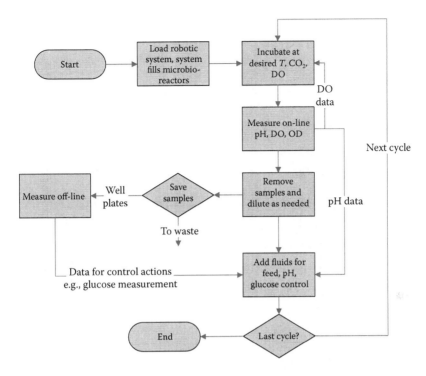

FIGURE 12.5   Cycle begins when the first reactor array is removed from the incubator. The array is measured in the sensing module, then sampled into well plates in the sampling module as programmed, and finally receives feeds and pH adjustment in the liquid-dispensing module before returning to the incubator.

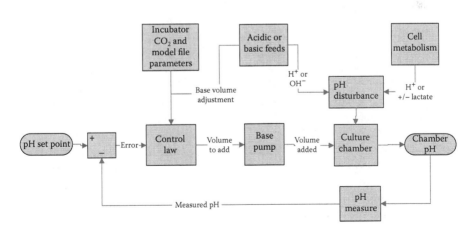

FIGURE 12.6   Block diagram of pH control loop. Measured variable is chamber pH level; manipulated variable is the base volume added in response to comparison versus set point. Note feed-forward action to compensate for acidic/basic feeds.

This challenge was met by developing a template-driven program that allows the user to describe a given cell culture experiment (usually the center point of a factorial design) as a series of measurements, liquid additions, and sampling actions. Once a single "baseline" culture condition is described in this way, it is automatically multiplexed to create a family of related conditions and corresponding descriptions that differ according to the user's desired factorial design. This family of related culture protocols is then automatically translated into a program executable by the robotic system. Using this formalism, a program for an experimental campaign featuring several hundred cell cultures under more than a hundred distinct conditions can be quickly and easily developed in less than 2 h.

12.4 CHALLENGE THREE: EVALUATING THE MODEL AND INTERPRETING THE DATA

Given that the purpose of a scale-down model is to generate data, and ultimately process understanding rather than products for sale, its utility is determined by weighing the quality of experimental data generated against the effort required for experimentation. Because process development represents truly multivariate, multi-objective optimization, the quality of the scale-down model should not be judged on the basis of results at a single set of conditions, and a good scale-down model will show the same linkages between changes in process inputs and process outputs as the larger scale system of interest. With this in mind, a multifactorial experiment was used by Legmann et al. (2009) to evaluate the SimCell microbioreactor as a scale-down model of a popular bench-scale bioreactor. The effort convincingly demonstrated that the microbioreactor predictions of growth, titer, purity, and incomplete glycosylation levels can agree well with the predictions of a 3-L-scale bioreactor.

Because these models are, by nature, empirical, it is natural to wonder if the microreactor-based modeling approach is generalizable. In the absence of first principles models, the robustness of the modeling approach itself can only be established by following established scale-down heuristics and verifying empirically. In a sense, the only "proof" of a model's validity is obtained by exhaustive comparison to established models. To this end, 132 distinct head-to-head comparisons between SimCell microbioreactors and established scale-down models are shown in Figure 12.7. The data were obtained using cell lines from 18 different companies and represent a wide variety of products and processes. Each data point represents the titer obtained using the SimCell system regressed against a conventional scale-down model such as a flask or bioreactor. The correlation is again very high, and the residuals are distributed roughly evenly about the 45° line. This suggests that while not always exact, the model is fundamentally sound.

Of course, it is also natural to wonder if the microbioreactor is more like a shaken flask than a true bioreactor. Segmenting the shake flask model results of Figure 12.7 yields a "bioreactor only" correlation (Figure 12.8). This figure shows that the microbioreactor is capable of modeling bioreactors as well as it models flasks. The system can be a good "mimic" of either model by adjusting the mode of operation.

The demonstrated agreement with diverse model systems reflects the flexibility of the microbioreactor approach. The microbioreactor model results can be made "flask-like"

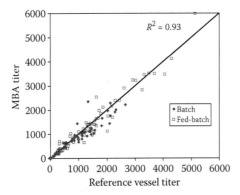

FIGURE 12.7 Since 2006, more than 132 comparisons between microreactors and conventional flask and bioreactor scale-down models have been made. Data set includes bioreactors run in batch and fed-batch modes, as well as flasks run in both batch and fed-batch modes. Predictions of titer showed excellent agreement, with an R^2 of 0.93.

with infrequent control actions when a flask-like process is followed, whereas results can be made more "bioreactor-like" by increasing the frequency of control action. The only drawback in obtaining bioreactor-like data is that this improved data quality comes at the cost of reduced throughput, as more measurement and liquid handling are required. However, it is important to note that even at this level of reduced throughput, more than 300 microbioreactors can be executed in parallel.

Although the high degree of correlation may be surprising, it is logical that the microbioreactor can be used to model other systems with similar accuracy as long as three conditions are met. First, the scale-independent variables such as pH, T, and DO must be matched between the two systems. Second, the larger scale system must be homogenous, with liquid and gas headspace in near equilibrium, as the homogenous microreactor could not be expected to represent different zones in a very large or poorly mixed system. Third, the cells should not interact strongly with the reactor walls (i.e., they should be well adapted to suspension culture). The first two restrictions of course would apply to any scale-down model, and it could be the

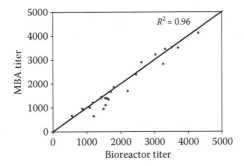

FIGURE 12.8 Data of Figure 12.7 with bioreactors only, that is, flasks excluded. Again correlation is very high, and residuals show little, if any, systematic bias.

case that in the absence of strong surface effects, it is easier to use a microreactor as a model of a bench-scale bioreactor than it is to use a bench-scale reactor as a model of a larger, heterogeneous system. The third restriction arises from the fact that the surface-to-volume ratio, S/V, is about 1 mm^{-1}, whereas it is roughly 100-fold greater in a bench-scale bioreactor. Although surface interactions are typically negligible even in the smallest conventional reactors, their impact was observed in our system with cholesterol-independent NS0 cells. These cells generally grew poorly in our system, as they often do in polymeric vessels unless special care is taken in the selection of materials or treatment of the polymer surfaces (Okonkowski et al. 2007). In general, it is proposed that the microreactor can be used to model any larger scale system, as long as the system does not display significant heterogeneity arising from mixing or surface effects. Said differently, if the system is truly homogeneous, then the scale-independent variables should be sufficient to specify the culture environment.

12.5 CHALLENGE FOUR: ASSEMBLING A COMPLETE SOLUTION

Although microbioreactor technology can enable multifactorial experimentation of much larger scope than has previously been practical, the ultimate utility of the technology is impaired without the use of complementary off-line or at-line analytics, as well as a multivariate cell culture data management system.

Most, if not all, relevant analytical measurements can be performed today using platforms with the throughput and volume requirements that are complementary to scale-down bioreactor models. Glucose and other metabolites can be measured using the Nova BioProfile series or YSI 7100 multiparameter bioanalytical system with sample volumes of less than 1 ml. Additionally, glucose and other metabolites can be measured on conventional absorbance/fluorescence plate readers using colorimetric enzymatic assays that require even smaller volumes. Cell counts and viability can be obtained using Trypan blue-based instruments such as the Cedex from Innovatis or Vi-CELL from Beckman Coulter, or using capillary-based instruments such as the Guava EasyCyte.

Product titer can be measured using standard high-performance liquid chromatography methods, enzyme-linked immunosorbent assay, or with the Octet series of instruments from ForteBio. The Caliper GXII system, a microchip capillary electrophoresis–sodium dodecyl sulfate (CE-SDS) platform, can also provide titer data in addition to size-based quality data (Chen et al. 2008). In addition to capillary-based SDS separations, the ProteomeLab PA 800 Protein Characterization System from Beckman Coulter can perform glycan separations and isoelectric focusing. The GlycoScope platform from ProCognia also provides high-throughput glycoanalysis using lectin arrays. Mass spectrometry for other product quality-related measurements is routinely conducted using extremely small sample volumes.

Complementary analytical systems fall into two major groups, depending on how their results are used. If the results are to be used to inform a feedback control loop within the cell culture system, such as feeding back in a glucose or lactate measurement, then it may be beneficial to automate the complementary analytical system as the schedule for feedback control actions will be dictated by the needs of the cells rather than the user. If the results are not part of a feedback control loop, or are "terminal" analyses, then automation may be advisable from a labor-saving point of view, but is certainly not necessary. In fact, automation may not even be desirable for these assays once the resulting increases in complexity are considered.

The small working volumes of these microreactor platforms also present a challenge for downstream purification process development. Several technologies are currently available to enable process development at these scales and throughputs. GE Healthcare offers PreDictor 96-well plates prefilled with chromatography media. PhyNexus provides PhyTip chromatography columns that can be packed with a variety of resins and operated manually or with automated liquid handlers. AssayMAP cartridges from BioSystem Development can be operated as spin columns or interfaced with liquid handlers for high-throughput purification process development. ATOLL RoboColumns have been interfaced with Tecan Freedom EVO liquid handlers for high-throughput separations.

Regardless of the choices made regarding which complementary off-line/at-line measurements to include, and the appropriate degree of automation, data sets from off-line and on-line measurements must be brought together in the context of the original factorial design if the linkages between process parameters and process results are to be analyzed systematically. Furthermore, the cell culture data should not be confined to an "island" with respect to the other activities in process development, and so the system must be designed to readily support bidirectional communication with other instruments, analysis packages, or laboratory information management systems.

Figure 12.9 depicts the data management system developed to support multifactorial cell culture experimentation. The system is constructed with a four-tier architecture and features a web services application programming interface (API) to enable bidirectional communication with the outside world. The system can also incorporate data obtained from other cell culture systems such as flasks and bioreactors.

Because all the data sets are linked in context of their respective factorial designs, the database becomes continually richer and more valuable as more experiments are

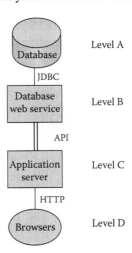

FIGURE 12.9 The data management system allows any user with a browser to view multivariate data sets in the context of the original factorial design. Bidirectional communication is supported with a web services API.

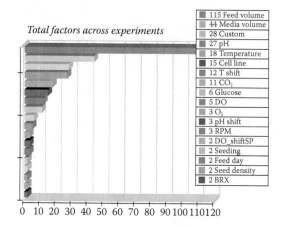

FIGURE 12.10 Factors from all experiments in the database ranked by frequency of impact. A similar analysis could be used to examine factors most often impacting a particular quality attribute or to perform sensitivity studies.

completed. At present, the database at Seahorse Bioscience contains results from over 20 different collaborations and 10,000 individual cell cultures. The database thus presents exciting opportunities for analyses particularly when combined with information from other databases or off-line analyses in the event that the primary database is not all-encompassing. Figure 12.10 shows a trivial example of retrospective analysis, showing the frequency with which certain factors were found to impact integrated viable cell count.

12.6 APPLICATION AND NEXT STEPS

With the advent of a high-throughput process development capability, opportunities to develop new process design paradigms appear. One of the most exciting possibilities is to use the new experimental bandwidth to execute experiments in parallel that have historically been done serially. Process development today frequently features a workflow similar to that shown in Figure 12.11, with clones screened against a baseline process and a limited set of media formulations in flasks, and then a reduced set of clones and media formulations is moved forward for further experimentation in bioreactors.

High-throughput experimentation would permit a "dynamic" clone ranking strategy, in which each clone could be subjected to an identical factorial design. These factorial designs could be 10- to 20-fold larger than those in use today, allowing more variables to be considered and more interactions to be resolved. This is useful not only for revealing the true optimum but also for providing much greater information about process sensitivities. This information could then be used to develop dense factorial designs to map responses, inform failure mode effect analysis studies, and indicate the likely boundaries of the design space for confirmation in bioreactors and positioning later efforts for success. The use of a microbioreactor model rather than flasks for early process development work might also allow PAT to be deployed at an

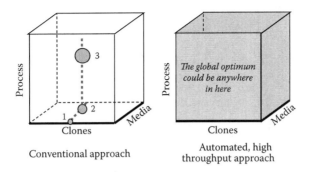

FIGURE 12.11 Serial approach for process development requires less experimental bandwidth but lengthens process development timelines and can miss interactions between parameters.

earlier stage than is common today. Clearly, process design decisions such as clone choice, media choice, and process operating ranges represent key opportunities for designing quality into a bioproduction process, and application of high-throughput technologies to support these decisions could be a significant step toward achieving the goals of the FDA's QbD initiative in biomanufacturing.

Ultimately, a process development effort that begins a cycle of continuous improvement can be envisioned. Process development could begin with dynamic clone ranking and screening experiments to determine the most important variables and interactions for the most promising cell lines. Process design decisions would be supported by these data and confirmed in validated scale-down models. Later, this information could also be used to develop model-based process control strategies and guide deployment of PAT at larger scales. Lastly, through integration with companion analytics and informatics, a continually improving body of knowledge is developed and shared throughout the organization, informing a "kaizen" effort toward continuous quality improvement and reduction of variability.

ACKNOWLEDGMENTS

The authors thank the following for their contributions: Benjamin Alexander, Brian Benoit, Scott Miller, Mohamed Shaheen, Fan Zhang, Tim Johnson, Bernardo Aumond, Rachel Legmann, and Andrey Zarur for their contributions to developing the SimCell system; Dr. Bahram Fathollahi of Caliper Life Sciences for his kind assistance on microchip CE-SDS techniques; and Pfizer for providing the cell line and the associated feeds used in some parts of this work.

REFERENCES

Anderson, D.C. and D. Reilly. 2004. Production technologies for monoclonal antibodies and their fragments. *Curr. Opin. Biotechnol.* 15: 456–462.
Betts, J.I. and F. Baganz. 2006. Miniature bioreactors: Current practices and future opportunities. *Microb. Cell Fact.* 5 (25): 5–21.

Butler, M. 2005. Animal cell cultures: Recent achievements and perspectives in the production of biopharmaceuticals. *Appl. Microbiol. Biotechnol.* 68 (3): 283–291.

Byrne, B., G.G. Donohoe, and R. O'Kennedy. 2007. Sialic acids: Carbohydrate moieties that influence the biological and physical properties of biopharmaceutical proteins and living cells. *Drug Discovery Today* 12: 319–326.

Chen, A., R. Chitta, D. Chang, and A. Amanullah. 2009. Twenty-four well-plate miniature bioreactor system as a scale-down model for cell culture process development. *Biotechnol. Bioeng.* 102: 148–160.

Chen, X., K. Tang, M. Lee, and G.C. Flynn. 2008. Microchip assays for screening monoclonal antibody product quality. *Electrophoresis* 29: 4993–5002.

Chun, C., K. Heineken, D. Szeto, T. Ryll, S. Chamow, and J.D. Chung. 2003. Application of factorial design to accelerate identification of CHO growth factor requirements. *Biotechnol. Prog.* 19: 52–57.

Cogdill, R.P. and J.K. Drennen. 2008. Risk-based quality by design (QbD): A Taguchi perspective on the assessment of product quality, and the quantitative linkage of drug product parameters and clinical performance. *J. Pharm. Innov.* 3: 23–29.

DeJesus, M.J., P. Girard, M. Bourgeois, G. Baumgartner, B. Jacko, H. Amstutz, and F.M. Wurm. 2004. TubeSpin satellites: A fast track approach for process development with animal cells using shaking technology. *Biochem. Eng. J.* 17: 217–223.

Dowd, J.E., I. Weber, R. Beatriz, J.M. Piret, and K.E. Kwok. 1999. Predictive control of hollow fiber bioreactors for the production of monoclonal antibodies. *Biotechnol. Bioeng.* 63 (4): 484–492.

Farid, S.S. 2007. Process economics of industrial monoclonal antibody manufacture. *J. Chromatogr., B: Anal. Technol. Biomed. Life Sci.* 848: 8–18.

Harms, J., X. Wang, T. Kim, X. Yang, and A.S. Rathore. 2008. Defining process design space for biotech products: Case study of *Pichia pastoris* fermentation. *Biotechnol. Prog.* 24: 655–662.

Isett, K., H. George, W. Herber, and A. Amanullah. 2007. Twenty-four well-plate miniature bioreactor high-throughput system: Assessment for microbial cultivations. *Biotechnol. Bioeng.* 98 (5): 1017–1028.

Jain, E. and A. Kumar. 2008. Upstream processes in antibody production: Evaluation of critical parameters. *Biotechnol. Adv.* 26 (1): 46–72.

Kunkel, J.P., D.C. Jan, M. Butler, and J.C. Jamieson. 2000. Comparisons of the glycosylation of a monoclonal antibody produced under nominally identical cell culture conditions in two different bioreactors. *Biotechnol. Prog.* 16: 462–470.

Legmann, R., H.B. Schreyer, R.G. Combs, E.L. McCormick, P.R. Russo, and S.T. Rodgers. 2009. A predictive high throughput scale down model of monoclonal antibody production in CHO cells. *Biotechnol. Bioeng.* 104 (6): 1107–1120.

Marques, M.P., J.M. Cabral, and P. Fernandes. 2009. High throughput in biotechnology: From shake-flasks to fully instrumented microfermentors. *Recent Pat. Biotechnol.* 3 (2): 124–140.

Micheletti, M. and G.J. Lye. 2006. Microscale bioprocess optimization. *Curr. Opin. Biotechnol.* 17 (6): 611–618.

Montgomery, D.C. 2009. *Design and Analysis of Experiments*, 7th edition, p. 656. Hoboken, NJ: John Wiley & Sons.

Moran, E.B., S.T. McGowan, J.M. McGuire, J.E. Frankland, I.A. Oyebade, W. Waller, L.C. Archer, L.O. Morris, J. Pandya, S.R. Nathan, L. Smith, M.L. Cadette, and J.T. Michalowski. 2000. A systematic approach to the validation of process control parameters for monoclonal antibody production in fed-batch culture of a murine myeloma. *Biotechnol. Bioeng.* 69: 242–255.

Okonkowski, J., U. Balasubramanian, C. Seamans, S. Fries, J. Zhang, P. Salmon, D. Robinson, and M. Chartrain. 2007. Cholesterol delivery to NS0 cells: Challenges and solutions in linear low density polyethylene-based bioreactors. *J. Biosci. Bioeng.* 103 (1): 50–59.

Puskeiler, R., K. Kaufmann, and D. Weuster-Botz. 2005. Development, parallelization, and automation of a gas-inducing milliliter-scale bioreactor for high-throughput bioprocess design (HTBD). *Biotechnol. Bioeng.* 89: 512–523.

Rao, G., A. Moreira, and K. Brorson. 2009. Disposable bioprocessing: The future has arrived. *Biotechnol. Bioeng.* 102 (2): 348–356.

Reis, N., C.N. Goncalves, A.A. Vicente, and J.A. Teixeira. 2006. Proof-of concept of a novel microbioreactor for fast development of industrial bioprocesses. *Biotechnol. Bioeng.* 95 (4): 744–753.

Swalley, S.E., J.R. Fulghum, and S.P. Chambers. 2006. Screening factors effecting a response in soluble protein expression: Formalized approach using design of experiments. *Anal. Biochem.* 351: 122–127.

Willoughby, N. 2006. Perspective scaling up by thinking small: A perspective on the use of scale-down techniques in process design. *J. Chem. Technol. Biotechnol.* 81: 1849–1851.

Yang, J.D., C. Lu, B. Stasny, J. Henley, W. Guinto, C. Gonzalez, J. Gleason, M. Fung, B. Collopy, M. Benjamino, J. Gangi, M. Hanson, and E. Ille. 2007. Fed-batch bioreactor process scale-up from 3-L to 2,500-L scale for monoclonal antibody production from cell culture. *Biotechnol. Bioeng.* 98 (1): 141–154

Zhang, Z., N. Szita, P. Boccazzi, A.J. Sinskey, and K.F. Jensen. 2006. A well-mixed, polymer-based microbioreactor with integrated optical measurements. *Biotechnol. Bioeng.* 93: 286–296.

13 Real-Time Monitoring and Controlling of Lyophilization Process Parameters through Process Analytical Technology Tools

Feroz Jameel and William J. Kessler

CONTENTS

13.1 INTRODUCTION

As the number of biomolecules for potential therapeutic use is increasing, the need to understand their structural complexities and stabilization is also increasing. The lyophilization process still remains as one of the most preferred stabilization methods relative to other drying technologies for the simple reason that it is a low-temperature process and allows processing of biological solutions that are otherwise susceptible to damage (Pikal 2002; Franks 1990). However, lyophilization is a complex process and can cause in-process and storage instabilities if it is not properly designed. The lyophilization process consists of three phases: (1) freezing, (2) primary drying, and (3) secondary drying. The freezing phase involves the conversion of water into solid ice by exposing the solution to temperatures $\leq -40°C$, followed by primary drying where sublimation of ice is carried out with the application of heat and vacuum. At the end of primary drying, depending on the composition of the formulation, there will still be a significant amount of water left that will be desorbed using elevated temperatures during the secondary drying phase.

With the recent roll-out of the quality-by-design regulatory initiative based on the process analytical technology (PAT) ICH Q8/Q9/Q10 guidelines, it is expected that product and process performance characteristics should be scientifically designed to meet specific objectives, rather than empirically derived from performance of test batches (FDA 2002, 2004; ICH 2005a,b, 2008, 2009). This requires definitions of the target process based on target product profile and critical process parameters (CPPs) based on critical quality attributes (CQAs) and tools to control them.

The design of a target lyophilization process requires in-depth understanding of material science, the multiple processes of lyophilization, the effect of independent/dependent variables, and the challenges associated with manufacturing operations. These challenges stem from the differences in environment (e.g., effect of particle-free environment), differences in load size (scale-related issues), differences in equipment (dryer) design, and time and procedural differences between laboratory-based lyophilization and production. From the commercial manufacturing point of view, the manufacturing process should be short (i.e., economically viable and efficient), operate within the capabilities of the equipment with appropriate safety margins, and efficiently, consistently, and reproducibly utilize plant resources within the established "design space."

Once the initial process design is developed, the next step is the characterization of the process to understand its robustness and to establish the "design space." This is done by using prior-knowledge and risk-based assessment to identify all of the parameters that have the potential to influence process performance and product quality attributes. They typically fall into four categories: (1) freeze-drying process operating parameters (shelf temperature, chamber pressure, ramp rates, and hold-times); (2) formulation parameters (protein concentration, excipients and their concentrations, vial configuration, stopper design, and fill volume); (3) equipment (capabilities and limitations, batch load/size, and scale effects); and (4) component preparation and devices. The next step is to design multivariate experiments supported with the accelerated stability studies to determine the degree of impact each parameter has on the CQAs. The evaluation of which parameter will have impact on CQA may be based on the statistical significance, and the process parameters that significantly impact CQAs are categorized as CPPs. This is the key to process understanding and an expectation by regulators. The CPPs driving the variability of the CQAs must be identified and understood during process characterization, so that they can be measured and controlled in real time during the manufacturing process. Thus, the measurement and control of the critical parameters should be enabled using a broad spectrum of analytical technologies interfaced to production plant control networks and assimilated into standard procedures (Jameel and Khan 2009).

The FDA defines PAT as a system for designing, analyzing, and controlling manufacturing through timely measurements (i.e., during processing) of critical quality and performance attributes of raw and in-process materials and processes, with the goal of ensuring final product quality (FDA 2004). In the absence of PAT, processes are generally designed empirically without a thorough understanding of the relationship between critical product qualities and process parameters. In commercial manufacturing, the value of the product processed in the freeze dryer may exceed several million dollars for each batch. Thus, an approach that does not use PAT places high value-added pharmaceutical product at risk for loss due to unanticipated process deviations and a lack of knowledge of how these deviations may affect product quality. The application of PAT enables science-based process design and continuous process monitoring and control to produce a product with predictable and consistent quality.

13.2 DEPENDENT VARIABLES/CRITICAL PROCESS PARAMETERS OF FREEZE DRYING

13.2.1 DEGREE OF UNDERCOOLING

The degree of undercooling, which is defined as the difference between the equilibrium freezing point and the nucleation temperature at which ice crystals first form in the solution, determines the morphology of ice. This in turn impacts the subsequent process performance and product quality attributes. Ice nucleation is a random process and variability in the design of the vials, contact with the shelf, heat transfer coefficients, and level of particulate matter in the product solution may contribute to variability in the degree of undercooling, which in turn may contribute to the heterogeneity in the product from vial to vial and batch to batch (Rambhatla et al.

2004). Hence, monitoring and control of ice nucleation is essential to the control of the product heterogeneity.

13.2.2 PRODUCT TEMPERATURE

In order for a product to look pharmaceutically elegant without collapse, the dried product must possess the following qualities: low residual moisture content, short reconstitution time, in-process retention of activity, and adequate shelf life. The pharmaceutical product must be dried below the maximum allowable product temperature, which is the collapse temperature for a predominantly amorphous system or eutectic melt for a crystalline system (Pikal and Shah 1990). Hence, product temperature is a critical parameter that needs to be accurately measured, controlled, and monitored during the process to control the product quality.

13.2.3 SUBLIMATION RATE

The ice sublimation rate is another dependent variable that needs to be monitored and maintained below a certain level and is dependent on the water vapor capturing capacity of the freeze-dryer condenser. Control of the sublimation rate through shelf temperature and pressure control will avoid condenser overload and a "choked flow" condition. The choked flow point is defined as when the velocity of water vapor traveling through the duct that connects the chamber to the condenser approaches the Mach one speed of sound limit. Under these conditions, the flow velocity no longer increases and the water vapor pressure within the chamber increases, leading to a loss of pressure control. The rise in chamber pressure leads to increased heat transfer to the product vials, a further increase in the sublimation rate, and a positive feedback runaway condition. Condenser overload occurs when the rate of incoming water vapor is faster than the rate with which the refrigeration system is able to remove heat from rapidly condensing water vapor and yet maintain the condenser coil temperature below the point at which the vapor pressure of ice on the condenser is greater than the chamber pressure. Both the conditions are characterized by a loss of chamber pressure control (Searles 2004). The determination of the ice sublimation rate enables process comparison during scale up and technology transfer.

13.2.4 PRIMARY DRYING ENDPOINT DETERMINATION

The primary drying time is directly related to the ice sublimation rate and is a dependent variable affected by the independent variables such as chamber pressure, shelf temperature, heat transfer coefficient of vials, fill volume, and product resistance. A method that precisely determines when all of the ice within the product vials is sublimed is important not only for maximizing the throughput of a process, but also from the product quality perspective. Advancement of the drying process to the secondary drying phase through an increase in the shelf temperature without the completion of ice sublimation carries the risk of product collapse, degradation in product quality, and the risk of a batch failure and monetary loss.

13.3 PAT FOR FREEZE-DRYING PROCESS MONITORING AND CONTROL

There are numerous commercially available analytical tools that have been used to monitor critical lyophilization process parameters. These tools can be separated into single-vial and batch process monitoring technologies. The batch monitoring techniques have the advantage of providing information related to all of the vials within the dryer. Batch monitoring techniques are also preferable because vials chosen for single-vial monitoring may not be representative of the entire batch, and in some instances, the measurement technique itself may influence the drying profile of the chosen vial. A single PAT tool will not provide all of the process information required for adequate process monitoring and understanding, and the choice of which technique or techniques used is dependent on not only the target process parameters but also the available resources. Independent of the specific tool, there are a number of instrument attributes that are preferable for successful use and acceptance as an on-line monitoring tool, including the following: (1) provides a measurement representative of the entire batch, not a single vial; (2) provides an absolute, quantitative measurement; (3) measurement capability for monitoring the slowest drying vials within a batch (see Table 13.1); (4) compatible with the process procedures and flow, for example, loading and unloading of trays/vials; (5) compatible with cleaning and steam sterilization; (6) does not compromise lyophilizer vacuum or sterility; and (7) scalable for use and integration with laboratory through a production-scale freeze-drying equipment.

In the following section, a brief review of a subset of the available single-vial and batch PAT monitoring techniques is presented. This chapter focuses on those techniques most commonly used and techniques that have the highest potential to significantly enhance monitoring of laboratory- and product-scale drying. A recent peer-reviewed publication by Patel and Pikal (2009) provides a comprehensive review of available process monitoring devices that have been used to measure CPPs. Many PAT techniques are well established and have been used to monitor freeze drying for decades. One of the newest batch monitoring techniques that has been applied to freeze drying is tunable diode laser absorption spectroscopy (TDLAS). A detailed description of this batch monitoring technique is provided. TDLAS may be applied to all scale dryers and it holds great promise for providing information that can be linked to numerous parameters that affect product quality.

13.4 SINGLE-VIAL METHODS

The single-vial methods are not representative of the entire batch and are primarily used to obtain information during laboratory-scale process development. Single-vial methods have limited use for commercial monitoring and control of lyophilization because of their incompatibility with the equipment and/or inability to provide information representative of the entire batch. The following section reviews some of the single methods that are currently available for obtaining process and product information.

TABLE 13.1

Advantages and Disadvantages of Process Analytical Tools for Monitoring and Control of Freeze-Drying Processes

Techniques	Batch Method	Single-Vial Method	Commercially Viable	Product Temperature	Sublimation Rate	Primary Drying Endpoint Indicator
TEMPRIS	No	Yes	Yes	Yes	No	Yes
Thermocouples/RTD	No	Yes	Yes	Yes	No	Yes
Pirani/capacitance differential pressure	Yes	No	Yes	No	No	Yes
MTM	Yes	No	Yes	Yes	Yes	Yes
Dew point	Yes	No	Yes, possible	No	No	Yes
Lyotrack	Yes	No	Yes	No	No	Yes
Residual gas analyzer (mass spectroscopy)	Yes	No	Yes, but cannot be steam sterilized	No	No	Yes
TDLAS	Yes	No	Yes	Yes	Yes	Yes

13.4.1 THERMOCOUPLES

Product temperature is a critical parameter that needs accurate measurement to monitor the process progress and performance and to ensure that product temperature does not exceed the critical collapse temperature. Temperature sensors in the form of thermocouples or resistance temperature detectors (RTDs) of various gauges are manually placed in the vials at selected locations and are used to monitor the product temperature. The precision and accuracy of the probes are important, and the use of 30-gauge thermocouples is recommended (e.g., Omega: 5SRTC-TT-T-30-36 or a similar model with 30 gauge). Product temperature measurements provide a method that precisely determines when all of the ice is sublimed within a vial. This is because when ice sublimation is complete, evaporative cooling ceases and the product temperature approaches the dryer shelf temperature. Thus, product temperature measurements are important for maximizing the throughput of a process and for ensuring product quality because advancing to the secondary drying phase without the completion of ice sublimation carries the risk of product collapse.

The product temperature data obtained through thermocouples can be used as an indicator of the endpoint of primary drying; however, the use of *in situ* temperature probes has numerous drawbacks. First, the vials with probes behave differently than those without probes. The presence of the sensor changes the ice nucleation behavior, resulting in less super-cooling and more rapid freezing. This changes the ice structure, resulting in larger pore sizes within the product matrix, reducing the product resistance to drying, increasing the ice sublimation rate, and resulting in product temperature measurements that are not representative of the entire batch. This behavior is most important in the sterile production environment because the super-cooling bias between vials with and without sensors becomes more significant due to the particle-free environment within sterile manufacturing facilities. Thus, in a production environment, it is misleading to use thermocouple- or RTD-based product temperature profiles as an indicator of end of primary drying without the use of a 10%–15% primary drying time "soak period" to ensure that all vials completed primary drying.

Second, thermocouples are commonly placed in the front row in a production-scale freeze dryer to avoid the risk of contamination. However, the atypical front row vials facing the door of a freeze dryer are exposed to elevated heat transfer, which also increases the product temperature and mass flux. Lastly, thermocouple or RTD probes measure the product temperature at the bottom of the vial rather than at the sublimation interface. It is the temperature at the constantly moving sublimation interface that must be maintained below the collapse temperature to ensure product quality. Thus, vials containing thermocouples are not representative of the overall batch product temperature and the use of thermocouples to determine product temperature in manufacturing-scale freeze dryers is not appropriate. The use of *in situ* temperature probes is limited to laboratory-scale experiments and process development.

13.4.2 Infrared Spectroscopy

Infrared absorption spectroscopy has been used to determine the moisture content in the vial and thereby predict either the sublimation rate and/or end of the primary or secondary drying phase. It is based on using a light beam with a wavelength in the range of 1100–2500 nm to monitor changes in sample reflectance to determine moisture content. The measured values have been shown to be in agreement with other techniques such as Karl Fisher (Lin and Hsu 2002). Although the technique is nondestructive, it requires development of calibration curves, specific to the product, in conjunction with the commonly acceptable method such as Karl Fisher. The measurement technique and calibration factor need to be robust enough to accommodate variations arising from the formulation and manufacturing processes. The use of peak area analysis rather than peak height analysis gives more accurate predictions and agreement with the measured values. Near-infrared (NIR) reflectance spectroscopy is a single-vial technique, not representative of the entire batch and not particularly useful as a PAT for freeze-drying processes because it requires direct visibility of the vials within the freeze dryer. However, it may be used as an on-line technique to determine the moisture content of every vial after the end of a freeze-drying process as part of release testing.

13.4.3 Microbalance Technique

The microbalance monitor uses a gravimetric measurement technique to monitor the drying rate during the freeze-drying process by periodically weighing a single vial within the freeze dryer (Roth et al. 2001). The microbalance instrument is placed on the surface of the dryer with a single vial located within the reach of the balance weighing arm. The microbalance is programmed to lift and weigh the vial at user-defined time intervals. Although this technique may be useful for the determination of heat transfer/drying rate homogeneity across the shelf, its use is limited by its inability to provide representative data that capture the effects of surrounding vials (within the hexagonal array) and the effect of vial location on the dryer shelf (edge vs. center vials). In addition, the application of this technology is limited by barriers associated with integrating it into commercial freeze-drying equipment, including the requirement for compatibility with clean-in-place and sterilize-in-place (CIP/SIP) systems.

13.4.4 Wireless Temperature Remote Interrogation System

Temperature remote interrogation system (TEMPRIS) is a new invasive wireless sensor designed to measure the product temperature in the vial during the process of freeze drying. The novel sensor is powered via passive transponder excitation using an amplitude-modulated electromagnetic signal in the internationally available 2.4 GHz ISM band. Its performance has been evaluated as a function of fill volume and solid content, and the values were found to be in agreement with the values obtained through the use of standard thermocouples and the manometric temperature measurement (MTM) technique described below (Schneid and Gieseler 2008).

Although it has the advantage of being a wireless sensor and the same sensor can be used in laboratory and manufacturing environments, it still does not ease constraints associated with aseptic handling. It requires manual placement, it is a single-vial measurement, and its measurements are not representative of the entire batch as previously described. However, it will be a useful tool for shelf mapping studies and to determine the edge effects of both small- and large-scale freeze dryers.

13.5 BATCH PAT METHODS

Analytical tools that can be used for real-time, at-line, or on-line process monitoring and control, which provide information on the entire product batch, are preferable for use in controlling commercial manufacturing processes. Some of the output data from the controls of the freeze dryer and process parameters such as nitrogen flow measurement at constant pressure, Pirani vs. capacitance gauge data comparison, pressure rise, and partial water vapor measurements can be used in conjunction/complementary with the newly developed PAT devices to determine some CPPs. Some of the batch methods that are available as PAT are discussed below.

13.5.1 PIRANI GAUGE DATA

Pirani gauge pressure measurements are based on the transfer of heat from a hot wire, located inside the sensor, to the surrounding gases. Because the gauge output depends on the thermal conductivity of the gases as well as their pressure, all thermal conductivity gauges provide an indirect, gas-dependent, pressure reading. Traditional Pirani gauges provide useful pressure readings between 10^{-3} and 10 Torr. The thermal conductivity pressure gauges (Pirani pressure gauge) are generally calibrated against air/nitrogen, and because the thermal conductivity of water vapor is roughly 1.5 times that of air or nitrogen, the gauge outputs a higher pressure reading during primary drying when most of the gas within the lyophilizer chamber is water vapor. When the last piece of ice is sublimed and the gas pressure within the lyophilizer chamber is mostly nitrogen, the Pirani gauge reading approaches the actual pressure determined using a capacitance manometer (e.g., MKS Baratron gauge), indicating the end of the primary drying phase. The capacitance manometer is a device consisting of a metal diaphragm, typically Inconel, placed between the two fixed electrodes. One side of the diaphragm is evacuated, and the other side is exposed to the chamber pressure. The deflection of the diaphragm determines the force per unit area providing the absolute pressure in the range of 0–760 Torr with a variability of $<\pm 1$ mTorr.

13.5.2 MANOMETRIC TEMPERATURE MEASUREMENT

MTM is a technique during which the valve between the dryer chamber and the condenser is momentarily closed for 25–30 s and the pressure rise data are collected. The MTM equation (Equation 13.1) is fitted to the pressure rise data through nonlinear regression to determine the vapor pressure of ice at the sublimation temperature and the sum of product and stopper resistance (Milton et al. 1997; Tang et al. 2005, 2006):

$$P(t) = P_{ice} - (P_{ice} - P_0) \exp\left[-\left(\frac{3.461 \cdot N \cdot A \cdot T_s}{V(R_p + R_s)}\right)t\right] \text{Term } 1 + 0.0465 \cdot P_{ice} \cdot \Delta T$$

$$\times \left[1 - 0.811 \exp\left(-\frac{0.114}{L'}t\right)\right] \text{Term } 2 + Xt \cdot \text{Term } 3, \qquad (13.1)$$

where P_{ice} is the vapor pressure of ice at the sublimation interface (an output to be determined); P_0 is the set chamber pressure; N is the total number of sample vials; A is the internal cross-sectional area of the vials; T_s is the set shelf temperature; V is the product chamber volume; $R_p + R_s$ is the area normalized product and stopper resistance (an output to be determined); ΔT is the temperature difference across the frozen layer; L' is the thickness of the ice; and X is a constant (an output to be determined). The MTM pressure rise is due to contributions coming from three sources: firstly, the pressure rise controlled by dry-layer resistance and the ice temperature at the sublimation interface indicated by term 1 in the MTM Equation 13.1; secondly, the pressure rise caused by the temperature rise at the sublimation surface arising from the dissipation of the temperature gradient across the frozen layer indicated by term 2 in the MTM equation; and thirdly, the pressure rise due to the increase in ice temperature by heat transfer from the shelf during MTM operation. Once the vapor pressure of ice is determined using the MTM equation, the product temperature at the sublimation interface is determined using the following pressure–temperature relationship:

$$T = \frac{-6144.96}{\ln(P_{ice}) - 24.01849} \qquad (13.2)$$

Because MTM determines the vapor pressure of ice at the sublimation interface and because at the end of primary drying there is no ice, a sharp drop in the ice vapor pressure will be indicative of the end of primary drying. Additionally, based on the determinations of vapor pressure of ice at the product temperature and the total resistance to mass transfer, one can deduce additional valuable information such as the heat transfer into the product. that is, dQ/dt, ice thickness (L_{ice}), vial heat transfer coefficient (K_v), and sublimation rate (dm/dt).

One of the advantages of the MTM technique is that it gives the average product temperature of the entire batch as opposed to thermocouples that are biased and not representative of the variation across all of the vials within the dryer. During laboratory-scale drying, when the insertion of the thermocouple into the product vial has little effect on the ice morphology (due to the particle loading within the laboratory environment), the agreement between MTM and thermocouple-based measurements is within ±2°C. The other advantage of MTM-based product temperature is that it gives the value at the interface as opposed to thermocouples that read the value at the bottom of the vial, which will be ~2°C warmer than at the interface. One of the limitations of the MTM technique is that the determinations are only accurate up to two-thirds of the primary drying phase. This is due to the small pressure rise near the end of primary drying. There is also a potential for erroneous measurements for products that have low collapse temperatures because the lowest product temperature that can be reliably measured by MTM is −35°C. In addition, products that are predominantly amorphous in nature tend to reabsorb sublimed water vapor during the pressure rise when the valve between the condenser and chamber is closed resulting in erroneous vapor pressure data (Milton et al. 1997).

Pressure rise data have been historically used to approximate the endpoint of the primary drying phase at commercial-scale freeze dryers, but their use for MTM technique requires swift closure of the valve to monitor the pressure rise and meeting this requirement is a challenge within large-scale production dryers. Additionally, because the pressure rise rate is dependent on the product dry layer resistance, the ice sublimation area, and the chamber volume, large volume freeze dryers and small sublimation areas or small number of vials can significantly limit its applicability.

Thus, using MTM, one can obtain valuable information such as product temperature, the sublimation rate, heat transfer coefficient of vials, and the endpoints of the primary and secondary phases of the drying process. This information is useful for improved process understanding, development, technology transfer, and for obtaining critical process data during deviations and control strategies of freeze-drying process.

13.5.3 DEW POINT MONITOR

A dew point sensor can be used as a tool to determine the endpoint of primary or secondary drying phase based on the principle of changes in the capacitance of a thin film of aluminum oxide due to adsorption of water at a given partial pressure (Roy and Pikal 1989). These sensors not only work for aqueous systems but also for mixed systems where the removal of a mixture of organic solvents and water is required. During the primary and secondary drying phases, moisture continues to evolve from product vials and is indicated at the sensor as a steady dew point temperature, usually between −35°C and −65°C. The end of the primary or secondary drying phase is indicated when the dew point drops and all of the product ice has sublimed. The challenge associated with the use of these moisture probes is their inability to survive steam sterilization during SIP operation; however, new sensor models incorporate a method of isolating the probe with a special fixture and include a biological barrier that can be sterilized within the fixture. These improvements may enable increased application of these relatively inexpensive sensors.

13.5.4 GAS PLASMA SPECTROSCOPY (LYOTRACK)

Gas plasma spectroscopy can be used as an on-line monitoring device to measure the gas composition profile within the dryer product chamber and determine the endpoint of both primary and secondary drying phases (Mayeresse et al. 2007). The device consists of a plasma generator and an optical spectrometer. The plasma generator ionizes the gas present in the chamber, whereas the spectrometer analyzes the gas species based on the wavelength-dependent fluorescence emitted by the ionized gas. In addition to being an on-line monitoring device, it offers a high signal-to-noise ratio and corresponding high measurement sensitivity and is compatible with steam sterilization (Hottot et al. 2009). However, its broader applicability is restricted due to its creation of free radicals that can negatively impact the stability of the product through free radical-induced oxidation. This effect is especially important when drying protein products. This problem can be moderated or eliminated by installing the device in the duct that connects the chamber to the condenser instead of having it in the chamber. Because the gas composition profile of the Lyotrack is the same as the pressure profile measured by a Pirani gauge, its added advantage is questionable.

13.5.5 Residual Gas Analyzer (Mass Spectrometer)

Mass spectrometry has also been used as an on-line device to monitor the composition of the gas in the freeze-dryer product chamber to determine drying progress and the primary and secondary drying endpoints (Nail and Johnson 1992). In addition to the determination of the endpoints, it has also been considered for the detection of leaks and ingress of other gases and solvents arising from vacuum pump oils, heat transfer fluid, and solvents used for cleaning. It consists of a quadrupole mass spectrometer that analyzes the residual gases based on mass-to-charge ratio and quantifies its measurements into partial pressures that can be further correlated with residual water. This enables on-line determination of moisture content. Because the profiles of partial pressure of water obtained by Pirani gauge and the residual gas analyzer are comparable and the inflection points are the same, the use of the more expensive mass spectrometer is questionable as the same information can be obtained through the use of less expensive Pirani gauge.

13.5.6 Tunable Diode Laser Absorption Spectroscopy

Recently, TDLAS has been applied to monitoring the water vapor mass flow rate in the duct connecting the lyophilizer product chamber and the dryer condenser (Gieseler et al. 2007). Using NIR absorption spectroscopy, TDLAS provides direct measurements of the water vapor temperature (K), concentration (molecules/cm^3), and gas flow velocity (m/s) within the duct connecting the lyophilizer chamber and condenser. These measurements are combined with knowledge of the cross-sectional area of the duct to calculate the instantaneous water vapor mass flow rate, dm/dt (g/s). The mass flow rate is integrated as a function of time to provide a continuous determination of total water removed. The mass flow rate may also be combined with freeze-drying heat and mass transfer models and additional process measurements (e.g., product chamber shelf temperature) and process-specific parameters (e.g., vial cross-sectional area and heat transfer coefficients) to determine the batch average product temperature as well as product resistance to drying. In the remainder of this chapter, a detailed description of the TDLAS technique and its application to real-time, continuous lyophilizer monitoring is provided.

13.5.7 TDLAS MEASUREMENTS OF VAPOR MASS FLOW

In TDLAS, sensors rely on well-known spectroscopic principles and sensitive detection techniques to continuously measure the concentrations of selected gases. The quantitative absorption measurement is described by the Beer–Lambert law:

$$I_\nu = I_{o,\nu} \exp[-S(T)g(\nu - \nu_o)N\ell], \tag{13.3}$$

where $I_{o,\nu}$ is the initial laser intensity at frequency ν; I_ν is the intensity recorded after traversing a path length, ℓ, across the measurement volume; $S(T)$ is the temperature-dependent absorption line strength; $g(\nu - \nu_o)$ is the spectral line shape function, and N is the number density of absorbers (the water concentration). The line shape function describes the temperature- and pressure-dependent broadening mechanisms of the fundamental line strength. For the low-pressure conditions present during lyophilization, $g(\nu - \nu_o)$ is described by a Gaussian function. In addition, by scanning the laser frequency across the entire absorption line shape, any pressure dependency of the line shape function is removed from the number density measurement. Scanning the fully resolved absorption line shape also reduces the effect of broadband absorbers in the background gas and nonresonant scattering from any aerosols or particulates in the flow.

The water concentration, $[H_2O]$, in molecules cm^{-3} is calculated using

$$N = \frac{-\int \ln\left[\dfrac{I(\nu)}{I_o(\nu)}\right] d\nu}{S(T)\ell}, \tag{13.4}$$

where $d\nu$ is the laser frequency scan rate per data point (cm^{-1}/point).

The NIR 1.3925 μm water vapor absorption feature that arises from the $3_{03} \rightarrow 2_{02}$ rotational line within the $\nu_3 + \nu_2$ vibrational band was chosen to monitor water vapor due to the availability of robust, fiber-coupled telecommunications-grade diode lasers to probe the transition, its strong absorption line strength, and the relative temperature insensitivity of the transition. The line strength for this transition changes by ~2.7% per 10 K gas temperature change under conditions of interest during lyophilization (Rothman et al. 1994). The water absorption line shape is analyzed to determine the gas temperature and calculate the line strength during drying to correct for temperature fluctuations. A Gaussian line shape profile is given by

$$\phi_D = \frac{2}{\Delta\nu_D} \sqrt{\frac{\ln 2}{\pi}} \exp\left[-4\ln 2 \left(\frac{\nu - \nu_o}{\Delta\nu_D}\right)^2\right], \tag{13.5}$$

where $\Delta\nu_D$ (cm^{-1}) is the Doppler full width at half maximum (FWHM) given by

$$\Delta\nu_D = 7.162 \times 10^{-7} \nu_0 \sqrt{(T/M)}, \tag{13.6}$$

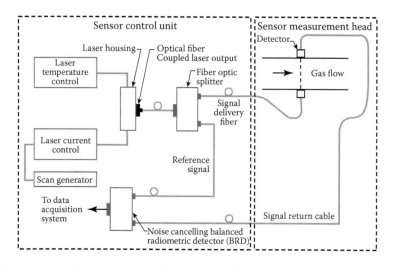

FIGURE 13.1 Schematic diagram for a TDLAS sensor.

where ν_0 is the line center frequency (7181 cm^{-1}), T is the gas temperature (K), and M is the molecular weight (g/mol) of the absorbing species (water vapor).

Figure 13.1 shows a schematic layout of a single-wavelength, NIR laser sensor configuration. This block diagram shows all of the major subcomponents of the sensor that are contained within one compact sensor control electronics unit. The diode laser temperature and injection current are controlled by an integrated diode laser controller. The laser may be temperature-tuned over approximately 20 cm^{-1} (~5 nm in the NIR spectral region) around the design wavelength and rapidly current-tuned over approximately 2 cm^{-1} (approximately 0.45 nm) about its nominal central operating wavelength. While monitoring the lyophilization process, the laser frequency is only tuned ~0.13 cm^{-1} (approximately 0.025 nm) to maximize the measurement sensitivity of the instrument.

To determine mass flow rate of the water vapor, the velocity of the target gas is also required (Miller et al. 1996). The velocity measurement concept is shown in Figure 13.2. The velocity is determined from the Doppler-shifted absorption

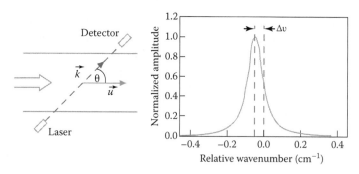

FIGURE 13.2 Schematic diagram showing the Doppler-shifted absorption spectroscopy velocity measurement concept.

spectrum that is shifted in wavelength or frequency with respect to the absorption wavelength of a static gas sample by an amount related to the velocity of the gas, **u**, and the angle, **θ**, between **u** and the probe laser beam propagation vector, **k**. Thus, a simultaneous measurement across the vapor path in the lyophilizer and through a sealed low pressure water vapor absorption cell using the same wavelength tunable laser source may be used to determine the water mass flow rate exiting the lyophilizer product chamber.

Equation 13.7 shows the relationship used to determine the gas flow velocity, u. c is the speed of light (3×10^{10} cm/s), Δv is the peak absorption shift from its zero velocity frequency (or wavelength) in cm^{-1}, v_o (7181 cm^{-1}) is the absorption peak frequency at zero flow velocity, and θ is the angle formed between the laser propagation across the flow and the gas flow vector.

$$u = \frac{c \cdot \Delta v}{v_o \cdot \cos \theta} \tag{13.7}$$

The velocity may also be determined using crossed measurement paths within the flow volume, one directed with the gas flow and one directed against the gas flow. In this configuration, Equation 13.7 is replaced by Equation 13.8 where θ_1 and θ_2 are the measurement angles with respect to the gas flow. We note that in either case, the same laser is used to produce both absorption line shapes and the frequency shift is determined by the shift in data points between the two absorption profiles and converted to absolute frequency using the diode laser frequency scan rate (cm^{-1}/point).

$$u = \frac{c \cdot \Delta v}{v_o \cdot \left(\cos \theta_1 - \cos \theta_2 \right)} \tag{13.8}$$

The mass flow rate (dm/dt, g/s) is calculated by the product of the measured number density (N, molecules cm^{-3}), the gas flow velocity (u, cm/s), and the cross-sectional area of the flow duct (A, cm^2) (and the appropriate conversion factors). This is shown by

$$dm/dt = N \cdot u \cdot A \text{ (g/s)}. \tag{13.9}$$

Figure 13.3 shows water vapor absorption data recorded in the spool of an FTS Lyostar II laboratory-scale freeze dryer during an ice slab sublimation test. The single line of sight TDLAS measurement across the lyophilizer duct was combined with a simultaneous measurement through a ~0.5 Torr reference absorption cell for the determination of the water vapor mass flow rate. The water vapor temperature was calculated from the absorption line shape FWHM, converting data points to frequency using the diode laser frequency scan rate calibration (cm^{-1}/point) and Equation 13.6. The temperature was used to calculate the absorption line strength, $S(T)$, which was used in combination with the integrated peak area, the absorption path length, and the laser frequency scan rate calibration factor to determine the water vapor density. The

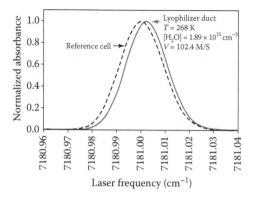

FIGURE 13.3 Sample water vapor absorption line shapes recorded using an NIR TDLAS mass flow rate monitor.

peak shift between the two absorption features was determined in data point units and converted to a frequency shift also using the diode laser calibration. The frequency difference was converted to a velocity using Equation 13.7.

13.5.8 INSTRUMENT REQUIREMENTS

Application of the sensor technology for monitoring water vapor mass flow during lyophilization requires an electronic sensor control unit (SCU) and an optical sensor measurement head (SMH) as indicated in Figure 13.1. Figure 13.4 shows a TDLAS LyoFlux SCU available from Physical Sciences (Andover, MA). The SCU contains an ultra-stable DC power supply, the NIR diode laser and diode laser controller, a pair of balanced ratiometric detection circuits and reference InGaAs photodiode detectors, a sealed, low-pressure reference absorption cell, and signal detector. The SCU is controlled by a computer outfitted with a 1.25-MHz data acquisition system. The SCU is connected to the SMH shown in Figure 13.5, used for measurement application within an FTS Lyostar II laboratory-scale lyophilizer.

FIGURE 13.4 TDLAS LyoFlux 100 mass flow rate SCU.

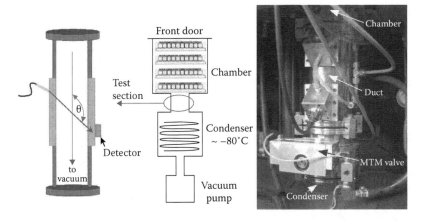

FIGURE 13.5 TDLAS water vapor mass flow rate monitor SMH installed in an FTS Lyostar II laboratory-scale lyophilizer.

13.5.9 SENSOR VALIDATION

Validation of the sensor measurement accuracy has been performed through a series of ice slab sublimation tests. These tests have been previously reported by Gieseler et al. (2007) and Schneid et al. (2009). The ice slab sublimation tests provide a direct comparison between the integrated TDLAS water mass flow rate (dm/dt) and the gravimetrically determined amount of water removed. The tests do not provide direct comparison with the three measurements made by the sensor, the average water vapor temperature, water concentration, and gas flow velocity along the optical line of sight through the lyophilizer duct, but do provide a standard method to evaluate the sensor mass flow rate measurement accuracy.

Sublimation tests within an FTS Lyostar II laboratory-scale dryer and an IMA Life LyoMax 3 pilot-scale dryer were conducted using "bottomless trays" made from stainless steel frames outfitted with thin plastic bags (0.003 cm thickness) attached to the frame to form the tray bottom (Schneid et al. 2006b). The three laboratory trays were placed on the lyophilizer shelves and filled with ~1500 g of pure water, whereas the four pilot shelves were each filled with ~7.5 kg of water, providing an ice thickness of ~1 cm on each dryer shelf. The shelf temperature was lowered to −40°C and held for 1 h to form the ice slabs. Before freezing the ice slab, fine wire thermocouples (Omega, CT) were placed in the middle of the tray water layer to monitor the slab "product" temperature. Following the freezing step, the chamber pressure was reduced using the dryer condenser and vacuum pump to the experimental set point pressure (between 65 and 500 mT). The shelf temperature was then ramped (typically 0.5°C–1°C/min) to the experimental set-point temperature, and approximately 50% of the ice slab product was sublimed under steady-state conditions. Following sublimation, the isolation valve located between the chamber and the condenser and downstream of the TDLAS optical measurement station was closed, ceasing water removal. The TDLAS data collection was simultaneously ended, followed by a ramping of the lyophilizer shelf temperature to melt the remaining ice. The water

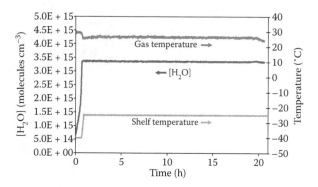

FIGURE 13.6 TDLAS-measured gas temperature, water vapor density (molecules cm^{-3}), and lyophilizer shelf temperature temporal profiles during a typical ice slab sublimation test.

remaining in the bottomless trays was removed and weighed, enabling a gravimetric determination of the total mass of water removed during the sublimation test. The gravimetrically determined mass balance was compared to the integrated TDLAS-determined mass balance to validate the measurement accuracy of the instrument. Alternatively, the average mass flow rates of each measurement technique may be compared. Figure 13.6 displays the TDLAS-measured water vapor concentration and gas temperature and the lyophilizer shelf fluid inlet temperature during a typical sublimation test. Figure 13.7 displays the measured velocity and the calculated water vapor mass flow rate. The data in Figures 13.6 and 13.7 both display the development of steady-state ice sublimation conditions following the initial shelf temperature ramp. Table 13.2 provides a summary of the laboratory-scale dryer measurement conditions and experimental results, including a ratio of TDLAS to gravimetric water sublimation rates. The ratio shows general agreement in mass flow rates with an average error of $<\pm2\%$. The measurement results for the pilot-scale dryer are compiled in Table 13.3 and were not as accurate as for the laboratory scale. This may have been due to the dryer geometry and nonaxisymmetric gas flow within the dryer

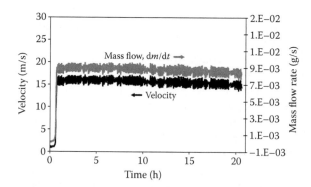

FIGURE 13.7 TDLAS-measured gas flow velocity and calculated mass flow rate, dm/dt (g/s) temporal profiles during a typical ice slab sublimation test.

TABLE 13.2
Summary of Laboratory-Scale Pure Ice Slab Sublimation Tests[a]

P (mTorr)	T_s (°C)	# Trays	Velocity (m/s)	Gravimetric Ave dm/dt (g/h)	TDLAS Ave dm/dt (g/h)	Ratio TDLAS/Grav
100	0	3	110	81.9	79.4	0.97
150	20	3	108	170.9	165.8	0.97
200	40	3	105[a]	191.4	185.7	0.97
500	40	3	51[a]	174.2	189.9	1.09
60	−30	1	15.9	18.5	18.8	1.02
100	−33	1	6.7	18.0	17.3	0.96
100	−27	1	12.4	28.6	28.5	1.00
100	−20	1	17.1	43.5	44.6	1.03
100	−5	3	94.5	200.0	206.5	1.03
100	0	1	29.7	77.3	79.6	1.03
100	40	1	73.1	147.8	152.2	1.03
150	−30	1	4.6	19.6	19.5	1.00
150	20	1	39.2	131.8	140.3	1.07
150	40	1	51.3	164.5	172.8	1.05

Source: Gieseler, H. et al., *J. Pharm. Sci.*, 96 (7), 1776–1793, 2007. Schneid, S. and Gieseler, H., *Eur. Pharm. Rev.*, 6, 18–25, 2009.

[a] Steady-state conditions not maintained due to loss of pressure control.

TABLE 13.3
Summary of Pilot-Scale Pure Ice Slab Sublimation Tests[a]

P (mTorr)	T_s (°C)	# Trays	Velocity (m/s)	Gravimetric Ave dm/dt (g/h)	TDLAS Ave dm/dt (g/h)	Ratio TDLAS/Grav
100	0	4	32	568.1	568.1	1.00
150	20	4	41	1016.7	1077.7	1.06
200	40	4	79	1188.0	1294.9	1.09
500	40	4	24[a]	1505.2	1670.8	1.11

Source: Gieseler, H. et al., *J. Pharm. Sci.*, 96 (7), 1776–1793, 2007.

[a] Steady-state conditions not maintained due to loss of pressure control.

spool. The standard TDLAS data analysis algorithm is based on the assumption of axisymmetric gas flow within the dryer spool.

13.5.10 SENSOR APPLICATIONS

TDLAS sensing technology may be applied to a wide variety of lyophilization monitoring and control needs (Patel and Pikal 2009; Schneid and Gieseler 2009) including lyophilizer operational qualification (OQ) (Patel et al. 2008; Nail and Searles

2008; Hardwick et al. 2008), determination of primary and secondary drying endpoints (Gieseler et al. 2007), vial heat transfer coefficients (Schneid et al. 2006a; Kuu et al. 2009), product temperature (Schneid et al. 2009), product residual moisture (Schneid et al. 2007), and freeze-drying cycle optimization (Kuu and Nail 2009). TDLAS sensing technology stands out from many other PAT tools applied to lyophilization because of to its ability to provide a direct measurement of the gas flow velocity in the duct, which can be used to determine the water vapor mass flow rate. In addition, the technique may be applied to laboratory-, pilot-, and production-scale lyophilizers. No other technology has been demonstrated to nonintrusively provide this measurement capability, which can be linked to not only dryer operation but also CPPs such as product temperature, which is linked to final product quality. It is the combination of continuous, real-time mass flow rates (dm/dt) with established heat and mass transfer models (Pikal 1985; Nail 1980; Rambhatla et al. 2006) that will drive application of the technology.

In the following section, a brief review of a few of the applications listed above will be provided to demonstrate the value of the measurement technology. This is not a comprehensive review nor is it meant to provide an in-depth analysis of any one of the applications. The reader is referred to the cited publications for additional information.

13.5.11 LYOPHILIZER OQ

Pharmaceutical companies are highly motivated by economic and regulatory forces to develop robust product formulations and lyophilization processes that maximize product throughput consistent with maintaining product quality. One aspect of maximizing throughput is the development of efficient drying processes that are consistent with the mass flow rate limitations of both the laboratory-scale process development dryer and the manufacturing-scale dryer that will be used to produce the drug product (Chang and Fisher 1995). The development of a process that can be transferred between lyophilizers requires knowledge of the maximum supported rate of mass transfer between the chamber and the condenser and the relationship between the ice sublimation rate (g/s) and the dryer shelf temperature, chamber pressure, and product temperature (Patel et al. 2007; Nail and Searles 2008).

The development of a family of sublimation rate curves versus chamber pressure and shelf temperature is needed to define the lyophilizer operational limitations. Traditionally, this information was gathered through a series of ice slab sublimation tests with each test providing a single data point at a single shelf temperature and pressure. The gravimetric determination of total water removed would provide the average sublimation rate during one experiment. Thus, a complete family of curves would require numerous experiments and a large investment of time and labor resources.

The TDLAS sensor technology enables the development of the required data set and the determination of choked flow conditions within a few experiments (Patel et al. 2008; Nail and Searles 2008). To accomplish this, ice slabs are formed on the dryer shelves and the chamber pressure is reduced to the set-point value (typically starting with the lowest values). The shelf temperature is then ramped to the lowest value of interest. The TDLAS instrument is used to monitor the sublimation rate (g/s) and to determine the establishment of steady-state drying conditions. During steady-state operation, the TDLAS-measured sublimation rate (dm/dt), the ice slab product temperature (T_b), the shelf temperature (T_s), and the chamber pressure are recorded. The chamber pressure is then changed and the sensor is again used to verify the establishment of steady-state operation (indicating that both the shelf temperature and the product temperature have stabilized). A new set of measurements is then recorded including dm/dt, T_s, T_b, and chamber pressure. After completing measurements over all pressures of interest, the shelf temperature is raised to the next setting of interest and the measurement process is repeated until a complete set of sublimation rates is measured for all pressures and temperatures. This process enables lyophilizer OQ to be completed in days rather than weeks, dramatically saving time and money.

13.5.12 DETERMINATION OF PRIMARY AND SECONDARY DRYING ENDPOINTS

TDLAS sensor technology has been used to monitor drying of numerous "product" drying cycles including mannitol, lactose, trehalose, sucrose, dextran, glycine, polyvinyl pyrrolidone, and bovine serum albumin formulations (Gieseler et al. 2007; Schneid and Gieseler 2009). Figure 13.8 displays a temporal plot of the TDLAS-measured water vapor concentration and the lyophilizer shelf temperature for drying a 5% w/w sucrose solution contained in 336 20-cc tubing vials. The vials were loaded onto three shelves of an FTS LyoStar II laboratory-scale dryer. The dryer pressure was set to 65 mTorr. The shelf temperature profile during primary drying is characterized by two-step changes at the start of primary drying and five-step changes near the end of primary drying to a temperature set point of −14°C. Throughout the primary drying phase of the cycle, the water concentration remained nearly constant. The spikes in concentration data correspond to lyophilizer isolation valve closings performed to enable MTM-based product temperature measurements throughout the drying cycle. Figure 13.9 displays the temporal profiles of the gas flow velocity measurement and the mass flow rate (dm/dt) determination. The mass

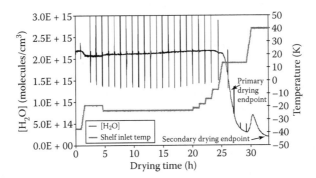

FIGURE 13.8 TDLAS water vapor concentration and lyophilizer shelf temporal measurement profiles during lyophilization of 5% w/w sucrose in a laboratory-scale dryer.

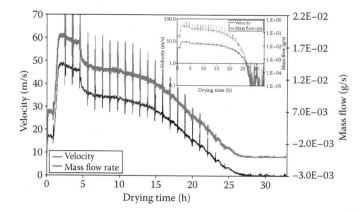

FIGURE 13.9 TDLAS vapor flow velocity and water mass flow rate temporal measurement profiles during lyophilization of 5% w/w sucrose in a laboratory-scale dryer.

flow profile closely follows the velocity profile throughout primary drying, including the obvious response to the shelf temperature adjustments during the early portion of primary drying. Careful inspection of the velocity and mass flow profiles also reveal response to the shelf temperature changes near the end of primary drying.

During secondary drying, the dryer shelf temperature was ramped to 40°C and the TDLAS sensor recorded a peak in the water concentration profile corresponding to the removal of bound water within the product cake. The logarithmic velocity and mass flow rate temporal plot inset within Figure 13.9 shows that there is a corresponding peak in the velocity and mass flow rate curves, both near the measurement sensitivity limit of the instrument. The primary and secondary drying endpoints are clearly indicated by the TDLAS water concentration measurements and mass flow determinations.

13.5.13 DETERMINATION OF VIAL HEAT TRANSFER COEFFICIENTS AND PRODUCT TEMPERATURE

During lyophilization, product temperature history is the most important characteristic of the pharmaceutical product, but its measurement has been problematic. The standard laboratory methodology has involved placing temperature sensors, usually thermocouples, directly in the product in a few selected vials. Product temperature determined by thermocouples represents the temperature at the bottom center of a vial, but does not directly measure the temperature of the product at the continuously moving sublimation interface. The temperature at the sublimation interface, however, governs the product quality. If the product temperature at the interface exceeds a critical temperature for the matrix, the product will undergo collapse, compromising product quality. Product temperature during drying directly affects cake appearance, residual moisture content, reconstitution time, and it may affect product stability and shelf life. Thus, the development of a widely applicable, robust measurement solution is an important industry goal, and temperature measurements during a process abnormality may prevent the loss of millions of dollars worth of product.

The previously described MTM pressure rise technique has been used to provide batch average product temperature during the first two-thirds of primary drying. Due to the requirement of a quick-closing isolation valve, this technique is generally only applied to laboratory-scale lyophilizers and does not provide a solution for production-scale temperature monitoring. In contrast, the TDLAS-based measurement technique described below may provide the needed measurement capability for all dryers.

Recently, it has been demonstrated that TDLAS-based mass flow rate measurements (dm/dt) may be combined with a steady-state heat and mass transfer model (Pikal 1985; Nail 1980; Rambhatla et al. 2006; Patel et al. 2007; Milton et al. 1997) to provide continuous, real-time determinations of batch average product temperature in a laboratory-scale dryer (Schneid et al. 2007). Due to the wide applicability of the TDLAS sensor technology, it is anticipated that this approach may also be applied to pilot- and production-scale dryers, providing a nonintrusive measurement solution that may be applied from process development through production.

As previously described, heat transfer during vial-based lyophilization can be described in terms of thermal barriers and temperature gradients. Heat is supplied to the frozen product from the drying chamber shelves through the bottom of the glass vials to compensate for the heat removed by sublimation. Heat flow from the shelves to the product is described by

$$dQ/dt = A_v \cdot K_v \cdot (T_S - T_b), \tag{13.10}$$

where dQ/dt is the heat flow (cal/s or J/s) from the shelves to the product; A_v is the cross-sectional area of the vial calculated from the vial outer diameter; K_v is the vial heat transfer coefficient (for a specific vial type at a specific pressure); T_s is the temperature of the shelf surface; and T_b is the temperature of the frozen product at the bottom center of the vial.

In steady state, the heat flow (dQ/dt) can be related to mass flow (dm/dt) by using the heat of ice sublimation, ΔH_s:

$$dQ/dt = \Delta H_s \cdot dm/dt, \tag{13.11}$$

where ΔH_S is given in the literature (650 cal/g) (Pikal et al. 1984). Equations 13.10 and 13.11 can be combined and rearranged to provide the product temperature in the bottom of the vial shown by

$$T_b = T_S - \left[\frac{\left(\Delta H_S \cdot (dm/dt) \right)}{A_v \cdot K_v} \right]. \tag{13.12}$$

In the laboratory, K_v can be separately determined using Equation 13.13 and by performing sublimation tests with pure water filled into vials instead of product.

$$K_v = \frac{dm/dt \cdot H_3}{A_v \cdot (T_s - T_p)} \tag{13.13}$$

Here, the average temperature difference, $(T_s - T_b)$, can be determined using thermo-couples in selected vials (bottom center) as well as adhesive thermocouples on the shelf surface during the experiments. Note that in the laboratory, temperature bias between vials containing thermocouples and vials not containing thermocouples is usually very small due to particulate contamination in the product fluid used to fill the vials. A_v is easily determined by measurement.

Mass flow can be determined either gravimetrically from the known initial mass of water and the remaining mass of water after a predefined time interval in primary drying (Tang et al. 2005) or by the TDLAS sensor. Figure 13.10 displays experimentally determined K_v from both gravimetric and TDLAS mass flow rate determinations as a function of chamber pressure (Schneid et al. 2007; Schneid et al. 2006a). As anticipated, increasing chamber pressure results in larger K_v values as the contribution from gas conduction to the vial heat transfer coefficient dominates over the shelf conduction and radiative heat transfer contributions. The TDLAS data may be used as either an integrated determination of average mass flow rate, TDLAS Ave (which may be directly compared to the gravimetrically determined value), or by using the instantaneous TDLAS dm/dt measurement during the steady-state sublimation, TDLAS Steady State. Figure 13.10 shows agreement between the K_v determination methods at low pressures and increasing differences at higher pressures. This is likely due to decreasing vapor flow velocities at higher pressures and limited instrument measurement sensitivity. This experiment used a single shelf of vials during the sublimation test. Alternative experiments using a full load of vials would result in approximately three times the velocity and mass flow rate and improved agreement throughout the pressure range of interest.

Following the determination of the weighted average vial heat transfer coefficient, experiments were performed with product-filled vials to demonstrate the use of TDLAS dm/dt measurements for the determination of batch average product

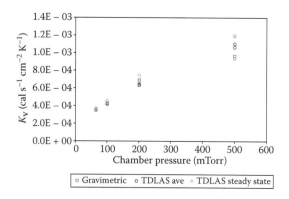

FIGURE 13.10 Gravimetric and TDLAS-based determinations of vial heat transfer coefficient, K_v, as a function of the laboratory-scale lyophilizer chamber pressure.

temperature (Schneid et al. 2007). The dm/dt measurements were combined with thermocouple-based shelf temperature measurements, the vial cross-sectional area, and the water heat of sublimation to determine the batch average product temperature using Equation 13.12. The TDLAS-determined bottom center temperature was compared to thermocouple-based product temperature measurements to assess the accuracy of the measurement technique. Experiments were performed using sucrose, glycine, and mannitol product formulations with the results of the 10% glycine drying primary drying experiment shown in Figure 13.11. The plot shows a clear difference between the center vial and edge vial thermocouple-based temperature measurements, with the edge vial product temperatures higher than the center vials due to radiative heat loading from the warm dryer walls and door. The TDLAS-determined batch average product temperature is initially biased to the center vial thermocouple-based measurements during early primary drying and then providing an average determination between edge and center vials in the later stages of primary drying. In addition to the TDLAS and thermocouple-based temperature measurements, the batch average product temperature was also determined using the MTM technique. Figure 13.11 shows that the MTM and TDLAS techniques agree very well during approximately the first half of primary drying before the MTM measurement techniques fails due to insufficient pressure rise.

Additional analysis may enable the determination of the product temperature at the sublimation interface, T_p, by using

$$T_p = T_b - \left[\frac{dQ/dt \cdot L_{ice}}{(A_v \cdot 20.52)} \right], \tag{13.14}$$

where dQ/dt is the heat flow, L_{ice} is the ice thickness, and A_v is the cross-sectional area of the vial. The value 20.52 in Equation 13.14 represents the thermal conductivity of ice (cal/h cm^2 K). L_{ice} may be instantaneously calculated from TDLAS mass flow rate measurements and the knowledge of the initial fill depth (Tang et al. 2005). This procedure has not yet been verified through experimental investigations.

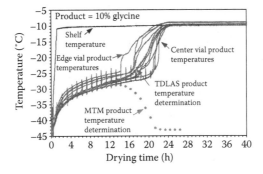

FIGURE 13.11 Product temperature temporal profile during 10% glycine primary drying as determined using thermocouples, TDLAS, and MTM measurement techniques.

13.5.14 TDLAS Summary

The TDLAS technique enables continuous, real-time, nonintrusive measurements of gas temperature, water concentration, and gas flow velocity based on fundamental principles of absorption spectroscopy. These measurements are combined with computational fluid dynamic modeling of the gas flow within the lyophilizer to interpret the line of sight measurement data and provide the mass flow or sublimation rate of water throughout both the primary and secondary drying phases of lyophilization. A number of measurement applications have been reviewed, including lyophilizer OQ, determination of primary and secondary drying endpoints, the determination of vial heat transfer coefficients, and finally the real-time, nonintrusive determination of batch average product temperature. These final two applications combine the TDLAS sublimation rate measurements with a steady-state drying heat and mass transfer model to provide temperature information that may be directly linked to product quality. Finally, we note that to date, the measurement technique has focused on monitoring water vapor during drying. TDLAS sensor technology may also be applied to additional solvents such as acetonitrile, provided that the solvent possesses narrow band spectral absorption features that may be probed using narrow spectral width, wavelength tunable semiconductor diode lasers. TDLAS measurement technology is a powerful tool for monitoring and future control of lyophilization processes and for linking measurements to product quality.

13.6 PAT SUMMARY

This chapter has provided a description of the motivation for the development and application of real-time, on-line process monitoring techniques for pharmaceutical lyophilization. A high level survey of both batch average and single-vial measurement techniques has been provided and is summarized in Table 13.1. Many of these techniques have been used for decades, but a few new measurement techniques have been recently introduced, including gas plasma spectroscopy (Lyotrack) and TDLAS (LyoFlux). A detailed description of the TDLAS technique was provided, including a description of the spectroscopy that underpins the development and application of the measurement technique. TDLAS measurements have been combined with established heat and mass transfer models to provide determinations of important in-process product parameters such as product temperature. Additional experimentation is required to demonstrate the application of this technique and other new techniques within manufacturing-scale dryers, allowing the use of PAT tools from laboratory-scale process development through scale-up to routine monitoring and control of product manufacturing. Continued application of PAT tools will result in improved product quality and should ultimately lead to improved efficiency and lower production costs.

ACKNOWLEDGMENTS

Partial financial support for the development and application of the TDLAS mass flow rate sensor to monitoring pharmaceutical lyophilization was provided by the

National Science Foundation and the National Institutes of Health, National Cancer Institute Small Business Innovative Research programs. The TDLAS sensor development and applications described within this chapter resulted from the dedicated work of numerous individuals at Physical Sciences Inc.: G.E. Caledonia, M.L. Finson, J.F. Cronin, P.A. Mulhall, S.J. Davis, D. Paulsen, J.C. Magill, K.L. Galbally-Kinney, J.A. Polex, T. Ustun, A.H. Patel, D. Vu, A. Hicks, and M. Clark; The University of Connecticut: M.J. Pikal, H. Gieseler, S. Schneid, S.M. Patel, and S. Luthra; and IMA Edwards: A. Schaepman, D.J. Debo, V. Bons, F. DeMarco, and F. Jansen.

REFERENCES

Chang, B.S. and N.L. Fisher. 1995. Development of an efficient single-step freeze-drying cycle for protein formulations. *Pharm. Res.* 12: 831–837.

FDA, Aug. 21, 2002. *Pharmaceutical CGMPs for the 21st Century: A Risk-Based Approach.* Rockville, MD: FDA.

FDA. 2004. *Guidance for Industry: PAT—A Framework for Innovative Pharmaceutical Development, Manufacturing, and Quality Assurance*, September 2004. Available online at http://www.fda.gov/downloads/Drugs/GuidanceComplianceRegulatoryInformation/Guidances/UCM070305.pdf. Accessed on March 10, 2011.

Franks, F. 1990. Freeze-drying: From empiricism to predictability. *CryoLetters* 11: 93–110.

Gieseler, H., W.J. Kessler, M.F. Finson et al. 2007. Evaluation of tunable diode laser absorption spectroscopy for in-process water vapor mass flux measurements during freeze-drying. *J. Pharm. Sci.* 96 (7): 1776–1793.

Hardwick, L.M., C. Paunicka, and M.J. Akers. 2008. Critical factors in the design and optimisation of lyophilisation processes. *Innovations Pharm. Technol.* 26: 70–74.

Hottot, A., J. Andrieu, V. Hoang, E.Y. Shalaev, L.A. Gatlin, and S. Ricketts. 2009. Experimental study and modeling of freeze-drying in syringe configuration. Part II: Mass and heat transfer parameters and sublimation end-points. *Drying Technol.* 27: 49–58.

ICH. 2009. Q8(R2): Pharmaceutical development. International Conference on Harmonization of Technical Requirements for the Registration of Pharmaceuticals for Human Use, Geneva, Switzerland.

ICH. 2005a. Q8: Pharmaceutical development, Step 4, Nov. 10, 2005. International Conference on Harmonization of Technical Requirements for the Registration of Pharmaceuticals for Human Use, Geneva, Switzerland.

ICH. 2005b. Q9: Quality risk management. International Conference on Harmonization of Technical Requirements for the Registration of Pharmaceuticals for Human Use, Geneva, Switzerland.

ICH. 2008. Q10: Pharmaceutical quality system. International Conference on Harmonization of Technical Requirements for the Registration of Pharmaceuticals for Human Use, Geneva, Switzerland.

Jameel, F. and M.A. Khan. 2009. Quality by design as applied to the development and manufacturing of a lyophilized protein product. *Am. Pharm. Rev.* (Nov/Dec Issue): 20–24.

Kuu, W.Y. and S.L. Nail. 2009. Rapid freeze-drying cycle optimization using computer programs developed based on heat and mass transfer models and facilitated by tunable diode laser absorption spectroscopy (TDLAS). *J. Pharm. Sci.* 98 (9): 3469–3482.

Kuu, W.Y., S.L. Nail, and G. Sacha. 2009. Rapid determination of vial heat transfer parameters using tunable diode laser absorption spectroscopy (TDLAS) in response to step-changes in pressure set-point during freeze-drying. *J. Pharm. Sci.* 98 (3): 1136–1154.

Lin, T.P. and C. Hsu. 2002. Determination of residual moisture in lyophilized protein pharmaceuticals using a rapid and non-invasive method: Near infrared spectroscopy. *PDA J. Pharm. Sci. Technol.* 56: 196–205.

Mayeresse, Y.V.R., P.H. Sibille, and C. Nomine. 2007. Freeze-drying process monitoring using a cold plasma ionization device. *PDA J. Pharm. Sci. Technol.* 61 (3): 160–174.

Miller, M.F., W.J. Kessler, and M.G. Allen. 1996. Diode laser-based air mass flux sensor for subsonic aeropropulsion inlets. *Appl. Opt.* 35 (24): 4905–4912.

Milton, N., M.J. Pikal, M.L. Roy, and S.L. Nail. 1997. Evaluation of manometric temperature measurement as a method of monitoring product temperature during lyophilization. *PDA J. Pharm. Sci. Technol.* 51: 7–16.

Nail, S.L. 1980. The effect of chamber pressure on heat transfer in the freeze-drying of parenteral solutions. *J. Parenter. Drug Assoc.* 34: 358–368.

Nail, S.L. and W. Johnson. 1992. Methodology for in-process determination of residual water in freeze-dried products. *Dev. Biol. Stand.* 74: 137–151.

Nail, S.L. and J.A. Searles. 2008. Elements of quality by design in development and scale-up of freeze-dried parenterals. *Int. Biopharm.* 21 (1): 44–52.

Patel, S.M., S. Chaudhuri, and M.J. Pikal. 2008. Choked flow and importance of Mach I in freeze-drying process design. Freeze Drying of Pharmaceuticals and Biologicals Conference, Breckenridge, CO.

Patel, S.M., T. Doen, S. Schneid, and M.J. Pikal. 2007. Determination of end point of primary drying in freeze-drying process control. Proceedings of AAPS Annual Meeting and Exposition, San Diego, CA, Nov. 11–15.

Patel, S.M. and M. Pikal. 2009. Process analytical technologies (PAT) in freeze-drying of parenteral products. *Pharm. Dev. Technol.* 14 (6): 567–587.

Pikal, M. 2002. Lyophilization. In *Encyclopedia of Pharmaceutical Technology*, edited by J. Swarbrick and J.C. Boylan, Volume 6, pp.1299–1326. New York: Marcel Dekker.

Pikal, M.J. 1985. Use of laboratory data in freeze drying process design: Heat and mass transfer coefficients and the computer simulation of freeze drying. *J. Parenter. Sci. Technol.* 39: 115–138.

Pikal, M.J., M.L. Roy, and S. Shah. 1984. Mass and heat transfer in vial freeze-drying of pharmaceuticals: Role of the vial. *J. Pharm. Sci.* 73: 1224–1237.

Pikal, M.J. and S. Shah. 1990. The collapse temperature in freeze drying: Dependence on measurement methodology and rate of water removal from the glassy phase. *Int. J. Pharm.* 62: 165–186.

Rambhatla, S., R. Ramot, C. Bhugra, and M.J. Pikal. 2004. Heat and mass transfer scale-up issues during freeze drying. II: Control and characterization of the degree of supercooling. *AAPS PharmSciTech* 5 (4): Article 58.

Rambhatla, S., S. Tchessalov, and M.J. Pikal. 2006. Heat and mass transfer scale up issues during freeze-drying. III: Control and characterization of dryer differences via operational qualification (OQ) tests. *AAPS PharmSciTech* 7 (2), Article 39.

Roth, C., G. Winter, and G. Lee. 2001. Continuous measurement of drying rate of crystalline and amorphous systems during freeze drying using an in situ microbalance technique. *J. Pharm. Sci.* 90: 1345–1355.

Rothman, L.S., R.R. Gamache, A. Goldman et al. 1994. The HITRAN database: 1986 edition. *Appl. Opt.* 33 (21): 4851–4867.

Roy, M.L. and M.J. Pikal. 1989. Process control in freeze drying: Determination of the end point of sublimation drying by an electronic moisture sensor. *J. Parenter. Sci. Technol.* 43: 60–66.

Schneid, S. and H. Gieseler. 2008. Evaluation of a new wireless temperature remote interrogation system (TEMPRIS) to measure product temperature during freeze drying. *AAPS PharmSciTech* 9: 729–739.

Schneid, S. and H. Gieseler. 2009. Process analytical technology (PAT) in freeze drying: Tunable diode laser absorption spectroscopy as an evolving tool for cycle monitoring. *Eur. Pharm. Rev.* 6: 18–25.

Schneid, S., H. Gieseler, W. Kessler, and M.J. Pikal. 2006a. Position dependent vial heat transfer coefficient: A comparison of tunable diode laser absorption spectroscopy and gravimetric measurements. Proceedings of CPPR Freeze-Drying of Pharmaceuticals and Biologicals, Garmisch-Partenkirchen, Germany, Oct. 3–6.

Schneid, S., H. Gieseler, W. Kessler, and M.J. Pikal. 2006b. Process analytical technology in freeze drying: Accuracy of mass balance determination using tunable diode laser absorption spectroscopy (TDLAS). Proceedings of AAPS Annual Meeting and Exposition, San Antonio, Texas, Oct. 29–Nov. 2.

Schneid, S., H. Gieseler, W. Kessler, and M.J. Pikal. 2007. Tunable diode laser absorption spectroscopy (TDLAS) as a residual moisture monitor for the secondary drying stage of freeze drying. AAPS Annual Meeting, San Diego, CA.

Schneid, S.C., H. Gieseler, W.J. Kessler, and M.J. Pikal. 2009. Non-invasive product temperature determination during primary drying using tunable diode laser absorption spectroscopy. *J. Pharm. Sci.* 98 (9): 3401–3418.

Searles, J.A. 2004. Observation and implications of sonic water vapor flow during freeze drying. *Am. Pharm. Rev.* 7 (2): 58.

Tang, X., S.L. Nail, and M.J. Pikal. 2005. Freeze-drying process design by manometric temperature measurement: Design of a smart freeze dryer. *Pharm. Res.* 22 (4): 685–700.

Tang, X., S.L. Nail, and M.J. Pikal. 2006. Evaluation of manometric temperature measurement: A process analytical technology tool for freeze-drying. Part 1: Product temperature measurement. *AAPS PharmSciTech* 7: E14.

14 Process Analytical Technology's Role in Operational Excellence

Fetanet Ceylan Erzen and Manbir Sodhi

CONTENTS

14.1 INTRODUCTION

As competition and cost pressures increase, pharmaceutical and biotechnology industries are placing an increased emphasis on operational excellence, which involves continuously improving a business, by increasing plant efficiencies and by cost reduction. Operational excellence focuses on both the processes and the performance, and this helps a company gain competitive advantage in the marketplace, enhances quality, and minimizes waste.

Companies adopt a variety of methodologies in pursuit of operational excellence. These include Lean, Total Quality Management (TQM), Six Sigma, or combinations of these. Lean methodologies focus primarily on waste reduction; Six Sigma directs attention toward reducing variability; and TQM reorients the companies' energy toward satisfying customer needs. Recently, process analytical technology (PAT) has been advocated as a methodology by which process and product specifications can be met or exceeded without adverse impact on productivity and minimizing waste by a quick response to unfavorable production circumstances. This chapter discusses how the implementation of PAT may result in a synergistic effect on the enhancement of operational excellence strategies used in the bio/pharmaceutical companies while also satisfying regulatory expectations. Instead of detailing the commonly used methodologies for enhancing performance (such as Lean), this chapter reviews PAT's role in the journey toward operational excellence in a biopharmaceutical manufacturing environment.

14.2 OPERATIONAL EXCELLENCE AND PAT

Operational excellence has been defined in various ways. Wikipedia states: "Operational Excellence is a philosophy of leadership, teamwork and problem solving resulting in continuous improvement throughout the organization by focusing on the needs of the customer, empowering employees, and optimizing existing activities in the process." Another definition is "Operational Excellence is when each and every employee can see the flow of value to the customer, and fix that flow before it breaks down" (Duggan 2011).

Companies adopt different strategies in their journey toward operational excellence. Some of the methodologies commonly followed include, either singly or in combinations, Lean, Six Sigma, and TQM. The common goal of any of the operational excellence methodologies adopted by companies is the continual improvement of quality and yield by seeking ways to eliminate different types of waste, by reducing variance, and by examining alternatives to reduce cost of goods of manufacturing. However, these methodologies focus on different aspects of the manufacturing system. Lean thinking focuses on enumerating the types of waste generated, thus leading to a reduction of waste. Six Sigma examines the variances present in the system, traces the origins of these variances, and recommends rooting out its causes, leading to predictable production transformations. TQM examines the activities of the company from a customer's viewpoint, redirecting effort away from work that does not add value to what the customer receives. However, even in the best of cases, a complete attainment of the goals of any of the individual focuses is unlikely. Waste cannot be eliminated entirely; variances will creep in because of a number of uncontrollable and/or unanticipated reasons; and maintaining a focus on the short- and long-term goals of a variety of customers is not always feasible.

In guideline published by the U.S. Food and Drug Administration (FDA 2004), PAT is defined as a system for designing, analyzing, and controlling manufacturing through timely measurements (i.e., during processing) of critical quality and performance attributes of raw and in-process materials and processes, with the goal of ensuring final product quality. Therefore, a desired goal of the PAT framework is to design and develop well-understood processes that will consistently ensure a predefined quality at the end of the manufacturing process. PAT is considered to be an enabler for achieving this common and key strategy, thus enhancing productivity and efficiency. Therefore, one of the primary visions for achieving operational excellence would require a PAT-enabled manufacturing environment.

Friedli et al. (2006) have defined an operational excellence reference model (Table 14.1), which is built on the elements of Lean manufacturing model [e.g., just-in-time (JIT) and total productive maintenance (TPM)] but also focuses on management practices. It is, therefore, a system containing two main subsystems: a technical subsystem and a social subsystem. The technical subsystem includes TQM, TPM, and JIT as well as standardization and visual management as basic elements that are shared by these three major practices. The social subsystem (i.e., management sys-

TABLE 14.1

Technical and Management Systems of Operational Excellence Model

Technical System		
TQM System	JIT System	TPM System
	Basic Elements: Standardization and visual management	
	Social Systems/Management System	

Source: Friedli, T. et al., *Operational Excellence in the Pharmaceutical Industry*, Editio Cantor Verlag, Stuttgart, Germany, 2006.

tem) as described by Loch et al. (2003) focuses on people involvement in continuous improvement.

The technical subsystem focuses on three major principles that support each other. In a JIT environment where the goal is to achieve one-piece flow (or single batch flow, as applied to biotech industries), TQM, which focuses on variance minimization, and TPM concepts, where the objective is maximum equipment effectiveness, become complementary (Friedli et al. 2006).

14.3 JUST-IN-TIME (JIT) SYSTEM

The focus in a JIT system is elimination of waste. The objective is to produce what is needed, when needed, and in the amount needed. Waste is defined as an activity that does not add value, and value itself is defined by the customer. Seven types of wastes in business or manufacturing processes are commonly identified: overproduction, waiting, transportation, overprocessing, excess inventory, motion, and defects (Liker 2003). Table 14.2 details the definition of the types of waste and shows examples from biopharmaceutical manufacturing operations of how the application of PAT can be used to reduce this particular type of waste.

Each of these types of waste and PAT solutions to remedy it can be elaborated further expanding on the specific examples from industrial applications.

Defects. By definition, producing defective parts resulting in repairs, rework, or scrap can be translated into a nonconforming intermediate or end product in the setting of biopharmaceuticals manufacturing. One example can be given from the drug product area where final formulation and dosage form are obtained. One common dosage form includes lyophilized (also known as freeze-drying) product in vials. Lyophilization is usually performed to increase the shelf life of biopharmaceuticals (such as protein-based products) that are susceptible to degradation when water is present. It is carried out in three main stages, namely, freezing, primary, and secondary drying. Precise determination of the primary drying end point before proceeding to the secondary drying is crucial because the increase in the shelf temperature may create a risk of product degradation and ultimately failing the batch. Conventional techniques are based on single-vial testing for primary drying end-point testing;

TABLE 14.2

Seven Types of Wastes Commonly Observed in Manufacturing Operations

Type of Waste	Definition	Traditional Methods Example	Future State Example: PAT-Enabled Manufacturing
Defects	Production of defective parts, which results in repairs, rework, or scrap	Large bottles or carboys containing product, placed in walk-in freezers, primary drying end-point determination based on single-vial methods used in lyophilization process for drug product vials	Batch PAT methods coupled with advanced spectroscopic techniques to monitor and control primary drying end point during lyophilization in real time
Waiting	Idle time created when material, information, people, or equipment is not ready	Waiting for in-process sample results to proceed to the next process step	Reduction of idle times for in-process sample results by use of on-line analyzers
Transportation	Moving materials, WIP, or finished goods long distances, into or out of storage or between processes	Transportation of samples from manufacturing to the laboratory. Example: long assay turnaround time for microbial testing	Reduction of transportation of samples between areas by the use of process analyzers at manufacturing floor operations via PAT-enabled CQAs testing. Example: Rapid PAT-based microbial testing methods
Inventory	Unnecessary raw materials in the warehouse, excess WIP, excess inventory of finished product causing longer lead times, damaged goods, transportation, and storage costs	Long cycle times in manufacturing, testing, and product disposition process. Example: long assay turnaround time for microbial testing	Reduced final product inventory due to shorter cycle time, reduced testing, and faster product disposition via PAT-enabled CQA testing. Example: rapid PAT-based microbial testing methods are available to expedite lot disposition to minimize inventory
Overproduction	Producing items more than the customer requirement	Excess buffer solution make-up taking a conservative approach in production planning	Predictive monitoring capabilities to estimate buffer demand ahead of time resulting in reduced buffer inventory and scrap
Motion	Any motion employees have to perform due to looking for, reaching for tools or parts. Walking from one place to another is also considered waste.	Motion waste resulting from taking, testing, and recording results for in-process samples	Reduced and ultimately eliminated motion waste due to automated sterile sampling
Overprocessing	Producing above and beyond the quality requirement, producing higher quality products than is necessary	Schedule-driven processing where processing time extends beyond the required performance end point	Biologics-driven processing where processing time is driven by the desired performance/quality end point

however, they are not representative of the batch (Pikal 2002). Batch PAT methods coupled with strong analytical tools (such as tunable diode laser absorption spectroscopy) offer real-time on-line process monitoring and control for this particular step and alleviate the aforementioned issues presented by conventional techniques, hence reducing the risk of defects or nonconforming products (Jameel and Khan 2009; Molony and Undey 2009; Low and Phillips 2009; Patel and Pikal 2009; Read et al. 2010a,b).

Waiting time. This is associated with idle time in the process due to an operational element (e.g., material, information, people, or equipment) not being ready, and it typically increases the turnaround time for a current good manufacturing procedures (cGMP) critical assay in a biopharmaceutical manufacturing process. It can be an assay that is required for further processing of an intermediate for instance and in the traditional approach, it is sampled by manufacturing personnel, submitted to quality control laboratories for off-line analysis, and results are waited on until proceeding to the next process step. A typical example may include an off-line sample taken for testing on product concentration by high-performance liquid chromatography (HPLC) in the QC lab, whereas it can be made by an on-line HPLC attached to the process equipment via sterile sampling device so that measurement and its result can be made readily available, significantly minimizing the waiting time for process decisions (Rathore et al. 2008; Molony and Undey 2009).

Transportation. This involves moving materials, work in process (WIP), or finished goods over long distances, into or out of storage or between processes, and the same example mentioned earlier for waiting times illustrates the waste due to transportation of the off-line samples to QC Laboratories for analysis, which require sample handling and transport procedures. It is also important to mention that how samples are handled and transported to the QC lab may have significant impact on the system productivity by increasing risks and variability, whereas automated sampling and analysis via PAT minimizes or eliminates this issue, as sampling and analysis can be made on the manufacturing floor in an automated manner. Industrial applications of this for biopharmaceutical manufacturing can be found in Derfus et al. (2010).

Inventory. Inventory can be excess raw materials in the warehouse, excess WIP, and accumulation of finished products before shipping, causing longer lead times, damaged goods, transportation, and storage costs. PAT-enabled manufacturing offers faster lot disposition, for example, by eliminating and automating some of the product-related release assays. A typical assay in biologics manufacturing with a long turnaround time is microbial testing. Conventional method typically involves an agar plate and a waiting period of 3–7 days for the results before a product can be released. A PAT-based rapid microbial testing assay, described in Miller (2005) and Greb (2010), significantly reduces the lot disposition cycle time down to 18–24 h, hence enabling lower inventories for a product.

Overproduction. This is usually the production of greater quantities than that required to meet customer requirement. Examples of overproduction are common in biopharmaceutical manufacturing and include making excess amount of buffer based on the theoretical maximum needed such as based on final titer in upstream. Excess buffer is made in advance in the absence of PAT tools because it is a required condition to avoid any shortage, which may end up as scrap if not needed at the end. However, PAT-enabled approaches such as real-time process models in the form of soft sensors present opportunities to accurately estimate how much buffer would be needed downstream so that the scrapped buffer amount can be minimized while also giving operations a lot more flexible production scheduling opportunities (Undey et al. 2010).

Motion. Generally, this relates to the action that employees have to perform when looking for or reaching for tools or parts; walking from one place to another is also considered waste. A typical example in the biopharmaceutical industry is again the sampling and transporting the sample to the QC or WIP laboratories back and forth as described above. Another example would be the automation of assays via PAT tools so that labor-intensive repetitive assays generating motion waste can be eliminated. Practical examples include using optical cell density probes for real-time cell counting instead of off-line sampling-based techniques such as using hemacytometers (Wu et al. 1995) and also using in-line PAT tools to determine total organic compounds (TOCs) for water used in bioprocess unit operations instead of off-line sampling-based approaches that involve motion waste (Molony and Undey 2009).

Overprocessing. This implies producing above and beyond the quality requirement and producing higher quality products than necessary. Based on this definition, this particular waste category may not appear to be very applicable for the biopharmaceutical manufacturing because higher quality products are always desirable. What is more applicable is the capability to monitor and control towards a process or product performance or quality end point. It is important to know when these end points are reached to stop processing versus continuing the process based on a prescribed duration resulting in overprocessing. One of the examples is in diafiltration purification operation based on a predetermined number of cycles for ultrafiltration/diafiltration step to exchange buffers to condition the product stream before the next step. This is typically done via a conservative approach by running excess buffer volume, whereas the required conductivity can be achieved in a shorter time (Rathore et al. 2006). Another example would be running a cell culture bioreactor longer due to scheduling reasons despite that it has reached its required acceptance criteria for the next step. PAT tools can be used in both cases to ensure that process end point is monitored in real time where possible and overprocessing is avoided so that process variability can also be reduced.

Based on the discussion above, it is clear that the implementation of PAT technologies will reduce all types of wastes that degrade the productivity of biopharmaceutical manufacturing.

14.4 TOTAL QUALITY MANAGEMENT

TQM focuses on increased customer satisfaction through continuous improvements, in which all employees actively participate (Dahlgaard et al. 2002; Dahlgaard and Dahlgaard-Park 2006). One of the main core elements of this subsystem is process management, which is defining the present process and determining customer needs, establishing measurements, analyzing, and making improvements toward minimizing the process variances as much as possible (Juran and Godfrey 1998).

Enhancing quality of a process requires very rigorous problem solving that is based on process understanding. Currently, due to regulatory requirements, problem-solving activities (such as investigation of nonconformances) must take place within a specified period of problem detection. However, the problems are not necessarily solved at the location where they have occurred. Even though the regulatory spirit is aimed at reducing the latency of recovery and corrective action, the lag time and distance can amplify the consequences of faults. Corrective action is essentially post-mortem. However, because PAT seeks to enable real-time action, this is not only a way to automate the process, but also process malfunctions can be identified immediately (sometimes before they lead to nonconformances) and allow real-time interaction with the process and performance improvement. The implementation of PAT therefore enables the manufacturing operations to grow in knowledge and skills. PAT is an enabler for employee development for a culture where problems are identified where they occur and corrected on a daily basis.

The major components of any system that deploys PAT are (1) process under-standing, (2) analytical tools, (3) quality system support, and (4) regulatory impact analyses. Construction of the design space that characterizes the process to facilitate a greater process understanding is key to PAT. This is typically carried out through designed experiments to determine proven and acceptable ranges of process inputs, including their interactive and nonlinear effects on process performance parameters as well as critical quality attributes (CQAs). A linear operating space is usually cho-sen for process inputs for practical reasons defining the design space while the actual acceptable operating region could be nonlinear. Greater process understanding leads to process models that can predict the process and the product performance. Based on the predictions, control schemes can be devised to ensure the same end quality at each production run.

Analytical tools and robust methods allow real-time decisions during manufac-turing and provide more frequent information about the progress of the process and also about the quality attributes of the product or of an intermediate. Advanced mod-eling, monitoring, and supervisory control systems may be considered in biophar-maceutical manufacturing as a self-adapting and learning from process data (Undey et al. 2004).

Conventional quality control is performed through extensive laboratory testing before intermediate transfer to the next step and also final product release to the market. In the PAT framework, quality systems should be developed such that they support real-time testing and decision-making during critical intermediate process-ing steps as well as during the final release. For instance, instead of testing products after each stage of production, a PAT-enabled system would develop standards for in-process testing of partially completed products. The range of acceptable specifi-cations that must be developed is clearly considerably more complex than for end-of-run testing, but the payoffs can be significant. In a similar manner, PAT approaches can be considered for improving the raw material monitoring and control toward variance reduction. This may also be implemented at the suppliers for improving the robustness of their processes to minimize the effect on the biopharmaceutical manufacturing plant.

14.5 TOTAL PRODUCTIVE MAINTENANCE

TPM is designed to maximize the performance of a production system by providing a systematic way for managing equipment over its life cycle (Hill 2007). PAT may complement this management philosophy. One such example can be given from the predictive maintenance approach. In a conventional preventive maintenance approach, which is common in the biopharmaceutical industry, equipment and measurement devices, for instance, are maintained based on predetermined schedule. Because this is a time-based approach, there may be certain cases where maintenance activities occur on schedule even though the equipment is perfectly functional, resulting in maintenance cost. In other cases, equipment is not maintained soon enough because its scheduled maintenance is not due yet. PAT-based approaches such as using vibration analysis of valves, or other model predictive real-time monitoring techniques help to predict time to failure (Aeppel 2002). In this manner, the maintenance schedule can be optimized and be based on actively measuring the equipment performance. Equipment issues are important from production standpoint as they may have an impact on process and product quality, and any unexpected downtime may result in impacting lot disposition or even loss of product. One way to capture equipment impact to the process performance is to calculate overall equipment efficiency (OEE) as given below.

Overall equipment efficiency = Availability × Performance × Quality

Although many different definitions are possible for a given process, production downtime due to breakdowns and the preparation of the equipment to production impacts the *Availability*, the *Performance* is impacted by the production run rate, and the *Quality* would be a process yield indicator including expected and unexpected yield losses. OEE can be tracked as a metric during the production campaign to identify improvement opportunities. PAT-based maintenance and manufacturing is expected to have a positive influence on this metric as it would improve equipment uptime, improve performance, and ensure quality. Another area under TPM can be seen as periodic monitoring of cleaning systems such as clean-in-place and steam-in-place. PAT offers unique and proactive solutions to this area. In-line TOC analyzers are used to determine when the cleaning end point is achieved. Additionally, parametric methods can also be considered via real-time monitoring of operating variable trends in real time such that any deviations can be detected in a timely manner enabling swift response. This will help improve the cleaning success rate, reduce contamination risks, and also reduce the rework, hence reducing manufacturing costs. Improved cleaning success rate leads to an improved uptime of given equipment, makes scheduling more predictable for the manufacturing operations, and hence increases the OEE metric. This becomes even more important at a multi-product high-run rate manufacturing plant.

14.6 BASIC ELEMENTS: STANDARDIZATION AND VISUAL MANAGEMENT

Standardization ensures that the process is always executed in a standard and repeatable manner. Once the standards are established, there needs to be a check in place to verify the conformance to standards resulting in process stability and action(s) in place to respond to the effects of the standards. In conventional approach, process inputs are controlled via automation and standardized manual operations; however, this does not always result in stabilizing the process end points. Therefore, PAT not only enables real-time monitoring of inputs to ensure that they are complying with the standards but also provides timely measurement of process end points to ensure more stable outputs toward those desired process end points.

The purpose of visual management is to make problems visible to create a common understanding and to improve problem solving. It provides timely information so that immediate action can be taken to correct the problem(s). Depending on the level of design, the operator interface for the PAT-enabled systems range from highly interactive monitoring and control to fully automated systems with indicator displays. As these interfaces become more accessible in the workplace and become a communication tool, they play a bigger role in visual management. This emphasis on visual management is not entirely new—it is essentially the paradigm used successfully in the control of cars, airplanes, and nuclear plants. The controller is given a dashboard of data relating to different measurements from the system and the environment, and he/she must make judgments on the state of the system and its performance in the future based on the current state. The goal of a PAT-enabled system is to give operators the essential information about operations in some visual form and train the operators to use and refine this information to optimize operations.

14.7 CONCLUSION

This paper has reviewed the potential synergistic effect of PAT in the enhancement of operational excellence strategies used in the bio/pharmaceutical companies while also satisfying regulatory expectations. Every company has a certain methodology for operational excellence (OpEx) and how PAT is incorporated. We see PAT as an enabler toward achieving OpEx endeavors, and it became clear that there are connections to other OpEx methods such as Lean, Six Sigma, TQM, and TPM. The key to organizational transformation is to be able to integrate and facilitate discussions between the various business units so that while PAT teams are enabling technology for advanced control and reduced variation, other Lean teams may help to best incorporate the technology within the biopharmaceutical manufacturing framework, such as setting the most optimal visual factory setting, tracking, and reducing variety of the waste as we described with examples. Culture of creativity and innovation via deploying OpEx in the biopharmaceutical manufacturing setting would involve PAT-enabled aspects that would lead to improved productivity and efficiency toward more flexible and cost-effective manufacturing.

REFERENCES

Aeppel, T. 2002. Workers aren't included in lights-out factories. *Wall Street Journal*, Nov. 19.

Cogdill, R.P., T.P. Knight, C.A. Anderson, and J.K. Drennen III. 2007. The financial returns on investment in process analytical technology and lean manufacturing: Benchmarks and case study. *J. Pharm. Innov.* 2: 38–50.

Dahlgaard, J.J., K. Kristensen, and G.K. Khanji. 2002. *Fundamentals of Total Quality Management*. London: Taylor & Francis.

Dahlgaard, J.J. and S.M. Dahlgaard-Park. 2006. Lean production, six sigma quality, TQM and company culture. *The TQM Magazine* 18 (3): 263–281.

Derfus, G.E., D. Abramzon, M. Tung, D. Chang, R. Kiss, and A. Amanullah. 2010. Cell culture monitoring via an auto-sampler and an integrated multi-functional off-line analyzer. *Biotechnol. Prog.* 26 (1): 284–292.

Duggan, K. 2011. Operational Excellence, Institute for Operational Excellence. Available online at http://www.instituteopex.org/cms/index.php?definition. Accessed on March 10, 2011.

FDA, US. 2004. *Guidance for Industry, PAT—A Framework for Innovative Pharmaceutical Development, Manufacturing, and Quality Assurance*, September 2004. Available online at http://www.fda.gov/downloads/Drugs/GuidanceComplianceRegulatoryInformation/Guidances/UCM070305.pdf. Accessed on March 10, 2011.

Friedli, T., M. Kickuth, F. Stieneker, P. Thaler, and J. Werani. 2006. *Operational Excellence in the Pharmaceutical Industry*. Stuttgart, Germany: Editio Cantor Verlag.

Greb, E. 2010. An overview of rapid microbial-detection methods. *Pharm. Technol.* 34: 46–50.

Hill, A.V. 2007. *Encyclopedia of Operations Management*. Eden Prairie, MN: Clamshell Beach Press.

Hussain, A.S., D.C. Watts, A.M. Afnan, and H. Wu. 2004. Process analytical technologies (PAT). *J. Process Anal. Technol.* 1 (1): 3.

Jameel, F. and M.A. Khan. 2009. Quality by design as applied to the development and manufacturing of a lyophilized protein product. *Am. Pharm. Rev.* (Nov/Dec Issue): 20–24.

Juran, J. and A.B. Godfrey. 1998. *Juran's Quality Handbook*. New York: McGraw-Hill Professional.

Liker, J. 2003. *The Toyota Way*. New York: McGraw Hill.

Loch, C.H., L. Van Der Heyden, L.N. Van Wassenhove, and A. Huchzermeier. 2003. *Industrial Excellence: Management Quality in Manufacturing*. Berlin: Springer.

Low, D. and J. Phillips. 2009. Evolution and integration of quality by design and process analytical technology. In *Quality by Design for Biopharmaceuticals: Perspectives and Case Studies*, edited by A.S. Rathore and R. Mhatre, pp. 278–281. Hoboken, NJ: Wiley Interscience.

Miller, M.J. 2005. The impact of process analytical technology (PAT), cGMPs for the 21st century and other regulatory and compendial initiatives on the implementation of rapid microbiological methods. In *Encyclopedia of Rapid Microbiological Methods*, edited by M.J. Miller, Volume I, pp. 195–215. PDA/DHI.

Molony, M. and C. Undey. 2009. PAT tools in biologics: Considerations and challenges. In *Quality by Design for Biopharmaceuticals: Perspectives and Case Studies*, edited by A.S. Rathore and R. Mhatre, pp. 228–230. Hoboken, NJ: Wiley Interscience.

Patel, S.M. and M. Pikal. 2009. Process analytical technologies (PAT) in freeze-drying of parenteral products. *Pharm. Dev. Technol.* 14 (6): 567–587.

Pikal, M.J. 2002. Freeze-drying. In *Encyclopedia of Pharmaceutical Technology*, edited by J.S. Swarbrick and J.C. Boylan, Volume 6, pp. 1299–1326. New York: Marcel Dekker.

Rathore, A.S., A. Sharma, and D. Chillin. 2006. Application of process analytical technology in biotech unit operations. *BioPharma Int.* 19 (8): 48–57.

Rathore, A.S., M. Yu, S. Seboah, and A. Sharma. 2008. Case study and application of process analytical technology (PAT) towards bioprocessing: Use of on-line high-performance liquid chromatography (HPLC) for making real-time pooling decisions for process chromatography. *Biotechnol. Bioeng.* 100 (2): 306–316.

Read, E.K., J.T. Park, R.B. Shah et al. 2010a. Process analytical technology (PAT) for biopharmaceutical products: Concepts and applications—Part I. *Biotechnol. Bioeng.* 105: 276–284.

Read, E.K., J.T. Park, R.B. Shah et al. 2010b. Process analytical technology (PAT) for biopharmaceutical products: Concepts and Applications—Part II. *Biotechnol. Bioeng.* 105: 285–295.

Undey, C., S. Ertunc, T. Mistretta, and B. Looze. 2010. Applied advanced process analytics in biopharmaceutical manufacturing: Challenges and prospects in real-time monitoring and control. *J. Process Control* 20 (9): 1009–1018.

Undey, C., E. Tatara, and A. Cinar. 2004. Intelligent real-time performance monitoring and quality prediction for batch/fed-batch cultivations. *J. Biotechnol.* 108 (1): 61–77.

Webber, K. 2005. FDA update: Process analytical technology for biotechnology products. *PAT—J. Process Anal. Technol.* 2 (4): 12–14.

Wikipedia. 2011. Operational Excellence. http://en.wikipedia.org/wiki/Operational_excellence. Accessed on March 10, 2011.

Wu, P., S.S. Ozturk, J.D. Blackie, J.C. Thrift, C. Figueroa, and D. Naveh. 1995. Evaluation and applications of optical cell density probes in mammalian cell bioreactors. *Biotechnol. Bioeng.* 45: 495–502.

15 Conclusions, Current Status, and Future Vision of Process Analytical Technology in Biologics

Duncan Low and Ray Chrisman

CONTENTS

15.1 INTRODUCTION

All signs seem to suggest that the future of process analytical technology (PAT) in biopharmaceutical process development and manufacturing is promising. This is based on the realization that to ensure efficient high-quality manufacturing, a deeper understanding of the details of the various process steps will be required. Thus, given the complexity and cost of many of these emerging biopharmaceutical processes, it seems clear that the industry will be doing everything possible to maximize throughput while reducing product variability. In addition, the need for better process understanding is also supported by the recent introduction of quality by design (QbD) by the US Food and Drug Administration (FDA) and the corresponding European International Harmonization activities. In contrast to this promise, however, is the current situation with PAT. It is felt by many that the industry is not embracing PAT, or more importantly, modern product quality and manufacturing approaches, with the enthusiasm and commitment that it should. In this concluding chapter, we shall explore both the current impediments and opportunities, real or perceived, and the future vision of PAT.

15.2 CURRENT STATUS—LACK OF INITIATIVE, INSIGHT, OR SIMPLY INERTIA?

Professional discussion groups on social websites and society blogs (e.g., LinkedIn and ISPE) are currently awash with discussions as to if and why the pharmaceutical industry has been slow, or even outright negligent in its adoption of PAT and/or QbD. There are almost as many reasons given as there are contributors to these discussion threads, and there is no clear single theme that overrides all the others. Equally, there are also a host of remedies, but because the cure depends on the ailment, it is likely to be highly situation dependent. The following sections are not meant to be an all-inclusive list of issues, but will give the reader a sense of the range of topics from environmental, technical, to business issues.

15.3 QBD VS. PAT

The FDA's PAT Initiative was the first of several documents that address quality and manufacturing in the 21st century (Low and Phillips 2009). Although the document points out that quality should be designed from the outset, the initial focus was toward applying superior manufacturing practices and in-process controls toward securing quality. This was followed by the ICH Q8 guidance for Process Development, which focused more on process and product design aspects, and emphasized the need for developing a significant degree of product and process understanding upfront. It is clear that good design should precede manufacturing implementation, but many in industry have taken a more QbD-centric approach to development which may have resulted in less focus on current manufacturing and more PAT opportunities for products in development. QbD stresses a science- and risk-based approach and the need to identify product critical quality attributes (CQAs) and link them to process parameters. This is not a simple task in biotechnology, where products and processes are highly complex and understanding the impact of subtle variations in product molecular structure may vary considerably from case to case. Case studies (e.g., A-Mab) have been developed around these issues to help with the discussion (ISPE PQLI 2009). However, until industry and regulators have consensus on what constitutes a CQA and a critical process parameter, it will be hard for manufacturers to develop PAT applications when it is not clear what must be controlled. This is by no means an excuse, and manufacturers must continue to explore multiple variables beyond commercialization in the certain knowledge that their understanding of processes and products will increase.

15.4 BIOTECHNOLOGY AND SMALL MOLECULES

As noted above, the complexity of biotechnology products and processes are often orders of magnitude greater than is the case for small molecules. Processes are expensive to run and can fail catastrophically (e.g., bioreactor contamination); even poor performance in product titer can have considerable impact because of the high level of fixed costs. As a consequence, many in-process analyzers have been in place for years, even if some of the elements of full PAT control are not present, and observers have noted that PAT approaches are more widely adopted by biologics manufacturers. Upstream analyzer-based controls focus on maintaining a favorable environment for cell growth, viability, and product expression, whereas downstream analyzers are deployed to ensure product quality, chiefly by removal of impurities. The use of UV monitors to detect the presence of product and to make fraction collection decisions is ubiquitous, but they do not distinguish between product and related impurities. Newer monitoring techniques and control strategies are being applied but there is a major opportunity for real-time sensors that could provide specific product quality information, for example, around immunogenic properties, which would result in rapid and broad uptake of PAT.

Judging from the volume of correspondence, a serious concern with small molecules in tablet form is dissolution and the consistency of dosage from one tablet to the next. Biotechnology products are often in liquid form and so homogeneity and dissolution are to some extent replaced by different concerns.

15.5   EMPIRICAL AND MECHANISTIC UNDERSTANDING

An interesting contrast can be drawn between the application of process models to upstream and downstream processes, as described in the preceding chapters. Upstream control strategies are based on maximizing productivity without compromising product quality, and controlling variability as far as possible. These controls are really about maintaining a favorable environment within the bioreactor, providing nutrients as required and allowing cells (bacterial, mammalian, or otherwise) to get on with the job of making product. Process monitoring strategies can correlate enormous amounts of data about the impact of raw materials and process parameters on process performance, but the models themselves are not necessarily fully understood at the level of fundamental biochemical processes. This means it is not always straightforward to build a solid business case for advanced monitoring systems because it is not necessarily possible to describe the problem to be solved. However, multivariate tools can uncover positive and negative correlations between parameters and product or process attributes, even if they are not always explainable. There is rarely any direct intervention in the metabolic pathways within the cells themselves, but similar to strategies used in the manufacture of industrial enzymes, there is potential to promote the activity of desirable pathways, suppress the activity of undesired pathways, and simplify subsequent downstream processes. This might have more to do with product design and the development of the cell line than advanced PAT control, but is nevertheless a potential opportunity for PAT.

The underlying theoretical basis of unit operations used in downstream processes, as discussed in the preceding chapters, are better understood and it is possible to build mechanistic models that capture the important variables in material attributes and process parameters. The post-bioreactor processes can often be described in terms of specific conditions and outputs and it is easier to build a financial justification for the proposed analyzer-based controls. Companies seeking to implement PAT may prefer to start with downstream processes because success with a few applications builds the case for PAT and gives management confidence to extend these practices further upstream.

15.6 ORGANIZATIONAL MATURITY

It has been suggested that QbD is a concept for big pharmaceuticals only because smaller companies will not have the resources to do the necessary work upfront to fully characterize the process, and by extension, this may also mean that application of PAT is limited. Whereas it is true that the resources required to develop a well-characterized and well-controlled process may seem daunting, there is still considerable value in doing so because smaller companies may well be dependent on transferring the process to a manufacturing partner for commercial-scale operation. Avoiding expensive failures or delays during process transfers can save in both lost profits due to greater costs as well as in lost revenues resulting from delays. Smaller companies may still prefer to take the risk, but it is likely that most contract manufacturing organizations will insist on certain minimal requirements being met in terms of procedures and controls. Laboratory-scale PAT-based approaches that were described in an earlier chapter, which can reduce this risk by providing the needed process information, should find growing use.

It is difficult to fully establish the full range of variability that can be expected at the manufacturing scale when only a limited number of lots are produced at small scale, which in its turn makes it difficult to fully define specific ranges and operating conditions. It is even harder to understand the significance of variability in, for example, product-related impurities, because clinical exposure and data will be lacking. This is not a problem that will be resolved by PAT, but it makes it harder to establish appropriate operating and control limits. Where PAT-based approaches would be helpful is the advanced monitoring (including real-time multivariate monitoring) during the commercial campaign at large scale to identify, track, and control variation. Learning at the large scale (such as the impact of raw material variation on the process) when monitored and the knowledge is captured appropriately can be very powerful during the product life cycle.

15.7 MANUFACTURING COSTS AND THE VALUE OF QUALITY

There has been discussion that the increase in competition from biosimilars will result in downward pressures on costs and improvements in manufacturing practices and a broader adoption of PAT. The impact on biologics will be felt to some extent; however, the costs of goods are still significant, and extensive clinical testing is still likely to be required because even though the efficacy of a molecule is not in question, the safety is. Moreover, innovative producers are more likely to respond as they always have, by innovating further. Molecules have been improved from partially humanized to fully human versions, molecular structure has been modified to improve potency, decrease adverse reactions, and increase the half-life in the body. What will drive innovators to embrace PAT more fully from a business perspective is the need to optimize yield, to improve consistency in manufacturing and eliminate waste, either as catastrophic failures or in inefficient processes.

An interesting proposal is to create an environment where there is a mechanism for manufacturers to compete on the basis of quality. The "customer" for a drug manufacturer can be the regulators, the insurance providers, the prescribing doctor, as well as the patient. The first three have considerable training and a vast array of data and expertise to help them make critical decisions about the safety and efficacy of drugs, whereas the vast majority of patients are not equipped to make analytical assessments of drug product quality at all. They can at least be informed of product consistency and variability, and a heightened awareness will assist them in making informed decisions as to which medications to prefer, and ultimately create market pressures that would favor manufacturers most willing to commit to meeting patient needs.

Again, the picture in biologics can be significantly different than for small-molecule drugs—because many are injectable materials that are presented as devices—resulting in a patient having a clearer perception of convenience and ease of use. Certain biologics, notably insulin, have reached a level of maturity where total patient care rather than simply providing a drug is required to be competitive. Surely, a treatment where the patient is directly monitored continuously and the drug is administered as required is the ultimate in the post-production application of PAT.

Joseph Juran—the father of QbD—pointed out that cultural resistance, that is, resistance to change, was the root cause of quality issues. If market pull for improved product quality is aligned with superior manufacturing technology, the future for PAT will be bright indeed.

15.8 FUTURE VISION

The first impacts of PAT, as noted in the book, are being seen in the manufacturing area as more companies seek to add analyzers to each unit operation in the process to maintain parameters within the design space of the step. This first stage of implementations usually takes the form of monitors that send an alarm when conditions begin to drift away from the optimum rather than the analyzer being integrated into a closed-loop control system. However, as has been seen in many processing industries, as new and reliable real-time data becomes available, the process control engineers will begin to incorporate this information into more sophisticated control strategies to gain higher throughput with reduced energy utilization while maintaining high-quality products.

The first units to be monitored will be the ones with the most variability and thus, the most to gain with enhanced control. This will be tempered somewhat by the technology available to monitor difficult to sample and analyze steps. The result will be that most early implementations of PAT will be on unit operations that are viable candidates using technology that can be readily adapted from other processing industries.

However, the need for better overall process control will drive the development of more bioprocess-specific PAT. Given that sampling is often the major source of problems, the first approaches to real-time analysis of difficult to characterize unit operations will be to utilize semi-automated analysis where the sample is first obtained and conditioned by the operations staff before introduction into the analyzer. In the case of protein therapeutics, the analyzers will be versions of systems that are utilized in the laboratories to characterize the complexity of these molecules.

The above scenario will drive the development of new sampling and analyzer systems because there will be a high likelihood that it will rapidly become apparent that this approach is fraught with problems. As was often seen in the chemical processing area, the use of operators for initial sample gathering and preparation will result in an additional and potentially unacceptable source of variability. In addition, the timeliness of the data is likely to not be well synchronized with control needs. In other words, new approaches will be needed to rapidly convert samples into the required form for analysis in a repeatable manner.

As a simple example, proteins that need to undergo enzymatic digestion for analysis will need streamlined methods to reduce the often hours-long step. Fortunately, Al-Lawati et al. (2006) have already shown that this step can be reduced to minutes. Thus, the expectation is that as the need for faster sample preparation times for steps such as lysing, solvent exchange, derivatization, and filtration develops, new techniques will appear. This expectation is based on the much more precise and faster control that is achieved in microscale equipment. This is the result of mass and heat transfer rates that are often orders of magnitude higher and the fact that most analyzers require very small quantities, which means much less material needs to be processed than is often done in standard laboratory manipulations.

Just as sampling and conditioning capabilities are being improved, analyzer capabilities will also experience significant growth. Although the current generation of laboratory-based techniques is now beginning to have the ability to characterize these often complex molecules, the instruments are not well suited for use in a production setting. For example, whereas mass spectrometers were some of the first on-line analyzers, the systems in current use for laboratory analysis are too complex, finicky, and expensive for significant use in process applications. However, a recent article by Ouyang et al. (2009) describing the state of the art in miniaturization of mass spectrometers suggests that the capabilities required for rapid on-line analysis of complex molecules is within the range of possibilities. Moreover, a recent review of column technology in liquid chromatography suggests that the resolution and speed required for this technique to be useful for these types of analysis is being developed (Nováková and Vlčková 2009).

Unfortunately, the speed with which PAT will be utilized will be somewhat hindered by the fact that there is relatively little research funding to evolve these techniques to the stage required for process utilization. Even though the impact on process economics can be rather significant, the market for analyzers is somewhat small and specialized, which means that companies supplying equipment do not have sufficient resources to develop needed hardware. In addition, government funding agencies do not normally fund equipment development research.

Once the equipment is finally developed and interfaced to various unit operations, a next level of advancement will occur, which will be the implementation of next-generation control strategies. As more understanding evolves as to cause-and-effect relationships within the unit operation, models will be developed that will enable much more precise control of the processing step. This will initially take the form of better precision of control for enhanced quality of materials. However, it will evolve to new more highly optimized conditions that will improve throughput and energy utilization. Additional outcomes can be expected such as reduced environmental impacts as well as better process safety.

An interesting parallel effort will occur in the development of real-time process analysis for laboratory-scale reaction screening. As was described in Chapter 12 of this book, equipment is becoming available, which enables a wide range of reaction conditions to be explored in small-scale equipment within the laboratory. As was demonstrated, the ability to utilize the data for process scale-up is quite good. This will mean that more and higher performance analyzers will be coupled to the small-scale reactors for information gathering such as correlating protein production with growth phase or with components in the raw material feed.

This trend will be quite important because it will provide a test bed for sampling system and analyzer development. As is typical with laboratory work, there are more skilled operators available to keep the analyzers running plus the conditions within the laboratory are less demanding and long term compared to process operations. These differences provide an opportunity to try new approaches in a breadboard fashion without the need to completely engineer all aspects of the sampling system and analyzer operation. In addition, the inevitable failures are much more easily corrected and they do not have any impact on the operations personnel's confidence in the reliability of data.

Laboratory-scale on-line analysis will also enable the control engineers to develop control models much more efficiently because significant parameter space can be explored safely and efficiently.

ACKNOWLEDGMENTS

The authors thank Ali Afnàn and Gawayne Mahboubian-Jones for their helpful and insightful comments.

REFERENCES

Al-Lawati, H., P. Watts, and K.J. Welham. 2006. Efficient protein digestion with peptide separation in a micro-device interfaced to electrospray mass spectrometry. *Analyst* 131: 656–663.

ISPE PQLI. 2009. A-Mab Case Study. Available online at http://www.ispe.org/PQLI_A_Mab_Case_Study_Version_2_1.pdf. Accessed on Sept 5, 2010.

Low, D. and J. Phillips. 2009. Evolution and integration of quality by design and process analytical technology. In *Quality by Design for Biopharmaceuticals*, edited by A.S. Rathore and R. Mhatre, pp. 255–286. Hoboken, New Jersey: John Wiley & Sons, Inc.

Nováková, L. and H. Vlčková. 2009. A review of current trends and advances in modern bioanalytical methods: Chromatography and sample preparation. *Anal. Chim. Acta* 656: 8–35.

Ouyang, Z., R.J. Noll, and R.G. Cooks. 2009. Handheld miniature ion trap mass spectrometers. *Anal. Chem.* 81: 2421–2425.

Index

Page numbers followed by f and t indicate figures and tables, respectively.